# PUBLIC ADMINISTRATION AND SOCIETY

## CRITICAL ISSUES IN AMERICAN GOVERNANCE

## RICHARD C. BOX

*M.E.Sharpe*
Armonk, New York
London, England

Copyright © 2004 by M.E. Sharpe, Inc.

**Library of Congress Cataloging-in-Publication Data**

Box, Richard C.
  Public administration and society : critical issues in American governance / Richard C. Box.
    p. cm.
  Includes bibliographical references and index.
  ISBN 0-7656-0824-3 (cloth: alk. paper) — ISBN 0-7656-0825-1 (pbk.: alk. paper)
    1. Administrative agencies—United States—History. 2. Bureaucracy—United States—History.
3. Public administration—United States—History. I. Title.

JK411.B69 2004
351.73—dc21

                                                                        2003050599


Printed in the United States of America

The paper used in this publication meets the minimum requirements of
American National Standard for Information Sciences
Permanence of Paper for Printed Library Materials,
ANSI Z 39.48-1984.

BM (c)  10   9   8   7   6   5   4   3   2   1
BM (p)  10   9   8   7   6   5   4   3   2   1

# PUBLIC ADMINISTRATION AND SOCIETY

# CONTENTS

# PREFACE

This book is a supplemental text for the introductory course in Master's in Public Administration programs. Many who teach the introductory course find that students have little knowledge of the *historical, social, and structural basis* of the American system of government and its effects on key issues facing today's public administrators. This is the specific area of knowledge addressed by the book.

Instead of the emphasis on management topics found in introductory public administration texts, this book provides a *macro-level view* of the cultural, political, institutional, and economic context of public administration, within a framework that is *historical and critical*. This view is grounded in the premise that future leaders in the field may significantly influence governmental policy and its implementation, so they should have broad understanding of matters beyond carrying out micro-level management tasks. The framework is historical because contemporary values and practices have been shaped by past actions; it is critical in bringing to the reader's attention the interplay of opposing ideas through time. The book deals with American society and public administration because that is its intended audience and because the context of the public sector in the United States is quite different from that of other nations.

Though the importance of nonprofit and private-sector organizations and citizen efforts in nongovernmental, "civil society" is recognized, the focus here is on *public governance*, goal-setting, and action potentially affecting all people in a defined geographic area. There is a unique character to a setting in which people act jointly as a nation, state, region, local public agency, or neighborhood, a joint character that makes it unlike other human activities. Given this focus, attention can be directed to understanding factors that shape the work of public organizations and how students and practitioners of public administration can contribute to constructive change.

People become accustomed to their personal and work lives, often doing what needs to be done without thinking too deeply about its meaning. We take for granted the complex web of human organizations, physical structures such as streets, utility systems, and buildings, commercial networks that supply such basics as food, clothing, and an almost infinite variety of other products, and the laws, procedures, and expectations about correct behavior, specific to our national culture, that make all this work as well as it does. On occasion, an unusual event, conversation, media presentation, or written material will jolt us out of the routine of daily habit, prompting reexamination of our lives and the society around us.

Another factor leading to unthinking acceptance of the familiar is pressure to conform to the norms of society, making it difficult to question "the way things are." Sometimes scholars, the media, or advocates of change bring to our attention fundamental *contradictions* between our beliefs about such matters as fairness, democracy, or protection of the natural world, and the way organizations and society actually function on a day-to-day basis in relation to treatment of human beings and the physical environment. Though it can be risky for public professionals to call attention to such contradictions, as they advance in their careers and take on broader responsibilities, they may sometimes find it necessary to question the status quo.

An assumption made in this book is that some of its readers are, or will be later in their careers, in a position to influence the course of public-service practice or research, with constructive impacts on the lives of citizens. A crucial skill for those who want to make a difference in the world is *critical thought*. The term "critical" is not used here to mean negative or judgmental, but probing and analytical, identifying contradictions between commonly held beliefs and the practices and perspectives we discover in the course of learning. Narratives and readings in this book are intended to facilitate critical, skeptical thought about the uses of power and resources in public-sector governance, giving readers tools for improving citizen self-determination in a democratic society.

Though these materials introduce questions about economic, political, and social systems that may cause some readers discomfort, the objective is not to provoke cynicism about the public sector or a defensive rejection of views that challenge the status quo. Instead, the intent is to give readers a glimpse into the richness of the unique American experiment in *democracy and self-governance* and opportunities to personally contribute. This experiment is continuous and sometimes appears chaotic, harsh, or unjust, but this open-endedness and constant debate over what is right and good are its strengths. At the core are a set of beliefs about human freedom and the relationship of the individual to society that were revolutionary at the time of the nation's founding, beliefs that today are often admired in principle but can be difficult to achieve. The challenge for students of public administration is to understand the nature of this society, clarify our value commitments as people in public service, and find ways to carry them out in daily practice.

The reader looking for clear answers to complex problems of public action may be disappointed by this book, because questions are raised here as well as answered. Public administration is not a science, in which many problems have one clear answer, so the best we can do is offer a framework for understanding and to provoke creative reflection. It is hoped that teachers and students will use the book as a starting point to link together other readings and discussion of the societal environment of public administration, encouraging active thought and questioning.

In one book it is impossible to fully develop the themes presented here or to offer all perspectives on each theme. The field is vibrant with contending descriptions of what seems to be happening in society and in public organizations at all levels. There is diversity of thought on preferred courses of action and possible outcomes, as well as academic disputes about how to research and write about these phenomena. The reader may find it helpful to think about the following three questions while using this book—they are central to the field and to the materials that follow.

1. *What ideas and events have shaped contemporary public administration?* This question assumes the present is better understood with knowledge of the past and this knowledge may help make a better future. It also assumes there are a variety of perspectives on events that have shaped public administration and that knowledge of the past is a matter of interpretation rather than searching for a single, "correct" explanation.

2. *What is the interaction between American society and the action alternatives available to public-service practitioners?* Public employees are not free to do whatever they wish; funds to operate their organizations come from a public that has expectations about appropriate public action in society.

3. *As public-service practitioners, whom do we serve, and for what purposes?* Is it current political and economic leaders, or a larger public whose voices are not often heard in public decision-making settings, or some conception of the long-term interests of all the people? Answers to this question in a given situation can make quite a difference in how decisions are made and what actions are taken by public agencies.

In search of useful ways to approach these questions, the book is organized into five parts. In Part I, three chapters set the stage conceptually for later readings. Chapter 1 describes the scope and content of public administration as a field of study and practice, Chapter 2 explores the societal environment of public administration over time, and Chapter 3 discusses concepts of citizenship, democracy, and ways people structure units of government to serve particular values or goals.

Parts II–V focus on crucial, often controversial thematic areas that are especially important for understanding contemporary public administration. It would be impractical to cover the full sweep of history and development of issues that affect contemporary affairs. Instead, these four parts are designed to address perceived gaps in the knowledge of public administration graduate students in these thematic areas. Each part begins with a narrative outlining key concepts in the thematic area, followed by readings offering a broad overview of issues that have significant impacts on public-administration theory and practice. Each reading is rich with interesting and useful ideas, so the introductory narrative highlights only a few key concepts. Readers are encouraged to identify and discuss additional points of interest in the materials.

The four thematic areas are: events that created the American governmental system and continue to shape its operation today; impacts of power, politics, and economics on the relationship between individuals and the larger community; the tension in public administration between desire to promote fairness in the economic system and desire to make government efficient, as we imagine the private sector to be; and contrasting views of the roles of public-service practitioners as they search for a balance of public service values in their daily work lives.

# PART I

## INTRODUCTION TO THE CENTRAL ISSUES

### CONTEXT, CHANGE, AND DEMOCRACY

# CHAPTER 1

# SCOPE AND CONTENT OF PUBLIC ADMINISTRATION

People in contemporary society are surrounded by information about aspects of public life. The media—including newspapers, magazines, radio, television, and the Internet—offer many stories every day about politics and government at the international, national, state, regional, and local levels. It would seem that people would be saturated with information about the public sector. Readers of this book have by definition more than an ordinary interest in public-sector matters, so we might expect them to be quite knowledgeable about government.

Instead, what we often find with the public and even with students of political science and experienced public-sector professionals, is knowledge of scattered, seemingly unrelated facts about various parts of the public sector, but only sketchy awareness of governmental history, systems, structures, and how government (the legislative, administrative, and judicial parts of the public sector) works. Public-administration ethicist Terry Cooper notes that people come to the university with "a wooden, over-simplified conception of the way public policy is formed and implemented" (1998, 62).

This is often true of graduate students as well, even those with significant experience in the public sector. Several years ago, an MPA student in the author's class used a vivid metaphor to describe this phenomenon, as she said her elementary, high school, and college undergraduate course work in American government left her with a "pastel, pink-and-blue" view of the nation's governance systems, a view that was simple and did not convey a sense of the American governmental experience as an exciting, often passionate, on-going debate over fundamental issues.

Many public-service practitioners play a significant role in guiding public agencies, policies, and programs, impacting the lives of real people on a daily basis. If these practitioners do not have a broad understanding of their institutions, what can be said about the outcomes of their work? Without such knowledge, on what basis do they make decisions about their public-service roles in a democratic society? In the absence of this knowledge, results of actions taken by public-service practitioners may be based on doing the same thing that has always been done, or on intuition about what might be best, or on selection of alternatives that seem most acceptable to those in positions of power. Any of these can be useful guides to action in a particular situation, but without greater breadth of knowledge, it is difficult to sort through the options in an informed, meaningful way.

["Public-service practitioner," instead of "public administrator" or "public professional," may be used to indicate that not all public employees are administrators or in occupations usually considered professional. "Administrator" suggests a person who plans and supervises the work of others, and a "professional" is someone who applies a recognized body of knowledge to daily practice. Also, the people referred to by use of these terms are career public employees selected for their jobs on the basis of job-related education and experience rather than personal affiliation with, or loyalty to, a political party or leader. They are not to be confused with elected officials/politicians, or their political appointees (such as cabinet members, department heads, or appointed subordinates) who are in office during the term of a particular elected leader and are chosen, at least in part, on the basis of acceptance of the beliefs of the leader or party.]

The paradox of limited knowledge of the public-sector in the midst of a flood of information suggests problems with the media and the educational system, but that is outside the scope of this book, which is about public-sector governance. It is assumed here that people involved in career public service have a greater responsibility to understand the nature of our society, democracy, and government than do most citizens, in part because such knowledge can improve professional practice. But beyond this practical value, to keep a society intact someone needs to carry the knowledge of how it came to be what it is, passing it on to others. People who do the work of public service are well positioned to play this role.

## INSTRUMENTAL AND CONTEXTUAL PUBLIC ADMINISTRATION

Many think of public administration as a largely technical, applied field of study and practice, all about how to do budgeting and manage personnel, send the Social Security checks out on time, or keep the potholes filled. This view of public administration is *instrumental*, that is, it views public administration as a tool used by elected representatives of the citizenry to accomplish public goals. This is not inaccurate, because a significant part of public administration can be reasonably described in this way, but it is incomplete. The instrumental view leaves out the sometimes difficult and conflictual process of identification of public purposes and how public employees interact with citizens and elected officials in helping shape governmental action. It assumes that millions of public employees wait quietly at their desks for orders to be handed to them from a mysterious other place, and that these orders are fully developed by knowledgeable people who comprehend the social world and public needs in their full complexity.

Of course, that is not how things work. Instead, describing public purposes and taking action toward fulfilling them in the daily "lifeworld" of citizens is a dynamic, repeating process. In this process, public professionals interact with peers, elected representatives, and citizens in formulating and reformulating programmatic goals and plans for action, putting them into effect, and revising them during implementation. New circumstances can emerge at any point, causing perceptions of the initial problem or the public action to change, for reasons that may operate singly or in combination, such as:

1. New information emerges about the problem (an example would be discovering there are many more homeless people than first thought when a program to house and feed them in cold weather was created);
2. The program produces on-the-ground results suggesting the need for modification (continuing the homeless example, maybe few homeless people want to participate and a significant number are still freezing to death on the streets);
3. The program causes peoples' perceptions of the situation or public action to change (some political leaders and media personalities say the program is pointless because the homeless would rather freeze, but a citizen's advocacy group thinks the program should be expanded by picking people up and taking them to shelters); and/or,
4. The broader social context of the problem has changed such that the program should be modified (for example, the economy weakens and even more people are living on the streets).

The process of identifying a problem, discussing potential solutions, finding resources, putting the plan into action, and in each phase accommodating new information, ideas, and perceptions, can be top-down, operating at the level of political and economic leaders and higher-level administrators. It can also take place from the bottom up, with citizens and public employees generating ideas from their daily experience, and top-down and bottom-up activity may occur simultaneously.

As public employees engage in these processes, they move outside the common role expectation of the value-neutral tool used to achieve predetermined ends. They are the people with detailed knowledge of techniques and practices and they also have considerable knowledge of social conditions and the needs of the population. This knowledge of public needs may differ from that of elected representatives because it is gained in forms of contact and interaction that are different. Nevertheless, it is important in shaping actions taken by public legislative bodies and tailoring services to the people who receive them.

So, commonly held views about public administration may not take in the full reality of the field. Public administration is not only instrumental—public-sector decisions and actions are often complex, involve multiple possibilities, and change with time; and public-sector practitioners are involved in determining what government does in addition to how it does it. Public employees work in political, economic, and social environments shaped in part by past events and accumulated ideas, values, and cultural preferences about the purposes and operation of the public sector. Those who know something about these environments may have a greater chance of succeeding because they are better able to craft options and alternatives appropriate for their circumstances. As a bonus, they may find that deeper understanding of the broader society leads to greater satisfaction in their careers and as citizens outside the professional role.

Providing a portion of that deeper understanding of the *social context* of public administration is a focus of this book. The reader will find that standard management subjects in public administration are not covered here. This is because the book is not about management techniques for motivating employees, designing or evaluating a public agency program, understanding financial relationships between the national government and state and local governments, and so on. Instead, the book is about *macro*-level aspects of American society that influence how management of public agencies takes place, how history and practice have brought us to this point, and what options there may be within this context for future action. This is not dry history or settled fact to be memorized. It is a complex, tumultuous story, full of intrigue, interesting lives, grand purposes, and failures and successes, stretching through historical time and across several scholarly disciplines. Given this scope it is possible here to highlight only a few areas, emphasized because of their particular importance to the public sector. Readers are encouraged to use the book as a starting point, identifying events or ideas not included that they think should be discussed as well.

## THE IMPORTANCE OF PUBLICNESS

People often have trouble conceptualizing the idea of public administration as a whole, aside from specific tasks or functions such as issuing drivers' licenses. This is not surprising, because public administration covers such a broad range of activities that its boundaries and contents can easily seem unclear. If we visualize ourselves in an introductory public administration course early in the semester, the professor may ask people to introduce themselves. Around the room, students in turn talk a little about their education, work history, and future career interests. In a typical introductory course it is common to find, for example, a captain in the military, a program manager in a nonprofit social service organization, a police officer or firefighter, a person who has recently graduated from college with a degree in majors such as political science, sociology, communication, a state highway engineer, a child welfare caseworker, a land use planner for a local community, a wildlife manager for the national government, and a person who has worked for years in the private sector and wishes to move into public service.

Some of these class members are in the course because they want a degree to advance their career potential; some are there because they want a credential that will allow them to change careers, and some simply want further education. Some think of work in the public or nonprofit sectors as just a job, and others feel a sense of commitment to a greater public good, a sense of changing the world for the better. It is generally believed this is a time of change and uncertainty in the world, a time of rapid technological advancement, realignment of political boundaries, and disintegration of certainties about things like moral values, the purposes of government, the nature of the family, and so on. Few areas of human activity are unaffected by this perception of change, and many organizations with a public-service mission are experiencing serious pressures to adapt to new demands.

Today, the boundaries between the three sectors of institutional and economic endeavor—private, public, and nonprofit—can be shifting and indistinct. This is due to several trends, including: the public sector contracting out work to nonprofit and private organizations, growth in the nonprofit sector, and increased emphasis in the public sector on businesslike efficiency. In an attempt to provide some clarity, public administration scholars have identified ways the public sector is different from the private sector. For example, Denhardt and Grubbs (1999, 5–9) suggest that public-sector management is different from that of the private sector because of: greater ambiguity of purposes in the public sector as compared to bottom-line profit in the private sector; pluralistic decision making in a public-sector organization involving many people instead of a few in a private organization; and visibility of public-sector management, the feeling of operating in a goldfish bowl of public scrutiny. These things make up an overall public-private comparative difference that Denhardt and Grubbs call *publicness*.

Our imaginary introductory course includes students representing experience in all three sectors—this cross-sectoral interest in education for public-sector management has become common in MPA programs. In many areas of management, concepts and techniques carry across sectors. For example, not only are there elements of leadership, motivation, accounting, human resource management, and information technology that cross sectoral boundaries, much of the innovation in these areas comes from the private sector and is adapted for use in the public sector. These similarities, and the cross-sectoral usefulness of certain management techniques, have led some to believe that management is management, regardless of sectoral location. However, a feature of public administration that is unique, that differentiates it from the private and nonprofit sectors, is that it is *public*, which indicates it involves every person in a defined geographic area. This simple word carries with it implications at the level of management practices—for example, the problem of potential intrusion of partisan politics into administration of personnel recruitment and hiring, or requirements for published, independent audits of the financial practices of public organizations.

Though the effects of the public character of organizations on management are important, it is at the broader, macro level of organizational purpose and the relationship of organizations to the people they serve that publicness becomes not just important, but crucial. Those who argue that the sectors are becoming more alike and that management is management are partially correct when they consider technical aspects of management practice. But when attention shifts to organizational purpose and relationships with those served by public administration, we find clear differences between sectors. These differences appear in areas such as identifying problems to be addressed and who is involved in problem solving, procedures for decision making and who the decision makers will be, to whom and in what ways organizations are accountable for their performance, whose interests shape organizational goals, and what appropriate roles are for practitioners.

The ways citizens, elected officials, and public employees involved with a particular organization approach these issues are shaped by American expectations about government, expectations formed by more than 200 years of dealing with balancing demands for open, democratic governance and desires for efficient, effective management. The American attitude about the relationship of the public sector to the broader society is quite different from that in many other countries. For example, there are countries (such as Scandinavia and some in Europe) in which people more readily accept the prominence of government in social affairs and government is considered a legitimate, permanent entity with a life of its own. In some Asian countries, the relationship of the public sector to society is framed within the cultural expectation that contributing to the collective well-being of society is more important than the interests of the individual. In the United States, the size, functions, even the right to exist of the public sector may be questioned, and government is often viewed as a threat to the freedom of individuals. In this environment of emphasis on protecting the rights of the individual from government, governmental action that to people in other countries might seem constructive, may instead appear inappropriate to an American.

## COERCION AND GOVERNANCE

Government cannot act without resources, and a large percentage of public-sector operating revenue comes from taxation of individuals. Taxation is often a focal point for concern about how large government has become and the extent of its intrusion into individual lives. This concern is expressed by politicians (such as Republicans in Congress in the 1990s who attempted to cut back or eliminate many public agencies and programs), it appears on television and radio news and talk shows, and it is reflected in public opinion polling about citizen mistrust of government. It motivates people who refuse to pay taxes, people who organize protests against the government, and in extreme form people like Timothy McVeigh, who was executed for destroying the Federal Building in Oklahoma City with a bomb, killing more than 160 people. In the United States, the role of the public sector in society is always in dispute, making it somewhat unstable and changeable, a work in progress. This situation can be thought of as an inefficient waste of time, or as a sign of a democracy that is constantly revitalized.

This discussion leads to identification of a key feature of the public sector that distinguishes it from the private and nonprofit sectors. For everyone in a specific geographic area, participation in the public sector is *mandatory and enforced coercively*. It is not necessarily specific services that are coercive, though some are, such as being arrested on suspicion of having committed a crime, or having one's child taken away by a government agency because of child abuse. People attending a particular publicly funded state university are not coerced to do so because there are other choices available in higher education. However, if a person decides that the public sector should not fund higher education and refuses to pay a portion of his or her income taxes equivalent to the percentage of the state budget used for that purpose, he or she can expect there will be consequences, including financial penalties and/or imprisonment.

In this way, citizens are coerced into participating in the support of public institutions, complying with the will of the majority, whether or not they agree and whether or not they personally benefit from a particular governmental service or program, on the grounds that it is good for all the people (in the case of state-funded higher education, an argument would be that an educated citizenry benefits everyone). The word "coercive" usually carries a negative feeling of something that is forced upon people against their will. But in public governance in a democratic society, at least in theory people agree to participate in the collective activity of the whole. If they wish, individuals may join together with others to change the way government is operated and the purposes it serves. At the subnational level, they may also choose to move to a different unit of government (state, county, city). The public sector must in this sense be mandatory, coercive, because there is no other way to operate government. It would fail if people were able selectively to withhold support for specific functions.

In the private sector, each customer chooses whether or not to buy a product or service. We can debate whether these choices are freely made or whether an economy that depends on growth in consumption trains people from an early age to want an endless supply of consumer goods. Nevertheless, there is a range of choices in a private economy and people choose what they want. The nonprofit sector includes associational, professional, and business organizations (examples: a neighborhood association formed to improve the quality of life in an area within a city; the American Medical Association; a city's Chamber of Commerce), mission-oriented organizations (philanthropic, charitable, social service, arts and culture, religious training, or education), and a variety of other organizations, including recreation and leisure nonprofits such as celebrity fan clubs, sports organizations, and so on. Some of these organizations are difficult to distinguish from those in the private sector, since they have similar purposes but differ only in the lack of profit accumulation for their owners. However, they all share with the private sector the central characteristic that their members participate voluntarily—the centrality of this difference from the public sector is noted by frequent use of the term *voluntary sector* to describe nonprofits.

Again, the mandatory, coercive nature of the public sector is how it must be to function. The intent here is not to portray this basic characteristic negatively, which is the way people join together to create the sort of community, state, and nation they want. The intent is to highlight the importance of this collective characteristic, both to distinguish the public sector from the nonprofit and private sectors, and to emphasize what this means for the relationship of public-service practitioners and citizens.

Public-service practitioners who are aware of the coercive nature of government have the opportunity to be particularly sensitive to ways decisions are made and how open or closed the process is to those who might want to participate. They can also take into account the impacts on people's lives of implementation of policies and programs, working to soften negative impacts and make government more responsive to people's preferences. In these ways, public-service practitioners recognize the special responsibility that comes from participation in an endeavor that includes, by definition, everyone within a defined area.

## SCOPE AND CONTENT OF AMERICAN GOVERNMENTS

Let us move from describing the public sector by contrasting it with other sectors, to examining the scope and content of public administration. As an applied field of study, grounded in delivery of public services by governmental agencies, the study and practice of public administration draws concepts and techniques from other academic disciplines and areas of professional practice. Examples include the academic disciplines of political science, philosophy, history, economics, and sociology, and the professional fields of planning, social work, criminology, engineering, accounting, and many more. It was common throughout the development of American public administration in the twentieth century for academicians to characterize this diversity negatively as a lack of focus rather than regarding it as open and vibrant. Some wish for clearly defined substantive content and rigorous quantitative research methods that might make public administration more respected in the academic world alongside disciplines such as economics or political science. Others worry about the fragmentation of the field into specializations with their own journals, organizations, and meetings (i.e., ethics, human resources, policy, financial management, critical and postmodern theory, intergovernmental management) as the public sector grows larger and more complex.

Despite this desire for clarity and academic respectability, public administration, both as academic study and practice, keeps growing, fragmenting, and drawing ideas and inspiration from whatever sources are helpful for understanding problems and solutions in public service. This is natural enough in a field that involves so many people in so many settings and occupations. Consider the size and complexity of the American public sector. Though it is common to think of the public sector as mostly the huge national government supplemented by a variety of state and local governments, concentrating on the national level is misleading. There are fifty one state and national governments, but more than 87,000 units of local government. These include school districts (13,726), cities (19,372), counties (3,043), special districts (34,683), and townships (16,629) (U.S. Census Bureau 1997).

In addition to the quantity of units of government, the number of people working in these organizations is impressive as well. In 1997 there were a total of 2,807,077 national government civilian employees (including the postal service and civilian defense employees), and 3,986,680 state employees. At the same time, there were 10,227,429 local government employees. Of these, 5,480,206 worked in elementary and secondary education, so the number of local government employees not in K–12 education was 4,747,223. Thus, national government civilian employment, at 2.8 million, was about 59 percent the size of local government employment outside of K–12 education, at 4.7 million. Adding state government employment to local (not including K–12) yields 8,733,903 employees; national civilian employment was about 32 percent the size of this figure.

[It is not uncommon for people to be surprised to hear that schools are units of local government. Their status is confusing because: school districts are administrative arms of the state formed to deliver education services; they receive state funding; to varying degrees their activities are controlled by the state; some are dependent on *general-purpose* units of local government such as cities for their financing or governance; and their organizational structures and boundaries can be changed by the state. Despite all this, running the schools has been a jealously guarded area of local control reaching back into pre-Revolutionary America and school districts are indeed *single-purpose* units of local government. They are governed by locally elected boards that levy taxes and they hire and fire administrators, teachers, and staff.]

All these public-sector workers are not necessarily in organizations with similar functions, goals, values, relationships with the surrounding society, and technical/professional employee backgrounds. On the contrary, the public sector is amazingly diverse in functions and professional specializations. A typical local government, for example the city government in a city of 50,000 people, may have employees with education and experience in areas such as (to list only a few): public works, including specializations in sewer, water, and street system design and maintenance; parks, including specializations in park acquisition, design, construction and maintenance, recreation programs, and programs for senior citizens; police services, including crime investigation, traffic patrol, communications and dispatch, community relations, and crime laboratory services; economic development, including public information, financial incentives for new business, and construction of industrial development areas; financial services such as accounting, computer systems, utility billing, and debt management; land-use planning, including long-range planning, development plan and subdivision design, review of architectural designs and proposals for modification of historic buildings, and zoning review of setbacks, signs, and variances from standards; fire suppression and inspection of buildings for fire safety; review of building plans for proposed structures and inspection of materials and practices as they are built; human resources, including recruitment, classification, health insurance and promotion, and labor contract negotiation and management; legal services; libraries; and so on.

County, special district (taxing authorities providing one or a few services, such as a school district or drainage district), and state agencies may also serve this typical city and its surroundings in functional areas such as: prisons and probation, child and adult safety and welfare, school systems, hospitals, transportation planning and provision of roads and mass transit, air and water quality, sales tax administration, property taxation, and so on. Overlaid on these governmental structures and systems are a range of national government services related to the aged and disabled, veterans, safety of food, air, water, vehicles, toys, workplaces, and he like; protection of wildlife and the natural environment; and other functions including those that are traditionally and uniquely national: defense, foreign relations, and regulation of interstate and international trade.

To make the overall picture a little more complicated, though the lists above allocate functions neatly to local, state, and national agencies, many services can be provided by different levels of government or by more than one level in the same geographic area. Examples include water and sewer service, various aspects of law enforcement, corrections and probation, transportation, and programs for seniors. Further, much public-sector work is actually carried out by people employed in nonprofit and private organizations that contract with public agencies. This intersectoral complexity has led to the blurring of boundaries noted by many authors in the public administration literature, some of whom believe there is, as a result, not much difference today between the public sector and other sectors.

It seems almost hopeless to search for common ideas and practices that stretch across this complexity of history, purposes, and perspectives on society and its problems, but this is what the field of public administration attempts to do. The common thread is service to the public, to all the people in a defined place, and the responsibility for sensitivity to societal values that comes with it.

## REFERENCES

Cooper, Terry L. 1998. *The Responsible Administrator: An Approach to Ethics for the Administrative Role*. San Francisco: Jossey-Bass.

Denhardt, Robert B., and Joseph W. Grubbs. 1999. *Public Administration: An Action Orientation*. Fort Worth, TX: Harcourt Brace College Publishers.

U.S. Census Bureau. 1997. *Governments Integrated Directory:* www.census.gov/govs/www/gid.html (April 13, 1999).

# CHAPTER 2

# TIME AND CHANGE

## The Environment of Public Administration

Most of us think of our circumstances in current terms first, that is, by concentrating on things that press on us at this moment instead of thinking about the past and factors that may have brought us to the present situation. The present, though, is composed of more than recent events and issues that concern us now. It is a complex web of relationships built on shared understandings of reality that have been constructed by many people over long periods of time. An example from organizational practice would be the conduct of negotiations between an employer and employees about conditions of employment such as wages and benefits. These negotiations can be complex to the point of being confusing, and the people involved are likely to be caught up in the many details of working with bargaining team members to establish positions, discuss disagreements with those on the other side of the negotiations, and so on.

At some point, questions may arise about the best positions to take that do not have answers within the current framework of issues on the table. To find solid answers, participants may need knowledge of the practices of other organizations, the development of labor law over an extended period of time, the history of the organizations involved, and the opinions or policies of board members or elected representatives responsible for approving a settlement. In addition, it would be helpful for bargaining representatives to be aware of the context of labor relations in the United States. In the nineteenth century and into the twentieth century, workers struggled with corporations and the government for the right to organize and bargain for wages and benefits, and they advocated laws regulating working conditions. The courts and police were used to stop labor actions and break up workers' gatherings, sometimes violently. Today's labor laws reflect the current balance of legal rights between labor and management that developed from these events.

Public-sector bargaining has a short history. The sort of labor relations and bargaining practiced in the private sector for much of the twentieth century did not begin in earnest in the public sector until the 1960s. In part this was because of the earlier belief, still held by some, that public-sector employees work for all the people, the people are sovereign and can create whatever sort of government they want, and their employees should not have the right to bargain for conditions of employment.

The balance between labor and management in the private and public sectors, in national, state, or local government, and in any particular organization, is not settled, static, or value-free. Every bargaining situation, legislative debate about public policy in relation to labor, and arbitrator's decision or court case resolving a labor dispute is another entry in the on-going story of the labor-management relationship in American society, a society in which the status of labor in relation to management is not assured or secure. This summary of the American labor-management story is just that, a summary, brief and incomplete, and to become knowledgeable in this area would require a significant investment of time. The options open to bargainers in each situation are in part determined by this story and the bargainer who understands there may be an advantage in framing proposals that will be useful for all parties, including mediators, arbitrators, and judges.

## PREDISPOSITIONS AND FREEDOM OF THOUGHT

The bargaining example presents just a quick sketch of the application of knowledge about history and society to one area of organizational practice. A list of all such applications in the public sector would include many thousands of entries. Each one is a story, a narrative, about how people have constructed the practices we are familiar with today. These socially constructed narratives reflect a degree of current common understanding, understanding that includes a range of opinions, ideas, and disagreement, and the potential for constant change as new situations, information, and interpretations develop. As Robert Bellah and colleagues noted in the book *Habits of the Heart*:

> Narrowly professional social science has given us valuable information about many aspects of contemporary society, but it often does so with little or no sense of history. . . . Yet what we need from history, and why the social scientist must also, among other things, be a historian, is not merely comparable information about the past, but some idea of how we have gotten from the past to the present, in short, a narrative. Narrative is a primary and powerful way by which to know about a whole. In an important sense, what a society (or a person) is, is its history. (1996, 302)

We live in a society shaped by the experience of emigrating from Europe to find new ways of living with greater opportunity and freedom. This story is more complex today because of the success of the United States as a place that attracts people from all over the world. The concepts of citizenship, governing, and governmental structure that appealed to European immigrants over three hundred years ago do not always fit contemporary situations or cultures, and contemporary events are reshaping our earlier understandings. Even so, the story of the creation of governmental systems in the United States has a powerful influence on today's public-sector systems and practices and on public expectations about the work of public employees. When, for example, a practitioner believes program changes should be made and wants to suggest new policies or expenditures, the likely reception she or he will receive depends on the attitudes of organizational superiors or elected officials. These attitudes are shaped to some extent by enduring American debates about the role of government and its career employees in society.

Attitudes toward government and citizenship do not form out of nowhere, suddenly appearing in each individual as if by magic. Attitudes vary from person to person, but there is an identifiable core of American thought. Most people are only dimly aware of the origins of their attitudes and do not consciously make a connection between historical events and current beliefs and opinions. The relationship between citizens and government is a constant feature of our collective social environment. Discussions about levels of taxation, whether to provide social services such as health care through public programs or mandates, the extent to which the use of private property or money is controlled by government regulation, what role citizens can or should play in governing themselves, and thousands of other matters large and small revolve around the question of the appropriate role of the public sector in relation to private citizens. This was a central part of founding-era American debates about the new Constitutional form of government. It is central to public life today, as individual freedom and the needs of the collective community are in constant tension, subject to debate and shifting balances of compromise.

We cannot be sure whether or not history is developmental, that is, whether human society is moving progressively toward a "better" future, whatever that could mean. There are recurring patterns to human history and certain things, such as science and technology, are clearly becoming more complex and sophisticated. There is no way for us to know, however, whether we are moving toward some desirable condition (such as a peaceful, humane world), an unpleasant condition (disorder, institutional disintegration, war) or simply experiencing relatively random events. Though it is useful to know the past to understand the present, people have many choices before them today, choices that are not fixed or determined by the past.

History does not appear to be determinative, that is, it does not seem that current events must take one particular form because of what preceded them. This is the case even though we are the product of our personal experiences, our experiences have been influenced by the ideas of those around us, and our ideas are influenced by people of earlier generations. One only has to study or travel to areas where people have cultural beliefs different from our own to understand the depth and strength of the effect of a specific cultural/historical background. A person's beliefs can include, among other things, a sense of identification based upon national, regional, or local citizenship, a particular generation, or socioeconomic class. A person is also likely to have beliefs about the appropriate relationship of citizens to government and about how public organizations and institutions should function.

These beliefs can function as *predispositions*, affecting how we respond to questions about the roles and behavior of citizens, political leaders, and public employees. Despite the strength of these predispositions, people are able to make choices that express independence of thought and vary from patterns of the past. Awareness of predispositions allows people to build upon the past to create a different future, choosing from a broader range of options than they would have without this knowledge.

## RECURRING THEMES

An important assumption of this book is that people who understand their societal and historical context are better prepared to act in ways that fit the needs of the present. There are recurring themes in public administration theory that can in part be explained, even anticipated, by reflecting on underlying ideas in society. Often, earlier ideas are reintroduced in updated form to deal with current circumstances, and some of what is thought of as new today consists of repackaging of values and practices familiar some time ago. Recognizing repeating themes is one part of understanding the context of public administration.

An example is the current idea that government should be "run like a business." Americans have sought businesslike efficiency in government for some time, alongside competing values such as democratic governance and social equity. Values may conflict: a program to prepare poorly educated low-income people for good jobs may be inefficient on a cost-per-person basis, but it may serve the value of social equity, assisting those who do not do well in a market economy. The idea that government should be efficient has been a primary reason given for many "reforms," including creating the national government's civil service system in the late nineteenth century, the strong mayor, commission, and council-manager plans for local government organization in the early twentieth century, restructuring the office of the president in the 1930s and 1940s; and so on through actions being taken today at all levels of government, such as contracting out service production to nonprofit and private companies and putting greater emphasis on measuring government performance.

Almost no one thinks it is a good idea for government to be intentionally inefficient, but a narrow focus on efficiency can have both intended and unintended effects. The movement to reform local government, making it more efficient, is a case in point. One of the ideas proposed during this movement in the late nineteenth and early twentieth centuries was the council-manager plan. It is similar to the administrative system used in school districts and other special local and regional districts. The concept is a unitary chain of command instead of the separation-of-powers model used in the national government, state governments, and many local governments. In the council-manager plan, the city council is structurally equivalent to the board of directors of a business and the appointed professional city manager is responsible for much of the daily operation of the city, a position parallel to that of the chief administrative officer or general manager in a private business. This contrasts with separated legislative and executive powers (used in many local governments in variations on the mayor-council model), in which the branches (council and mayor in cities) operate somewhat independently with defined spheres of authority and the chief administrative officers are elected.

The council-manager plan, based on the structure of private firms, was created to bring efficiency to American cities (Stillman 1974). This plan was intended to shift the emphasis in cities from political responsiveness to cost-effective professional management. It was part of an effort by business and professional leaders to counter the influence of local *bosses* who created powerful political organizations, often supported by groups of recent immigrants to the United States. These political *machines* used governing techniques the professional/business class thought were corrupt and not in the best interests of all community residents. Thus, the value of efficiency became a tool in a conflict between people with different visions of what government is for—professional and business leaders favored what they saw as efficiency, and some politicians favored responding to the needs of specific groups and individuals in their communities. The original council-manager plan included electing each city council member by citywide vote instead of by districts, to make it harder for machine candidates to be elected from ethnic neighborhoods. Mayors were appointed by councils from among their members instead of being separately elected for the position; this tended to reduce the mayor's strength as a leader.

There are some interesting themes in this story. Efficiency, of course—it was central to the people who proposed and promoted the council-manager plan and other reforms. They also believed that reforms could promote active citizenship and democracy (Schachter 1997), but the core idea was that government could be made efficient if organized, structured, in a certain way. The use of governmental structure in the service of specific values such as efficiency, political responsiveness, or protecting against or strengthening executive power appears and reappears in different forms throughout the history of the American public sector. This structural perspective, however, is something of an abstraction, a summary that overlays the human concerns and actions that it describes. At the level of human concerns and actions, the council-manager plan may be viewed as an effort by the upper-middle class to counter increasing influence of people of lower socioeconomic status, including recent immigrants. Elements of the plan had the effect of dampening open public debate on issues of importance, leading to successful efforts in the past two or three decades to reverse parts of the system. This is an instance of a value—democratic openness to the public—being reasserted to create a new balance in the relationship of government and citizens.

Calls for greater efficiency in government are not new. Sometimes they mean just what they say, government should be more efficient, and sometimes the underlying message is a conflict over power, money, and control. We are at a height of interest today in "running government like a business"; the origins of this recent trend toward efficiency as a dominant value stretch back several decades, but the impact has been felt in the United States most intensely in the last two decades of the twentieth century and into the early twenty-first century. This phenomenon will be discussed later in the book and it is not especially useful just now to label it "bad" or "good," but rather to observe that it has advantages and disadvantages as well as lessons to teach about American attitudes toward the relationship of government and the private sector.

The point is that history matters, that knowledge of past events and ideas helps public-service practitioners in a practical way to do a better job in the present. Beyond such direct usefulness, knowledge about the founding of the nation, ideas of citizenship and democracy, and development of specific practices are central to the role public employees play in conveying knowledge of the governmental sector to the public. There are many examples of the importance of knowing about the past for interpretation of the present and actively shaping the future.

## A PICTURE OF CHANGE

American public administration is set in a complex and fascinating environmental context of social, economic, and political systems. There are public employees who are not often affected by this context or trends and events in the broader society. They perform professional/technical tasks largely free from concerns about public opinion, adequacy of program funding, or their role in the process of creating public policy. But for many others, sound decision making depends on knowing about the nature of the environment and using that knowledge to improve their effectiveness.

At a general level, the times in which we live are sometimes called postmodern. The concept of postmodernism can be used in different ways, but for our purposes the key idea is the assertion that ideas about people and society that have been accepted as "real" for some time are coming apart, potentially leaving us "in a state of relativity reduced to normlessness and a conclusion that 'anything goes,' because no one has a basis for claims to moral rectitude and obligation" (Cooper 1998, 34). From the Age of Enlightenment in the eighteenth century to the Industrial Age of the twentieth century, modern thinking has attempted to apply science and reason to solve the mysteries of human nature, the natural world, and social life, including the structure and operation of institutions and organizations.

Today, despite the success of the modernist paradigm in creating systems and technologies that have transformed society, questions are raised about the validity of the quest for certainty and objective truth. Evidence given in support of the idea that many people question the certainties of the past includes:

- *Diminishing trust in science and government.* Only a few decades ago, people tended to think most problems could be dealt with through rational methods and technology, and there was faith that government was a good way to improve social conditions. People were optimistic that government could eliminate enduring problems such as poverty, environmental abuses, inadequate education, and joblessness, but the complexities of social and economic systems have made these goals difficult to achieve, or even to clearly define. Though we still think that technology can create useful physical tools and solve concrete, natural-science problems such as curing disease, many no longer trust it to solve social problems, and government is often viewed as the servant of special interests rather than a means of addressing issues of public concern. People at one time may have believed there is a single objective "Truth" to guide the way, but that belief is less common today.

- *Social fragmentation.* The old certainties of East versus West and communism versus capitalism have changed into ethnic and tribal conflict in the context of the dominance of global capitalism. Within the United States, people identify to a lesser extent with traditional party affiliations or loyalty to institutions and are more interested in the expanding variety of communities of interest available through technology and associations at the neighborhood and community level.

- *Vanishing norms.* Social fragmentation and greater choice bring with them uncertainty about the values that underlie morality and decisions about ethical conduct. A disturbing sense of relativism seems to be common, fed by images in popular culture of people whose values would have seemed outrageous and unacceptable in earlier times. There is less agreement on the nature of the family, what sort of education is best, which occupations are most desirable, the role of science and technology in human life, and so on. People with seemingly extreme views react to what they perceive as the threat of disintegration of moral control with proposals to impose their view of morality on others. The part of this agenda affecting the public sector involves limiting and prescribing actions of public-service practitioners such as school administrators and teachers, child-care workers, librarians, medical researchers, and so on.

- *Skepticism about authoritative answers.* Public skepticism toward unambiguous, single-answer explanations includes portrayals of history, institutions, and events (Rosenau 1992, chap. 4). The creation of the Constitution is a good example. It has become more common to encounter questions about these events, such as the morality of failing to eliminate slavery and grant citizenship status to women, and whether those favoring adoption of the new Constitution did so from a desire to promote individual liberty or to limit public access to a government run by a wealthy elite. Another example is debate about the role of women, since the Progressive Era in the late nineteenth and early twentieth centuries, in making the public sector more democratic and responsive to the needs of lower-income and disadvantaged people. Have women advocated a different view of the relationship of the public sector to the plight of individuals than have men (Frazer and Lacey 1993; Stivers 1993, 2000)? If we no longer believe there are clear and effective scientific answers to human problems, we are likely to question people with authority and expertise. For public administration, the consequence is dealing with a disillusioned and disinterested citizenry; some of those who remain involved examine and criticize administrative actions in detail.

- *Small and local is preferred to large and distant.* Given social fragmentation and skepticism about scientific solutions and those in authority, people naturally feel most comfortable with what is closest to them. This can mean paying less attention to the barrage of media coverage of national and international events, instead focusing on private, daily challenges associated with work and family. For some it can mean becoming more involved in local associations or community affairs.

- *The paradox of globalization.* Despite the preference for things that are closer and thus easier to understand and control, people worldwide are subject to the impacts of economic globalization. Globalization is the spread of the corporate businesses of developed Western countries across national boundaries, sometimes affecting local cultures in ways that people resist. It leads many to abandon traditional values and beliefs of their societies, in favor of a lifestyle of consumerism (Barber 1996).

This is quite a picture of change and uncertainty, of localized values, commitments, and action, including a shift from interest in "public things" to private interests and pursuits, and pressure to join a global economy. To many of us, it seems that culture, politics, and technology are radically different today from the way they were in the recent past and that the rate of change is accelerating. On the other hand, change, uncertainty, and questioning of commonly accepted values can be found at many points in history, along with periods of calm and stability.

## HISTORICAL PECULIARITY

In some ways the past does seem peculiar, not simply how the present would be if there were fewer people and less technology, but really different. A brief look at America during the founding era of the 1780s highlights some of the differences between that era and the present. At the time of the debate over the proposed United States Constitution, the nation was largely rural, most people were farmers, and government was mostly an insignificant and distant part of everyday life. Frederick Quinn (1993, 3) offers this description:

> The long coastline was fair prey for foreign invaders. Roads were few, muddy when it rained, dusty otherwise. Transportation was slow and irregular, most dependable by water. The potentially prosperous, primarily agrarian economy was stagnant, owing to the recent eight-year war, and entrepreneurial people were not sure how it would improve. Scattered insurrections flared, and the prospect of angry mobs or unschooled peasants taking the law into their hands threatened whatever form of government the newly independent states selected. The central government was powerless, lacking authority to raise funds or an army, or to administer justice. Politicians debated at length whether the existing government should be patched up, or if there should be a strong president, a president and council with shared powers, or a legislature with most power vested in it; but the discussions went nowhere.
>
> The confederation's thirteen isolated states were in infrequent contact with one another, except for commerce along the main maritime arteries. Spanish, French, English, and other metallic coins still circulated long after the war; the Continental Congress's money was valueless. "Not worth a continental" was a popular expression. The wartime military leader, George Washington, wrote state governors in 1783 that he feared "the union cannot be of long duration, and everything must very rapidly tend to anarchy and confusion." Thomas Jefferson, then Minister to France, said, "We are the lowest and most obscure of the whole diplomatic tribe." A British cleric said Americans were "a disunited people till the end of

time, suspicious and distrustful to each other, they will be divided and subdivided into little commonwealths, or principalities."

When the Constitution was ratified and the new national government was formed in 1789, the thirteen states had a total population of approximately 4 million (roughly as many as today live in, for example, Kentucky, or South Carolina, or Colorado), spread out along the Atlantic coast. The size of all the land in the new United States was equivalent to the combined land area of France, Italy, Spain, Germany, Britain, and Ireland (Rossiter 1966, 24). Most people lived outside the larger cities, in "towns and settlements but just as many in isolation. Only twenty-four places had more than 2,500 inhabitants, only five cities had more than 10,000: Philadelphia (45,000), New York (33,000), Boston (18,000), Charleston (16,000), and Baltimore (13,500)" (Rossiter 1966, 25). This contrasted with the populations of European cities at the time, such as Paris (600,000 in the late eighteenth century, about 9 million today) or London (950,000 in the late eighteenth century, about 7 million today). Communication between Europe and America took weeks by boat and transportation in America was difficult, since "roads were bad, bridges few, ferries leaky, rivers whimsical, stagecoaches cranky, and inns ill-kept" (Rossiter 1966, 25). Incidents and delays due to these conditions could happen to anyone, as described by Leonard White:

> On October 25, 1794, the President of the United States [George Washington] was returning from Bedford, Pennsylvania, to the seat of government. While crossing the Susquehanna River, his coach became lodged between two boulders in midstream, and there he was forced to sit in the rain until it could be extricated. (1956, 1)

Several years earlier Washington had assumed office as the first president, in 1789, and the government consisted of: "a foreign office with John Jay and a couple of clerks to deal with correspondence from [ambassadors] John Adams in London and Thomas Jefferson in Paris . . . a Treasury Board with an empty treasury . . . a 'Secretary at War' with an authorized army of 840 men . . . a dozen clerks whose pay was in arrears, and an unknown but fearful burden of debt, almost no revenue, and a prostrate credit" (White 1956, 1).

To this picture of physical and governmental differences from contemporary society, we could add social and cultural differences. There was no general education in that era and only a relatively small number of people were familiar with the writings of European authors on politics and government. A number of leading figures in the Revolution and the new government, such as George Washington and Thomas Jefferson, were slaveholders. Active participation in the public affairs of the day, and citizenship rights such as voting, were limited to white male property owners. Attitudes toward citizenship and the relationship between classes of people were very different from those of today.

Despite these dramatic differences between the founding period and today, those in the founding generation with the knowledge and desire to do so were active in discussing matters that seem quite contemporary, such as keeping government from interfering in the lives of individuals and at the same time making it effective. It was their ideas of liberty and a government that would preserve order and simultaneously allow some sense of self-governance that we live with today, as citizens and as public service practitioners. Society, government, and interpretations of the Constitution have all changed significantly since the time of the nation's founding. Thomas

Jefferson (thought of as a leading theorist of American democracy), Alexander Hamilton (credited with laying the foundation for contemporary administrative systems), and James Madison (leading political theorist and guiding hand in creating the Constitution) would probably be surprised at what their creation has become.

## HISTORICAL SIMILARITY

Though these founders might be surprised, they likely would find basic concerns of today to be similar to those of their own time, as contemporary Americans seek a balance between collective action through government and allowing as much space as possible for people to act outside the public sphere. Also, they might tell us that change and apparent disintegration of certainties in society are not unique to our time. An overarching theme of American thought in the formative period of the eighteenth century was the Age of Enlightenment desire to shed old constraints imposed by monarchical governments and religious beliefs, turning instead toward individual responsibility for reason and choice. One result was political thought that broke with tradition and created an experiment in governance that, though uncertain and risky at the time, has proven durable for many generations.

Through the nineteenth century, Americans worked to make their governments more democratic and they expanded the scope of government action. They also experimented with a variety of legal, institutional, and structural forms as they pursued democracy, self-interest, and efficiency. These things were often in conflict, as evidenced during the nineteenth century in heated political debate, violent confrontations in cities, war with England, and war between sections of the country. Another aspect of radical social transition was the phenomenon of people leaving home and safety to migrate in long, dangerous journeys westward to places about which they knew little, creating new communities literally from the ground up.

By the latter part of the nineteenth century, America had entered the modern era and the Industrial Revolution, which brought two changes, among others, on a mass scale that would permanently change society. One was the movement of populations from rural to urban areas in search of jobs, a movement that brought with it transformation in the lives of individuals and families. The other was the shift in work, from individual responsibility for a workshop or farm, to large-scale employment in factories and huge organizations (Rodgers 1974). These changes caused considerable human suffering and despair, as the conditions of the working class were often terrible, including long work hours at repetitive and boring jobs, child labor, hazardous materials and processes in the workplace, and ruthless business owners (Zinn 1995, chap. 13). But there was an optimistic side to the time, a belief that history was moving forward to something better. Historian Samuel Eliot Morison wrote of the turn of the twentieth century:

> Americans like myself who were so fortunate as to be born in the late nineteenth century and brought up in the early twentieth, often look upon the years prior to 1914 as a golden age of the Republic. In part, this feeling was due to our youth; in part to the fact that the great middle class could command goods and services that are now beyond their reach. But there was also a euphoria in the air, peace among the nations, and a feeling that justice and prosperity for all was attainable through good will and progressive legislation. (1965, 841)

This feeling of peace and prosperity could be attributed to the 1950s as well (though overshadowed somewhat by the Cold War). However, World War I, the Great Depression of the 1930s, and the periods directly before and after the 1950s (World War II in the 1940s and the countercultural revolution of the 1960s), were times of tremendous upheaval in personal and national life and the slaughter of hundreds of thousands of people in world wars.

Consider a writer whose perspective on social phenomena we might think of as characteristic of today's postmodern era. About people who wish for what they think of as the beauty of returning to an earlier time of certainty and order in life, he writes (the author's name and year are held until the reader absorbs the quotation):

> I don't see how this dream can succeed. Their solution is built on a wild impossibility, for in order to realize it they will have to abolish machinery and communication, newspapers and popular books. They will have to call upon some fairy to wipe out the memory of the last hundred years, and they will have to find a magician who can conjure up a church and a monarchy that men will obey. They can't do any of these things, though they can bewail the fact and display their grief by unremitting hostility to the modern world.
>
> But though their remedy is, I believe, altogether academic, their diagnosis does locate the spiritual problem. We have lost authority. We are "emancipated" from an ordered world. We drift.
>
> The loss of something outside ourselves which we can obey is a revolutionary break with our habits. Never before have we had to rely so completely upon ourselves. No guardian to think for us, no precedent to follow without question, no lawmaker above, only ordinary men set to deal with heart-breaking perplexity. All weakness comes to the surface. We are homeless in a jungle of machines and untamed powers that haunt and lure the imagination. (111–112)

The use of the male gender dates the piece to sometime before, maybe, the 1970s. However, this quotation is much older than that. It is taken from the book *Drift and Mastery*, by political writer Walter Lippmann, originally published in 1914 (reprinted in 1985). The point is that a feeling of drift, of being cut loose from the certainties of the past, is not unique to the beginning of the twenty-first century. We could no doubt find a similar sense of the crumbling of moral and social certainties if we examined the thoughts of Europeans during the Black Plague in medieval times, people emigrating from Europe to the American colonies in the seventeenth and eighteenth centuries, Jews and other minorities in Europe in the 1930s and 1940s, Western pioneers in the United States in the nineteenth century, Americans during the 1960s who believed the counterculture and Vietnam War protests could mean the end of civilization as we know it, and so on. The specific uncertainties experienced by these people are not our uncertainties, but clearly many people have faced the breakdown of values and beliefs in ways that rival or exceed anything we know today. The important thing for our consideration of the societal environment of contemporary public administration is that these are indeed trying and uncertain times, but we in the present are not unique or alone in having such experiences.

## CHANGE AND ECONOMICS

This longer view of history is one way to add some perspective to our thoughts about the current, supposedly postmodern era. There is another factor in the contemporary societal environment that runs counter to the postmodern view of fragmentation and loss of certainty. This is the globalization of economics and the spread of economistic thinking in the social sciences, including public administration.

The United States is a *liberal-capitalist* society. The term liberal is not used here in the common twentieth-century sense of favoring large-scale income redistribution through social welfare programs, but rather in the sense in which it was used in the Age of Enlightenment in the eighteenth century, into the nineteenth century. This use of liberal carries a concern for individual freedom from authority and has become associated with economic freedom to make profit and invest the capital in new or expanded business ventures (*capitalism*). Worldwide, existing economic and cultural systems have been affected by globalization of the products and techniques of Western capitalist societies, in some cases allowing improvements in living conditions and allowing greater freedom in life choices, in some cases causing environmental degradation or loss of cultural identity.

In the United States, public administration exists within the context of societal values in which individual freedom, including the freedom to make economic choices, is part of the cultural norm of emphasis on the individual. There has always been a countertrend in American culture toward limitation of extreme differences in income or ostentatious displays of wealth. In colonial and founding times, this trend toward equality of position was present in resistance to European-like titles of nobility and in the informality of relationships between people in different economic circumstances. Still, there was a wealthier economic class in the population and distinctions were made between people on this basis. With the coming of industrialization and the increasing urbanization of the nation in the later part of the nineteenth century, differences in wealth became even larger and more apparent. People with wealth often wielded political power as well, and the laws and institutions of the nation tended to protect this group against governmental action and demands from labor.

Today, American society is large and *pluralistic* (characterized by many groups with different interests), but there are also significant differences in wealth greater than those in many other fully developed nations, with the richest 1 percent of households owning almost 40 percent of the wealth, compared to 18 percent in Britain, for example (Bradsher 1995, A1, C4). In addition, the dominant American attitude toward the relationship of people to each other and to the broader community of citizens is one of negative rights, that is, of protecting the individual from interference by others. In this environment, government is regarded primarily as a means of obtaining individual rights and protections and only secondarily as a forum for citizens to debate and enact measures that serve some broader public good. The public sector is valued to the extent it provides an orderly structure of regulation that allows business to function smoothly (this is called a *mixed economy*, in essence a balance, or tension, between the private and public sectors), but there is often serious resistance to attempts to expand the role of the public sector into substantive questions of human or environmental well-being (McCollough 1991, chap. 4).

Whatever our perceptions of the nature of contemporary society, we may be able to agree this is a time of rapid and significant change, coupled with citizen demands for efficient and effective public services and voice in governance. In addition, fewer and fewer people today are able to count on lifelong job certainty and the clear career path their parents could expect several decades ago. The rate and depth of change in the environment of organizations shapes the way they use human resources. Many workers today must cultivate marketable skills rather than a sense of loyalty to a particular organization. They are expected to "behave nimbly, to be open to change on short notice, to take risks continually, to become ever less dependent on regulations and formal procedures" (Sennett 1998, 9).

Richard Sennett (1998) calls this phenomenon "flexible capitalism." He argues that workers in this environment develop a sense of confusion and anxiety and have trouble forming stable values or ideals. This sense of uncertainty has not yet affected the public sector to the extent it has the private and nonprofit sectors. In many parts of the private and nonprofit sectors, it has become the norm for managers to constantly scan their environment, seeking to adapt to a changing clientele and the initiatives of competitors. Marketing, image, and pleasing the customer are crucial. Though this model of turbulence and sensitivity to the environment has not yet affected the public sector as intensely as the private sector, it is nevertheless partially evident in many organizations and some have moved close to the private-sector model. In addition, because of the thrust to "run government like a business," as evidenced in trends such as contracting out, privatization, downsizing, public-private partnerships, and so on, we may expect continued pressure for public agencies to function more like private firms.

The expansion of economistic thinking in the public sector, citizen demand for greater accountability from public employees, and citizen desire for greater control over the public-policy process all generate pressures to change the way we view the professional role in public service. Bureaucratic control of public agencies is under attack from three directions: those who want public professionals to be more *entrepreneurial* in their methods, operating to maximize economic efficiency as do private-sector managers and staying away from policy making (this is the economistic, or market, model of the public sector); those who want public professionals to have more power and freedom to act, that is, greater *legitimacy* in the Constitutional system and *administrative discretion*, the ability to act independently; and those who would like public professionals to assist citizens to self-govern, through *discourse* or *facilitation*, processes of discussion and deliberation in which citizens decide policy matters for themselves. With challenges to public professionalism come challenges to the institutional structures that support it, such as civil service systems and the council-manager plan of local government organization.

## REFERENCES

Barber, Benjamin R. 1996. *Jihad vs. McWorld*. New York: Ballantine Books.

Bellah, Robert N., Richard Madsen, William M. Sullivan, Ann Swidler, and Steven M. Tipton. 1996. *Habits of the Heart: Individualism and Commitment in American Life*. Berkeley: University of California Press.

Bradsher, Keith. 1995. "Gap in Wealth Called Widest in West." *New York Times*, April 17, A1, C4.

Cooper, Terry L. 1998. *The Responsible Administrator: An Approach to Ethics for the Administrative Role*. San Francisco: Jossey-Bass.

Frazer, Elizabeth, and Nicola Lacey. 1993. *The Politics of Community: A Feminist Critique of the Liberal-Communitarian Debate*. Toronto: University of Toronto Press.

Lippmann, Walter. [1914] 1985. *Drift and Mastery: An Attempt to Diagnose the Current Unrest*. Madison: University of Wisconsin Press.

McCollough, Thomas E. 1991. *The Moral Imagination and Public Life: Raising the Ethical Question*. Chatham, NJ: Chatham House.

Morison, Samuel Eliot. 1965. *The Oxford History of the American People*. New York: Oxford University Press.

Quinn, Frederick, ed. 1993. *The Federalist Papers Reader*. Washington, DC: Seven Locks Press.

Rodgers, Daniel T. 1974. *The Work Ethic in Industrial America, 1850–1920*. Chicago: University of Chicago Press.

Rosenau, Pauline Marie. 1992. *Post-Modernism and the Social Sciences: Insights, Inroads, and Intrusions*. Princeton, NJ: Princeton University Press.

Rossiter, Clinton. 1966. *1787: The Grand Convention*. New York: W.W. Norton.

Schachter, Hindy Lauer. 1997. *Reinventing Government or Reinventing Ourselves: The Role of Citizen Owners in Making a Better Government*. Albany: State University of New York Press.

Sennett, Richard. 1998. *The Corrosion of Character: The Personal Consequences of Work in the New Capitalism*. New York: W.W. Norton.

Stillman, Richard J., II. 1974. *The Rise of the City Manager: A Public Professional in Local Government*. Albuquerque: University of New Mexico Press.

Stivers, Camilla. 1993. *Gender Images in Public Administration: Legitimacy and the Administrative State*. Newbury Park, CA: Sage.

———. 2000. *Bureau Men, Settlement Women: Constructing Public Administration in the Progressive Era*. Lawrence: University Press of Kansas.

White, Leonard D. 1956. *The Federalists: A Study in Administrative History*. New York: Macmillan.

Zinn, Howard. 1995. *A People's History of the United States, 1492–Present*. New York: Harper Perennial.

# CHAPTER 3

# DEMOCRACY, CITIZENSHIP, AND GOVERNMENTAL STRUCTURE

There has been lively discussion in America for more than 200 years about the relationship of people to their governments—national, state, and local. This involves questions about the freedom of people to act without interference from government, the duties of citizens, whether people are able to govern themselves, and tension between economic classes in society. Today we often think of these matters in the context of *democracy*, but during the founding era of the American government in the late 1700s, democracy was often thought to mean irrational mob rule. Instead of democracy, people in that era discussed *liberty*, *republicanism*, and *civic virtue* (Lakoff 1996, 26). Though they used words other than democracy, people during the founding era were, as we are today, interested in the extent to which citizens could determine for themselves the nature of the relationship between the individual and the larger society.

Today, we include many ideas and concerns within the broad definition of democracy. In situations where it is not explicitly the focus of attention, on closer examination it is found behind discussion of other issues. When Americans complain about high taxes, they may in part be saying they want more influence over how government uses their money. When they express concern about the size of government, in part they may mean the public sector seems to be moving into so many areas of the lives of individuals that they feel squeezed, constrained. When they talk about unresponsive bureaucrats, government "experts" who do not listen to citizens, or nonsensical regulations, they are worrying that somehow democracy, whatever the word means to each individual, is being damaged.

The concept of democracy is not well defined in common usage, but it expresses a wish for individuals to be heard in, and to participate in decision making about, public matters if they choose to do so. It often carries with it meanings that might better be termed *freedom*—the unrestricted ability to think and act—and *liberty*—the greatest possible freedom consistent with respecting the rights of others and doing one's duty toward the community. (A speaker in 1773 described liberty as "the happiness of living under laws of our own making" [in Wood 1969, 24].) Overall, democracy is often thought of as "the quest for autonomy" (Lakoff 1996).

A distinction may be drawn between *procedural democracy*, in which people have equal rights to participate in the economic and public life of society (this is the question of *equality of opportunity*), and *substantive democracy*, in which society considers to what extent it will allow inequalities of wealth, power, and privilege (this is the question of *equality of outcome, situation, or circumstances*). Democracy that involves elected officials making decisions for the people is called *representative democracy*, and a situation in which citizens personally participate in decision making is *direct democracy*.

## DEMOCRATIC CITIZENSHIP

As society moves away from authoritarian rule by one person or a small group of people, there are a number of alternatives available in the relationship of the individual to society. The concept of *citizenship* is central to understanding democracy because it provides perspectives on relationships between citizens, elected officials, and public employees.

Of the interesting aspects of democratic citizenship, we may highlight three that are useful here. One is whether the emphasis of a particular model of citizenship is placed on the community as a collective social group or on the individual. If the former, the interests of society are favored over those of the individual and citizens think of themselves as contributing parts of the whole, acting for the benefit of all. If the latter, there is much individual independence from the collective group, the interests of the individual are favored, and citizens think of themselves as autonomous, protected from the influence of society and free to act for their own benefit. Another aspect of citizenship is the level of commitment to procedural equality, that is, on giving each citizen full and equal access to the rights and responsibilities that accompany participating in public life. The third useful aspect of citizenship is the extent of commitment to substantive equality, that is, on ensuring that each citizen has adequate housing, food, education, career opportunities, health care, and so on, and that differences in wealth and power between people are not so great there are separate classes of people far removed from one another and inclined to treat some better than others.

The *classical republican* model of citizenship favors society over the individual. The term *republican* is used not in reference to a political party, but to a form of governing in which citizens believe that "ultimate authority is rooted in the community at large" and virtuous citizens demonstrate love of country, readiness to serve, preference for the public good over personal advantage, and belief that citizens achieve "moral fulfillment by participating in a self-governing republic" (Phillips 1993, 24). This model of citizenship has the virtue of fostering a sense of community and selfless service to a greater good. It has the disadvantage of emphasizing conformance to a particular sort of community and tends to exclude from full citizenship people who have backgrounds or beliefs different from those who are defined as citizens. In part, this exclusion comes from concern that broad, direct participation in public affairs may mean mob rule or the tyranny of the majority.

The *classical liberal* model of citizenship favors individual autonomy. As noted in chapter 1, the term *liberal* is used here in the way it was meant from the Age of Enlightenment in the eighteenth century until the mid-twentieth century, that is, freedom of the individual from the dictates of church and state, allowing each person to decide for themselves their beliefs and interests. (Note that use of the word *classical* does not necessarily mean a particular practice has disappeared, but is intended to avoid confusion with other, more recent uses of terms.) The strength of the model is the clarity of its vision of the individual as the center of thought and action, countering claims of institutions, the powerful, and tradition to restrict personal liberty. It has the disadvantage of potential concentrations of wealth and power that restrict public access to procedural and substantive democracy. The assumption that the public sector should stay out of private matters (as defined by those with the greatest stake in the status quo) means that many citizens must deal with powerful people and corporations on their own. Concepts of the classical liberal model were expressed by Thomas Jefferson in the Declaration of Independence, as he wrote of the individual's right to "life, liberty, and the pursuit of happiness." The founding generation read European authors on matters of history, philosophy, and politics. Jefferson's phrase was borrowed from English philosopher John Locke (1632–1704), who believed that each person was born with "natural rights," including life, liberty, and property.

The classical liberal and classical republican models of citizenship are the two most commonly discussed in the literature of the American public sphere and the relationship of citizens to government. However, there are likely several other models that could be useful for understanding the full range of options open in crafting democratic citizenship. One in particular we may call the *radical democratic* model of citizenship. It envisions an egalitarian society lacking great differences in wealth and power between members of society. Citizens of all backgrounds could take part on equal footing in the democratic process, with reasonable hope that public dialogue on issues of importance would yield a society that curbs excess wealth, redistributing some part of it to ordinary citizens in the form of improved education, health care, and other social benefits.

This model of citizenship and democracy may be found in American historical events such as the labor, civil rights, and other social movements. The nation moved closer to it during the Great Depression of the 1930s, and the model is illustrated later in this chapter by the English example of the Diggers. One author, Richard K. Matthews (1984), believes that radical democracy was Thomas Jefferson's vision for America's future. This model appeals to our sense of fairness and it captures the motivation behind events we now think of as having improved society. However, to imagine this as the primary model of citizenship is to imagine a country in which government exerts significant, constant control over private behavior. Many Americans would find this unacceptable, worrying that it would be an unfair intrusion on individual rights that would damage individual initiative.

## PRECURSORS OF AMERICAN DEMOCRACY

The city-state of Athens from the fifth to the third centuries B.C. is often referred to as the source of the classical republican model of citizenship. This is because a number of Athenian residents with rights as citizens (male property owners whose parents were also Athens natives), met in the policymaking Assembly more than forty times each year, and many would also be involved in public debates about philosophy, in dramatic presentations, and in legal contests. The purpose of life and sense of identity of Athenian citizens was closely tied to participation in the community, so that "The good life in the classical polis [the city as community] involved the citizen in discussion that was continuous, intense, and public" (Phillips 1993, 142). About 60 percent of citizens were farmers who worked small plots of land, or were "skilled craftsmen, shopkeepers, or casual laborers" (Lakoff 1996, 49). Some citizens performed public administrative duties, there were a few slaves who served as clerks, and some others who would be used on public works projects. This form of democracy is very different from public life today in the United States, in which citizens may follow the news about public events and vote but are otherwise not involved. A more active level of participation in public affairs can be found in some neighborhoods and local communities, but only for a few people does public life approach the intensity it held for the Athenian citizen.

During this time, Athens consisted of a central city and surrounding lands with a population of approximately 200,000 to 300,000 people (Phillips 1993, 142). Of these, about 40,000 were citizens, adult males with full rights to participate in public discussions, own land, and receive protection under the laws (Lakoff 1996, 49). Depending on estimates, possibly 60,000–150,000 people were slaves; some of these worked in farming and others in commercial enterprises, including mining, where conditions were poor (Lakoff 1996, 50). Women had no rights as citizens, could own little more than personal property such as jewelry, and could not buy or sell items worth more than a measure of barley sufficient to feed a family for a week (allowing them to serve household needs without being too independent and spending too much money). Girls would be married at about age fourteen to men of about thirty (young men were in military service), have children, and die relatively young, on average about nine years younger than men (Phillips 1993, 138–140).

The dilemma for those advocating a fully participatory, classical republican model of citizenship is posed clearly by ancient Athens. Athenian citizens vigorously debated the issues of the day but they were a relatively homogenous group with shared beliefs and political preferences. To achieve agreement on institutions and processes meant excluding from citizenship people unlike the male property-owning class, and this exclusion brought limitations on the individual freedom of the excluded. Some of the prominent philosophers of Athens were wary of an excess of democracy and preferred the seemingly more rational rule of well-educated, virtuous citizens.

The classical liberal model grew from the ideas of Enlightenment thinkers and the actions of people involved in movements to transform society, from monarchical governance to systems allowing citizens greater voice in governing. An example of people acting to transform society comes from the English civil war in the 1640s. There was a movement to enhance the power of Parliament against that of the king, involving many political leaders, armies, and armed clashes. Overall, the people in this movement were interested in "the limitation of executive power and the prevention of tyranny; they were therefore principally concerned with the redistribution of power among institutions, not among social classes" (Dow 1985, 25). This seemingly limited objective was quite radical at the time, since people had for centuries believed that kings were God's representatives on earth and had a divine right to rule. The movement to shift power from the monarch to Parliament reflected classical republican ideals: "whereas power originated in the popular element, in the actual administration of the state the 'aristocratic' element should prevail. In effect, classical republicans stressed that the people had delegated their power to the other elements in the state" (Dow 1985, 27).

Within this broader movement there was a group called the *Levellers* whose members believed "in the rights and liberties of the individual." They thought "the actions of government required the consent of the people, and that ultimate sovereignty resided in the people" (Dow 1985, 40). This is in essence the classical liberal position, claiming that governmental power ultimately comes from the people, with emphasis on the rights of the individual. Though there were differences of opinion within the movement, the Levellers advocated suffrage (citizenship rights) for most male adults, decentralization of much of the authority of government to the local level, accountability of elected officials to the citizenry, and full individual equality before the law. Some Levellers also favored social reforms such as creating schools and hospitals for the poor. There is evidence the Levellers were careful to say they were not proposing radical economic levelling, or redistribution of property (Dow 1985, 40–48), but it may be they assumed political equality (procedural democracy) would result in economic equality (substantive democracy) (Holorenshaw 1971, 85–86).

A radical fringe group of the Levellers called the *Diggers* believed the thrust toward democracy should begin with economic reforms. They openly advocated redistribution of property and elimination of much private ownership. They envisioned an agrarian society with common ownership of land and the buying and selling characteristic of commerce was to be replaced by exchange based on need (Dow 1985, 77–78). Small groups of Diggers briefly occupied open, unused lands in several areas, planting crops and hoping that many people would join them to create permanent communities, but few did and the Diggers were driven off.

In these narratives of the history of Athens and the English civil war, we can find the beginnings of contemporary conceptions of citizenship in democratic societies. This includes the classical republican model of service in the interests of the community by knowledgeable citizens of good reputation, the classical liberal model of individual rights and political equality, and the radical democratic model of economic equality. In different forms and contexts, these models also appear in the first two centuries of American history and remain present today.

## TOWARD A NEW AMERICAN GOVERNMENT

The people who emigrated from the Old World to the New sought *freedom from political and religious constraint*. They did not, in the weeks it took their boats to make the journey, miraculously shed the notions of social organization of their parent societies, including attitudes toward wealth, social status, power, and the relationship of the citizen to the community. It is sometimes assumed that people arrived in America and suddenly became completely free and unconstrained. Though parts of the new land offered considerable opportunities for freedom, in other places it would be more accurate to say that people used the large, unpopulated spaces as a way to create multiple communities that were in some ways different from one another, but quite homogeneous within. For example, New England communities in the seventeenth and eighteenth centuries often enforced rigid standards of conformance to religious principles. People who wished to become permanent residents of a town were subjected to oral examination to determine their fitness to join the community, and they were required to agree to obey the principles and leaders of the town. The social structure was hierarchical and included the assumption that "better" people would be at the top. In one Massachusetts town, residents

> held fast to the belief of their Puritan culture in the natural inequality of men. It was foreordained by God that some men should have both greater capabilities and virtues than others and should rise and prosper. It was equally fated that some men should be incompetents and sinners who would lag behind the rest. Nor was this without its social purpose, since obedience to men of high rank was the cement of an orderly society, while the needs of less fortunate souls kept men attentive to their duties of Christian charity. (Lockridge 1985, 11)

Relations of *class, wealth, and hierarchical power* were in evidence as they had been in the Old World. Immigrants often brought to local government the forms and practices of the English borough, which was governed by people of a certain class and status in society. As the eighteenth century progressed, planters in the South and merchants in the North increased their political and economic control of the colonies through "property qualifications for participation in government and representation disproportionate to their numbers" (Jensen 1940, 8). As a counterbalance to distinctions of social class and status, there was also widespread belief that differences in wealth and status should not be so great that some were wealthy far beyond their basic needs while many were impoverished.

Americans of the middle and late 1700s believed that large differences in wealth occurred in aging and overpopulated societies that relied on mass manufacturing to support large numbers of the dependent, working poor. Those who traveled to Europe brought back stories of the excesses, foolishness, and idleness of the rich, and in America it was upsetting to watch the behavior of obscure people elevated by officials of the English king to positions of authority over their social betters. The problem became so serious it was believed "the Crown actually seemed to be bent on changing the character of American society" (Wood 1969, 111). The spirit of the American Revolution had for a while put off fears that Americans would fall into the same habits of self-centered greed and misuse of power observed in other places and times, but by the 1780s leaders were fearful that it had not, in the words of Robert Livingston, "rendered them more worthy, by making them more virtuous, of the blessings of free government" (in Wood 1969, 424).

This fear of social corruption was associated with the hope that America could be forever different from the Old World, retaining a "republican" simplicity with virtuous citizens and a simple life close to the land. ["Corruption" in this sense meant threats to liberty, including for New Englanders loud, immoral, or blasphemous behavior or excessive exhibition of wealth, and for people throughout the American colonies, it included standing armies, priests and bishops, aristocrats, speculators and paper shufflers, and monopolists (McDonald 1985, 70–77).] This vision came from knowledge of earlier societies such as ancient Athens. It depended on continued availability of land, allowing the expanding population to move outward into new territory rather than concentrating in urban centers, and

> In its purest form, classical republicanism stipulated that republics had to be rather rude, simple, pre-commercial societies free from any taint of luxury and corruption. The essence of corruption was the encroachment of power on liberty, an insidious process most likely to occur in advanced, stratified societies where great wealth and inequality promoted avaricious behavior and dangerous dependencies among men. (McCoy 1980, 67)

The problem for this view of a peaceful America of landowning farmers was that by the later eighteenth century, the American economy was based in part on foreign trade. In 1785, George Washington wondered of trade with other nations, "whether the luxury, effeminacy, and corruptions which are introduced along with it; are counterbalanced by the convenience and wealth which it brings with it" (McCoy 1980, 102). Disturbing signs had appeared that America might be moving in the direction of Europe, including "unemployment, swelling poor relief rolls, an upsurge of crime and immorality . . . " (McCoy 1980, 105). To one writer of the time, David Ramsay, "Venality, Servility and Prostitution, eat and spread like a cancer." Ramsay noted that "we have sought to check the growth of luxury" by passing laws against displays of wealth, but that English officials governing the colonies had prevented them from doing so. Ramsay believed that if Americans had not revolted in the mid-1770s and freed themselves from Britain by the beginning of the 1780s, "our frugality, industry, and simplicity of manners, would have been lost in an imitation of British extravagance, idleness, and false refinements" (in Wood 1969, 110).

Wood (1969, 416) quotes an orator of 1783 as saying: "The great body of the people, smote by the charms and blandishments of a life of ease and pleasure, fall easy victims to its fascinations. . . ." Though this could be blamed on such things as the "increase of private and public credit and the paper money and debtor-favoring legislation stemming from it," according to Wood it was widely believed that these were only the symptoms, and the cause was that people were living "in a manner much more expensive and luxurious, than they have Ability to support," and they were "captivated" by "*an immoderate desire of high and expensive living.*" This description of the disintegrating moral character of citizens living beyond their means sounds much like contemporary concern about consumer credit debt and lack of personal savings.

Even so, a number of thinkers had concluded it made no sense to expect a nation to develop into a leading power on the basis of exporting farm yields and basic items made in small workshops, such as rough clothing or furniture. What was needed was to expand into larger-scale processing of agricultural products and production of other commodities, giving laboring people employment and reducing dependence on the advanced countries across the Atlantic. The question was how to achieve this while keeping intact a republican society based on some degree of social equality and a citizenry actively involved in public affairs, yet preserving an orderly society in which the poor did not threaten the property and position of those of higher social status.

In addition to pressure to shift the national economy to some extent from farming into manufacturing, there were concerns about the potential for democracy to spin out of control, with the mass of people confiscating the property of the wealthy and redistributing it in small parcels in a "levelling spirit" (McDonald 1985, 92). During the Revolutionary period, the property of many large landowners loyal to Britain had been confiscated and the sale of that land was still taking place when the Constitutional Convention was held in 1787. This confiscation and redistribution involved about one-tenth of the value of all developed real estate in America (McDonald 1985, 90–92). It was unlikely that land confiscation would be used against loyal American landholders, because "the vast majority of American families held a comfortable amount of land, and poverty of the depth that was common in Europe was all but unknown . . . " (McDonald 1985, 93). Nevertheless, there was a lingering fear this could happen if the people as a whole gained power over political and judicial systems.

Another indicator of potential instability in society was *citizen unrest and rebellion* against governing authorities in the colonies under British rule up to the mid-1770s, during the Revolutionary War, and in the states following the end of the war in 1781. This unrest often involved economic issues that divided the wealthy from the ordinary citizen in hard economic times, and it could become violent. Merrill Jensen (1940, 11) observes that the urge to seek freedom through rebellion was, among the urban masses and rural population, directed against the local wealthy and powerful as much as Great Britain. Clinton Rossiter (1966, 44) writes that incidents such as "the descent of an armed mob on the New Hampshire legislature in Exeter in September, 1786 . . .," and Shays' Rebellion (discussed later) showed that some people "were full of the animosity of distressed men toward their 'betters.' Beneath the veneer of deference there had always been malice and envy in the attitude of the 'lower orders' toward the gentry in America, and the pressures of the 1780s brought such feelings frothing to the surface."

In the 1760s and 1770s, groups of farmers in Western parts of the Carolinas protested lack of government in the backcountry regions of South Carolina and bad government in North Carolina, calling themselves "Regulators." In North Carolina their grievances included "unequal taxation, extortion by centrally appointed judges and corrupt sheriffs, greedy lawyers, uncertainty of land titles, scarcity of hard money to pay taxes, refusal of the assembly to provide paper money or to allow taxes to be paid in produce, consequent tax levies 'by distress,' and government taking over poor men's farms" (Morison 1965, 195). They took actions such as breaking up the Superior Court at Hillsboro and beating and wrecking the houses of unpopular lawyers. In 1771, about 2,000 Regulators fought a battle with half that many militia and the Regulators were defeated. Fifteen who were taken prisoner were tried for treason and six were hanged. This brought an end to the Regulator movement, which was put down by people who would later be Patriots during the Revolutionary War (Morison 1965, 195–196).

Another example comes from Western Massachusetts in the 1770s and into the 1780s. Gordon Wood (1969, 284–285) describes this situation:

> Throughout the entire Revolutionary era the Massachusetts towns west of the Connecticut River were in a state of virtual rebellion from the governing authorities in the East. Popular uprisings and mob violence were continual and extraordinarily effective. The courts were closed in 1774 and did not open again in Hampshire County until 1778 and in Berskshire County until after the Massachusetts Constitution of 1780 was put into effect, and even through the eighties mob outbursts periodically forced the courts to suspend judication.

By the fall of 1786 economic conditions had become bad for farmers. There was too much produce in the market, trade was slow, and "court judgments for debts or overdue taxes could in most cases be enforced only by stripping a farmer of his real estate, his cattle, and his furniture. In Worcester County alone, 92 persons were imprisoned for debt in 1785" (Morison 1965, 302–303). In some other states, legislatures had enacted relief measures to lighten the debt burden for farmers. These measures included tax relief and even excusing debtors from obligations to private creditors in addition to government. These measures were regarded by some as fiscally irresponsible and the Massachusetts legislature would not pass them. Its upper house was controlled by coastal merchants who blocked relief legislation proposed by representatives of the Western areas in the lower house, leading to rumors "that the wealthy men of Boston and Salem were trying to get all the land into their hands and convert the free farmers of Massachusetts into a dependent peasantry" (Morison 1965, 303).

In September 1786, the governor of Massachusetts, in response to groups of farmers who disrupted courts in four counties to prevent action against debtors, "issued a proclamation against unlawful assemblies and called out the militia to disperse them" (Morison 1965, 303). In January 1787, a group of about 2,000 farmers gathered to protest, led by Daniel Shays, a veteran of the Revolutionary War and an impoverished farmer. They attacked a federal arsenal hoping to better arm themselves, but the militia had artillery and forced the rebels to flee. They were pursued and several days later some of them were arrested and those remaining were dispersed. Fourteen captured leaders were sentenced to death but then given short prison terms or released.

*Shays' Rebellion*, as it came to be called, had quite an impact on national affairs. Though it was only one of many such events, it occurred at a crucial time in the affairs of the states. Its size and threat were magnified out of proportion by the superintendent of war for the Confederation of states, who wrote in a letter to George Washington "that the Shaysites had between twelve and fifteen thousand disciplined men under arms and that they intended to march on Boston, loot the Bank of Massachusetts, recruit additional rebels in New Hampshire and Rhode Island, and then march southward with the intention of redistributing all property" (McDonald 1985, 177). The content of this letter appeared in newspapers and was circulated in private correspondence. The Shays incident gave an extra sense of urgency to strengthening the central government and averting a possible rebellion that would overturn wealth, property, and the social order. Today, democracy is a good normative concept, something desirable to be defended, even if its meaning is ambiguous. During the founding era democracy could be a threatening idea, suggesting mob rule and anarchy. It was in this environment that representatives from several states gathered in Philadelphia in the summer of 1787 to strengthen the national government.

## CONTEMPORARY PUBLIC SERVICE INSTITUTIONS

From the sketch above of the social, economic, and political environment during the time of the constitutional founding we can identify significant divisions in society and ongoing disagreements about the relationship of citizens to government. These divisions and disagreements are somewhat different today because of the passage of two centuries. However, what is remarkable is not so much differences resulting from the passage of time, but the fundamental similarities in founding-era and contemporary concerns about the purposes of government and the citizen-government relationship. People express their preferences about the size and scope of government in part by how they structure it using constitutions, laws and policies created by legislative bodies, and the organizational and operational features of administrative agencies.

In Part II of this volume, we find a debate over the draft Constitution between people who viewed the proposed new governmental structure as a threat to individual liberties and state independence, and people who believed the new system was needed to create a coherent, economically viable nation. Though we often think of the founding era of the nation as a time when people gathered together quietly to create our current form of government, it was instead a time of intense debate over the relationship of citizens to government. Many people saw the proposed Constitution as a mistake, a move toward national, centralized rule by an elite who would rob citizens of their newly won freedoms. These people believed that states should be the focus of the governmental system, with a weak national structure to take care of the few matters states could not handle effectively. This was the idea behind the Articles of Confederation, the governmental form adopted by the states during the Revolutionary War. Some of those who wrote and spoke of these concerns are known as *Anti-Federalists*.

Those who successfully advocated for ratification of the Constitution (the *Federalists*) believed that people are often self-interested and seek power and economic advantage, though they also have the capacity to be virtuous and act for the good of society. Because people can be self-interested, using government to benefit themselves, government should be designed to counter abuse of power by dividing and balancing it between different parts, making it difficult for anyone to gain great advantage over others. The Federalists viewed the mass of citizens as incapable of governing wisely because they were largely uneducated, could be stirred by irrational political passions, and did not have sufficient property and wealth to guarantee allegiance to the existing order. Given this view of human nature and the public, the Federalists thought good national government would control the excesses of popular representation with the wisdom of stable, well-educated, property-owning citizens as the elected leaders. This also meant some shifting of authority from the states to the national government on matters such as trade, defense, foreign relations, financing of government functions, and currency.

People often express their ideas about the relationship between citizens and government by allocating authority and responsibility within organizations and between parts of government in certain ways. Structural features of governmental institutions are often important, mostly in relation to actions of political leaders, for example the president's authority to veto legislation from Congress. However, structural features can also be important for public administration; four of these which emerged from the debate over the Constitution are discussed below. They were important to the founding generation and they are important now, appearing in public discussions about citizenship, roles of the executive and legislative branches of government, the relationship between the national government and states, the size and scope of government, and the role of career public professionals in creation of public policy. Though the Federalists "won" the argument over adoption of the Constitution, discussion about its implementation and the effects of governmental structure continues today and probably always will, because these are timeless questions applicable to each generation.

### Republican Form of Government

The United States has a *republican* form of government. Use of the word republican here is not a reference to a political party, but to the form of representation of the public will. In a republic, power is vested in the citizens (the word republic comes from the Latin *res publica*, which may be translated as "the people's affair" [Lakoff 1996, 26] ) but it is expressed through an elite group or groups representing portions of the people, rather than all the people directly and individually. On a continuum of democratic government from nondemocratic to fully democratic, a republican form lies somewhere between monarchy and direct democracy. This was its advantage to the American founders, because a republic seemed to avoid the potential irrationality or excesses of direct democracy while incorporating elements of democracy. Though in earlier republics those who governed were not chosen by widespread election, the founding American generation linked the idea of election to the idea of republicanism, and American representative democracy came to be thought of as republican (Lakoff 1996, 25–29).

Republican government, or a *democratic republic* such as the United States, is different from a direct democracy, in which the citizens as a whole make decisions instead of elected representatives. It is difficult for the entire citizenry to take part in governance except in small communities, for example in a number of New England towns where it is the traditional form of government. Even in those towns, where people periodically meet as a whole to discuss and decide on public matters, they often choose a small group of their members to act for them on routine matters between general meetings.

Throughout the life of the nation there has been debate about the relationship between democracy, an idea cherished by Americans since they left Europe to find greater freedom from control by church and state, and the form of representation. In the era of the founding of the Constitution (the 1780s), the Anti-Federalists were worried that the national government proposed by the draft Constitution would take government too far away from the people. Today, those in favor of allowing unlimited citizen-initiated ballot measures argue this allows for greater democracy, while those in favor of limiting decision-making authority on public matters to elected representatives argue that greater citizen involvement subverts the normal and rational process of making public policy. Some people envision a new direct democracy, even at state and national levels, made possible by technology and the possibility of instant access to large-scale electronic voting on current public issues.

For public administrators, the question of whom they serve and for what purposes is related to the form of democratic citizen representation or involvement. Where citizens are far removed from the decision-making process (often the case in large public organizations), public service practitioners may look to elected officials, political appointees, and possibly advocates from interest groups for perspective on the needs and wishes of the public. Where citizens are more directly involved, for example through participation in neighborhood associations, or committees or boards associated with government programs or agencies, public administrators often must balance conflicting demands for their attention.

Public employees may perceive significant differences between public preferences and those of elected leaders about public services and priorities for allocating scarce public dollars. Sometimes, instead of differences between elected leaders and citizens, there may be such diversity of opinion among both groups that it is difficult to know what should be done. Since elected officials are responsible for supervising the work of public employees, it could be reasonable for public employees to follow their lead, assuming that a clear policy direction emerges from deliberation by elected leaders. At the same time, an argument can be made that public service is about service to the public rather than only to elected representatives, so public employees should also take into account the wishes of the public if they can determine what those are. There is no easy answer to this problem, which presents an ethical dilemma for many public-service practitioners in many different settings (Cooper 1991).

There are perspectives on public administration that attempt to resolve the dilemma of the source of policy direction for public administration. One of these, including work by a number of authors writing from several theoretical perspectives, advocates *administrative legitimacy*, the idea that public administration has become an essential and accepted part of the constitutional system though it is not explicitly described in the document itself (Lowi 1993; Rohr 1993; Spicer and Terry 1993a, 1993b; Stivers 1993; Wamsley et al. 1990; Warren 1993; Wise 1993). Another perspective, also including work approaching the policy dilemma from several theoretical perspectives, considers it appropriate for public administrators to exercise considerable *administrative discretion*, that is, independent judgment, in deciding what action is in the best interests of the public and how it should be implemented (Fox and Cochran 1990; Frederickson 1980; Harmon 1981).

## The "Vertical" Structure of Government

Because the Constitutional founders recognized the importance of the states to the people, but also wanted to create a stronger national government, the structure they designed gave powers to legislate and take action to both state governments and the national government. The structure they created is called "federal," which means in the American context that for a particular geographic area, certain powers are allocated to the national government and certain powers are allocated to state governments. (Local governments are not part of the Constitutional system; they are created by the people within each state.) Thus, *federalism* is about *vertical* relationships between *levels of government*. People in the founding era and for several decades thereafter took seriously protecting the states from the influence of the national government. This was reflected in a *dual* federalism intended to keep national and state governments distinct from one another, operating in their separate spheres of influence with relatively little interference or interaction.

Critics of the new form believed it was not truly federal, meaning a confederation of separate entities joining voluntarily to accomplish certain functions in common, but national or unitary, a system of central control with subordinate units of government (the states). This critique is reflected in the contemporary setting, in which the national government has grown much stronger and more active than during the founding era or the first century of the nation, in relation to states.

Despite the notion of dual federalism, there has always been cooperation between levels of government and over time the roles and functions of different levels of government have become intertwined (Elazar 1962). During the nineteenth and twentieth centuries, with territorial expansion, growth of population, and increasing complexity of society and government, the boundaries between national and state powers and responsibilities became increasingly blurred, and local government has become a significant partner in delivering government services. Though some governmental functions are performed exclusively by one level of government, for example the printing of currency, many others are shared, for example highway planning, construction, and maintenance; funding and administration of social welfare services; and education.

Until the twentieth century, local and state governments were by far the largest part of the public sector. At the end of the nineteenth century and into the early twentieth century there was public support for curtailing the excesses of private business, excesses such as use of child labor in difficult and hazardous conditions for long hours, production of unsafe food, monopolies in industries such as railroads, and so on. In the 1930s, conditions associated with the global economic depression led American political leaders to seek fundamental changes in the governmental role for taking care of the unemployed, infirm, sick, and elderly, and in regulating relations between labor and business management. In addition, the nation had to deal with two world wars plus smaller-scale conflicts, the need for policies and programs to address the rapid growth of the urban population, and the question of civil rights.

The challenges of these events and the public response as reflected in governmental action resulted in a much larger and more powerful national government than had been the case for over a century since the Constitutional founding. Today, we largely accept this enlarged role of the national government, though people often complain about the size of government and the taxes required to support it. Politicians such as Ronald Reagan and George W. Bush find this a good way to appeal to a public that thinks itself heavily taxed. However, taxation is relative and compared to most other developed countries, the level of taxation in the United States as a percentage of the overall domestic economy is low (Mikesell 1999, 280–281). Nevertheless, when the suggestion is made to cut government programs in response to perceived public desire for lower taxes, special-interest lobbying groups rush to urge cutting programs other than their own, and opinion polls reveal that citizens are not in favor of significant changes in government services. Thus, change is usually slow and incremental.

The vertical structure of government touches public administration in several ways. In many programmatic areas of practice, for example those related to the environment and law enforcement, public employees at each level of government work with agencies and policies at other levels. A variety of national laws and decisions of the federal courts apply to local and state public administrators and influence or determine their actions in specific areas—examples would include water quality standards and guidelines for police searches related to people suspected of violating laws. Choices available to local government employees and elected officials about which services to offer and how they are funded are limited by state laws and constitutions. Also, increased or decreased funding by one level of government (such as federal grants) for services delivered at another level of government can have significant impact on the jobs of public employees.

## The "Horizontal" Structure of Government

Another aspect of structure is distribution of authority within one level of government, the *horizontal* relationships internal to national, state, or local governments. Given the assumption of imperfect human nature, the founders thought it wise to split government, and therefore the powers of government, into branches—the legislative, executive, and judicial—and to further split the legislative branch into a "lower" house with many representatives, serving short, two-year terms, and an "upper" house (at that time chosen by state legislatures), with fewer members, serving six-year terms. This allowed one body to more closely represent the immediate will of the people, and the other body to, supposedly, represent a more detached, prudent, possibly virtuous view of the public-interest characteristic of wealthier, better-educated citizens.

This is the *separation-of-powers* model closely identified with the structure of government in the United States. As a generalization, the balance of power between the branches of the national government favored the legislature in the early years of the republic, but the executive and judicial branches have gained influence. The situation is complex at the local level. Here, some units of government have elements of separated powers (e.g., a mayor with significant executive functions, balanced by a council with power to approve or disapprove of mayoral actions), while others have *unitary* structures, designed much like private-sector organizations to maximize control and efficiency. Unitary forms have a legislative body (board or council) and an appointed, career professional general manager (examples: a general manager of a regional wastewater treatment district; a city manager) who supervises daily operations.

Governmental structures are an important way to express values about governing. Separation of powers is evidence of concern about potential concentration and abuse of power; adoption of unitary structures with a direct line of authority from top to bottom indicates a desire for efficient and businesslike administration. As a generalization, at all levels of government the separation-of-powers model was dominant from the founding era through the nineteenth century. At the local level, beginning in the mid-nineteenth century, school districts moved toward unitary structures, followed by some city governments in the early twentieth century.

Unitary structures with professional managers accountable to elected representatives are a departure from the traditional American model, derived from the experience of emigrating from Europe to escape authoritarian systems, of separated government administered by elected representatives of the people, or appointees of the elected representatives. In the late nineteenth century, the *civil-service* movement in the national government was intended to remove some of the influence of partisan politics from a separation-of-powers structure. This illustrates use of modifications to a separation-of-powers structure to shift emphasis from the value of political responsiveness to the value of efficient, technically competent work.

For the career public practitioner, the work environment can be quite different in organizations structured to emphasize action based on knowledge from a specific area of vocational practice (examples: park planning, hazardous materials management, public health nursing, neighborhood policing, military communications systems) than it is in organizations structured to emphasize responsiveness to politicians, interest groups, corporations, or powerful individuals. The perceptions of practitioners about the nature of the work environment can vary depending on their position in the organizational hierarchy; people working close to the top of a hierarchy may be more likely to experience the influence of values and demands from those outside the organization, while those in other parts of the organization may focus instead on largely technical matters.

Also, structure is not completely determinative. Examples of people who base their work primarily on vocational knowledge and norms of practice may be found in organizations with structures designed to facilitate responsiveness, such as a federal agency headed by a political appointee or a city with a charter granting chief-executive authority to an elected mayor rather than to a professional city manager. Examples of people who base their work primarily on responsiveness can be found in organizations with structures designed to facilitate practice based on vocational knowledge. As noted above, modifications within structures may create a shift in operating values, such as use of civil-service job protections to emphasize vocational knowledge and competent technical performance instead of political responsiveness. However, despite all the complexities and possible variations of practice associated with governmental structure, structures can make a significant difference in the daily work experiences of practitioners and also for experiences that citizens and elected officials have with administrative agencies (Abney and Lauth 1986). For example, though political responsiveness is part of the culture of cities with the professionalized, unitary council-manager structure, in many such cities there is an emphasis on continuing education, knowing about people and practices in other places, keeping up with written knowledge in an area of practice, and resisting the influence of those who want the practitioner to violate vocational norms of practice for their own benefit. This emphasis on what may be called *professionalism* is often stronger in places with structures created, or modified, to emphasize practice according to vocational knowledge and norms rather than practice according to responsiveness to pressures from outside the organization.

Practitioners may think the structure they are familiar with in their workplace, whether it includes elements of separation of powers or the unitary model, is "the way things are" everywhere. Instead, national and state governments and many local governments have some version of the separation-of-powers model and almost half of American cities, along with some counties and many special districts, including school districts, have a unitary structure.

## Politics and Administration

It is easy to confuse separation of powers with politics and administration. Separation of powers describes the distribution of authority and responsibility between branches of government headed by elected officials or judges, though as discussed immediately above it has implications for public-service practitioners as well. The term *politics and administration*, or *policy and administration*, signals a focus on the relationship between elected officials or political appointees at the head of public organizations and the larger body of career public-service practitioners who do much of the daily work of those organizations. (In the national government there are a few thousand political appointees and 2.8 million civilian employees; in a hypothetical medium-sized city with a mayor-council structural form there may be a total of three or four dozen elected officials and political appointees and 2,000 or so career public employees; in a council-manager city of the same size there may be only twelve to fifteen elected officials and political appointees.) Politics and administration is most commonly how this focus on the elected official/practitioner relationship is phrased, but policy and administration is used as well. It is used to draw attention away from partisan party politics, toward the process of discussion of issues of public importance, resulting in creation of policies implemented by public agencies.

The early national government was both led and administered by distinguished elected leaders and political appointees such as George Washington as president; Alexander Hamilton as Washington's secretary of the Treasury; Thomas Jefferson as Washington's secretary of State and later as President; and James Madison as Jefferson's secretary of State and later as president. As noted in Chapter 2, the national government was very small. These people did more than create policy and monitor the performance of units of government; they were involved in the details of daily departmental administration in a way that would be impossible for contemporary national leaders. At the local level in the early years of the nation, government consisted largely of elected governing body members and community volunteers. Beginnings of organized public services such as police and fire services, public health, and education emerged during the nineteenth century and were formed during the twentieth century into the organizations and practices familiar today.

This was adequate for the times, but as the nation grew and public agencies became large and complex, it became apparent that staffing them with supporters and friends of politicians who won the last election, or with volunteers, was insufficient. As administrative agencies grew and the tension increased between political responsiveness and the need for technical competence, Americans thought seriously about a problem that can be identified in a variety of ways, such as politics and administration, bureaucratic *accountability*, *controlling bureaucracy*, the allowable degree of *administrative discretion*, and whether administration has *legitimacy* within the American constitutional system. However it is approached, this issue is central to American public administration, to the question of whom the public administrator serves and for what purposes.

As new structural formats were tried in the nineteenth and early twentieth centuries to increase efficiency and effectiveness through professional expertise, people discovered an uncomfortable trade-off: gaining expertise meant shifting some portion of the control over policymaking and implementation away from elected representatives of the people and their political appointees, to full-time career employees. This created problems with defining the roles of elected officials and public employees, and concern that democracy could be damaged if too much of government were in the hands of a cadre of professional managers.

By the middle of the twentieth century, scholars had recognized that a clear split, or *dichotomy*, between politics and administration did not exist in practice. Citizens, politicians, and political appointees often are involved in aspects of administration as policy is put into practice following a policy decision, and public-service practitioners are often involved in providing information and giving recommendations to policymakers as they formulate policy, before a policy decision is made. However, it was also recognized that the relationship between practitioners and elected officials is an ongoing part of governance in public organizations, and that, as Dwight Waldo put it, "no problem is more central to public administration, the existential, real-world enterprise, and to Public Administration, the self-aware enterprise of study and education, than the relationship of politics and administration" (1981, 65).

Today, many citizens are accustomed to the bureaucratic nature and the size of government at all levels, which makes it often unresponsive and remote from the people. Even so, many would prefer to have government shaped by citizen preferences, government that demonstrates both openness to citizen involvement in formulating public policy, and direct accountability to citizens for the efficient and effective delivery of public services (King and Stivers 1998). Citizens of democratic nations tend to believe that government should ultimately answer to them. In particular, as a people, Americans retain a normative expectation, part of the culture of a country founded in rebellion against authority, that public employees should be subservient to the will of elected officials and citizens. However, the size and complexity of government means that much of the policy and most of the day-to-day decisions are made by unelected public practitioners, leading to the question of the relationship of public administration to the public in a democratic society (Gruber 1987). Does public administration have a role and responsibility in shaping the future, or should it wait to be told what to do by elected representatives? How much freedom to act, often called administrative discretion, are we willing to grant public-service practitioners? These are the important questions associated with the politics-administration relationship.

## TRANSITION

At this point, we move from outlining the field of public administration and its environment, to discussion and readings in four thematic areas important to public administration theory and practice. Each area is part of the broad, macrolevel setting of public administration, involving values and normative ideals that have always been a focus of debate and controversy in America. In Part II, attention turns to the founding era of the nation in the late eighteenth century, an era that shapes the present and future despite the passage of more than two centuries of time.

## REFERENCES

Abney, Glenn, and Thomas P. Lauth. 1986. *The Politics of State and City Administration*. Albany: State University of New York Press.

Cooper, Terry L. 1991. *An Ethic of Citizenship for Public Administration*. Englewood Cliffs, NJ: Prentice-Hall.

Dow, F.D. 1985. *Radicalism in the English Revolution, 1640–1660*. Oxford: Basil Blackwell.

Elazar, Daniel J. 1962. *The American Partnership: Intergovernmental Co-operation in the Nineteenth-Century United States*. Chicago: University of Chicago Press.

Fox, Charles J., and Clarke E. Cochran. 1990. "Discretionary Public Administration: Toward a Platonic Guardian Class?" In *Images and Identities in Public Administration*, ed. Henry D. Kass and Bayard L. Catron, 87–112. Newbury Park, CA: Sage.

Frederickson, H. George. 1980. *New Public Administration*. University: University of Alabama Press.

Gruber, Judith E. 1987. *Controlling Bureaucracies: Dilemmas in Democratic Governance*. Berkeley: University of California Press.

Harmon, Michael M. 1981. *Action Theory for Public Administration*. New York: Longman.

Holorenshaw, Henry. 1971. *The Levellers and the English Revolution*. New York: Howard Fertig.

Jensen, Merrill. 1940. *The Articles of Confederation: An Interpretation of the Social-Constitutional History of the American Revolution 1774–1781*. Madison: University of Wisconsin Press.

King, Cheryl Simrell, and Camilla Stivers. 1998. *Government Is Us: Public Administration in an Anti-Government Era*. Thousand Oaks, CA: Sage.

Lakoff, Sanford. 1996. *Democracy: History, Theory, Practice*. Boulder, CO: Westview Press.

Lockridge, Kenneth A. [1970] 1985. *A New England Town, the First Hundred Years: Dedham, Massachusetts, 1636–1736*. New York: W.W. Norton.

Lowi, Theodore A. 1993. "Legitimizing Public Administration: A Disturbed Dissent." *Public Administration Review* 53 (May/June): 261–264.

Matthews, Richard K. 1984. *The Radical Politics of Thomas Jefferson: A Revisionist View*. Lawrence: University Press of Kansas.

McCoy, Drew R. 1980. *The Elusive Republic: Political Economy in Jeffersonian America*. Chapel Hill: University of North Carolina Press.

McDonald, Forrest. 1985. *Novus ordo seclorum: The Intellectual Origins of the Constitution*. Lawrence: University Press of Kansas.

Mikesell, John L. 1999. *Fiscal Administration: Analysis and Applications for the Public Sector*, 5th ed. Fort Worth, TX: Harcourt Brace.

Morison, Samuel Eliot. 1965. *The Oxford History of the American People*. New York: Oxford University Press.

Phillips, Derek. 1993. *Looking Backward: A Critical Appraisal of Communitarian Thought*. Princeton, NJ: Princeton University Press.

Rohr, John A. 1993. "Toward a More Perfect Union." *Public Administration Review* 53 (May/June): 246–249.

Rossiter, Clinton. 1966. *1787: The Grand Convention*. New York: W.W. Norton.

Spicer, Michael W., and Larry D. Terry. 1993a. "Legitimacy, History, and Logic: Public Administration and the Constitution." *Public Administration Review* 53 (May/June): 239–246.

———. 1993b. "Advancing the Dialogue: Legitimacy, the Founders, and the Contractarian Argument." *Public Administration Review* 53 (May/June): 264–267.

Stivers, Camilla. 1993. "Rationality and Romanticism in Constitutional Argument." *Public Administration Review* 53 (May/June): 254–257.

Waldo, Dwight. 1981. *The Enterprise of Public Administration: A Summary View*. Novato, CA: Chandler and Sharp.

Wamsley, Gary L., Robert N. Bacher, Charles T. Goodsell, Philip S. Kronenberg, John A. Rohr, Camilla M. Stivers, Orion F. White, and James F. Wolf. 1990. *Refounding Public Administration*. Newbury Park, CA: Sage.

Warren, Kenneth F. 1993. "We Have Debated ad Nauseum the Legitimacy of the Administrative State—but Why?" *Public Administration Review* 53 (May/June): 249–254.

Wise, Charles R. 1993. "Public Administration Is Constitutional and Legitimate." *Public Administration Review* 53 (May/June): 257–261.

Wood, Gordon S. 1969. *The Creation of the American Republic, 1776–1787*. Chapel Hill: University of North Carolina Press.

# PART II

# DEBATE AND DECISION IN
# THE FOUNDING ERA

The readings in Part II explore controversy about the founding of the nation and the nature of government in a democratic society. The structure of government stretching over the states following the Revolutionary War was the Articles of Confederation. The Articles preserved the states intact as sovereign units of government, allowing the weak national legislative body to act only when the states would allow it. Frederick Quinn describes the situation of the Confederation under the Articles as chaotic, with a weak and dysfunctional central government, bankrupt state governments, and awkward economic relationships because many states had their own currencies and regulations for trade from outside their borders. Citizens tended to identify with their states rather than with the nation.

Not everyone thought this situation troublesome. People who would benefit from stable conditions in trade and finance were especially concerned, but many others believed it best to preserve the separate authority of the states and feared creation of a strong national government as a threat to liberty and self-governance. In 1786 there was a dispute between Maryland and Virginia over the oyster fishery in the Chesapeake Bay. Such disputes were common under the Articles. People representing several states met in Annapolis, Maryland, in the fall to discuss this problem, but they did not stop at settling this specific dispute. Instead, it was decided something ought to be done about the underlying problem, the absence of governmental structure that would allow for efficient commercial activity. This led to a Constitutional Convention in Philadelphia the following summer.

Shays' Rebellion (see Part I, Chapter 3) occurred soon after the Annapolis meeting, adding a sense of urgency to the upcoming convention. The convention began in May 1787 and continued until September. Twelve states named seventy-four delegates; fifty-five attended all or part of the convention, and thirty-nine were present when the draft Constitution was signed. The delegates were lawyers, merchants, and planters, professionals and property owners. They met in secret to allow discussion and action free from the pressure of public visibility (the only remaining record of the discussion, notes made by James Madison, was not published until 1840). The draft Constitution was given to Congress on September 20, 1787, and Congress moved on September 28 to send it to state conventions for approval.

State delegates voted on the new Constitution in ratification conventions. Delaware was the first state to accept the Constitution, in December 1787. With approval by New Hampshire in June 1788, the nine states required for ratification had approved. However, this left 40 percent of the population out of the ratification process, since Virginia, New York, North Carolina, and Rhode Island had not acted. Advocates of the Constitution did not want the new nation to begin without approval of large states such as New York and Virginia. After long debate, New York and Virginia voted in favor in the middle of 1788, North Carolina voted no in July 1788 but yes a year later, and Rhode Island approved the Constitution in May 1790, a year after the new government under the Constitution had begun operation in the spring of 1789.

Events of this era reflect American ideas about the purpose of government and the relationship of each person to society. People wanted to preserve the freedom won in the Revolutionary War, the ordinary person did not want the rich and powerful to become a dominant aristocracy, and there was disagreement about whether a strong central government would be a good thing for the economy or a bad thing for individual liberty and relative equality of social class.

The nation and the world are much different today, but the ideas underlying the debate over ratification of the Constitution are as important now as they were then. Contemporary policy debates about issues such as national security and military intervention in foreign countries, welfare, health care, retirement, free speech and privacy, powers of the national and state governments, and so on, are framed in terms that would be familiar to people from the founding era, though the national government is much more dominant in the American governmental structure than conceived at that time. For public-service practitioners, the constitutional debate reveals basic American values about government that affect daily practice, including choices made about the conduct of public programs, programmatic recommendations made to policy makers, and interactions between government and citizens. For all Americans in their role as citizens, applying or modifying these values is the essence of shaping the future of the nation as they participate in public decision making about policies, candidates, and issues.

It is hoped readers will experience heightened awareness of the conditions that shaped American government and public administration; a questioning and critical attitude toward people, events, and the institutions created during this period; and greater appreciation of the nature and importance of the American experiment in democracy. The relationship of citizens to governments and public-service practitioners is different in the United States from that in other countries, given the American emphasis on the individual and liberty formed by emigration from Europe during the Age of Enlightenment, a large and rich land mass, and relative geographic isolation. Upon completing this part, readers may find it helpful to consider whether, and in what ways, their perceptions have changed of the founding era and its meaning for citizenship and contemporary public administration.

## THE FEDERALIST PAPERS

In the first reading in this part, Frederick Quinn's description of the young nation is striking. It is not what we think of as a nation-state, but rather a poorly defended collection of largely independent areas (states), with a stagnant economy, crude and difficult transportation, and periodic uprisings of the lower classes against perceived injustices of governmental and financial systems controlled by the wealthy.

Quinn writes from a viewpoint of belief in the Constitutional system created by the founders, a viewpoint in which action needed to be taken to save the fledgling nation from disintegration. He emphasizes the coming together of a variety of conditions that favored creation of the new form of government at that moment:

A few years earlier, in the shadow of the war and the British crown, a proposal for a strong central government would not have found acceptance. Nor could such a concept have succeeded during the Age Of Jackson, just around the corner. Its republican features, which kept the people somewhat distant from the reins of power, would have been voted down. *The Federalist Papers* thus explain a revolutionary document that faced a hotly contested ratification battle for the political soul of a nation at a critical turning point in its history. (1993, 4)

Thus, the constitutional convention and the ratification debate that followed occurred at a time when circumstances favored stronger central government at the expense of democratic self-governance. Five or ten years earlier, Americans pleased with liberties won in the Revolutionary War would not have allowed creation of a strong central government. Quinn argues they would also not have supported it during Andrew Jackson's presidency (1829–1837), an age of strong democratic sentiment, but this may also be said of the time of the presidency of Thomas Jefferson (1801–1809). Jefferson's election ended the governmental leadership of the Federalists, slowing for a time the trend toward a stronger national government.

Quinn introduces several issues that appear in the *Papers*; for our purposes here, the reader is asked to note the sequence of events in the process of creating and ratifying the Constitution, and the issues of interest to the founders. The 85 *Federalist Papers* were published as newspaper articles in New York and later came to be thought of as the statement of founding philosophy for the Federalists. They are especially important to American government and citizenship because: they were written by Alexander Hamilton and James Madison (and John Jay), key figures in the national founding; they offer a concise statement of the intent of the "winners" in the ratification debate; and the concepts underlying the government they describe are present with us today.

The three *Federalist Papers* included here (Nos. 10, 17, and 51) have been chosen for current relevance to public-administration practice. Readers may initially find it difficult to understand the language of the *Papers* because the writing style is quite different from that of the twenty-first century. With some effort, the style becomes understandable, allowing us access to fundamental documents of American government and citizenship. As you read the papers, consider how these ideas have developed in the intervening years and how they influence current political affairs and administration of public agencies, not only at the national level, but also in states and local communities. *The Federalist Papers* were written by advocates of a stronger national government during a time when strong national government meant one that represented the nation in international affairs, collected duties on trade and taxes from citizens, operated a postal service, and maintained a very small military. Other governmental functions were left to the states and their local governments. This system of *federalism* remained largely intact until the late nineteenth century, when issues associated with the growing urban-industrial nation began to require a greater national presence.

The pattern of government described in the three *Papers* is crucial to understanding American government. It includes diluted influence of parties and special interests; national governance by a remote, philosophically inclined elite rather than the people; most governmental functions left to the states; and separation of powers between the legislative, executive, and judicial branches of government to limit abuse of power. These features respond to experiences of the founding generation in splitting from Britain, they are based on a somewhat gloomy view of human nature as self-interested, and they are designed to promote, at the national level, stability in the interest of private property and commerce, rather than democracy. Today, they affect how Americans think of citizenship, the structure and operation of national, state, and local public agencies, and the influence of the national system on state and local governments, including the public-service practitioners who work in them.

## FEDERALIST PAPERS 10, 17, AND 51

In number 10, Madison argues that the proposed national government would dilute the influence of what was then called *factions* and today are called parties and special interests. A particular faction would supposedly have less opportunity to dominate if there were many of them, making the national government safer from such influence than the state governments. In the opening paragraph, Madison writes that the proposed Constitution would help answer complaints from "friends of public faith" and friends of "public and personal liberty." By friends of public faith, he means people concerned that social instability and actions of state legislatures (such as failure to enforce contracts and debts, and printing an excess of paper money) threaten private property; many of these friends of public faith were Federalists.

Madison understands the most common reason for formation of factions in society to be the unequal distribution of property. Unlike some in the founding generation, he does not advocate reducing inequalities of wealth as a means of limiting accumulation of power. Instead, he would use the large geographical size of the proposed government to minimize factional conflict, since there would be so many different interests that it would be difficult for one or a few to dominate. He contrasts the form of this large government with that of government in smaller geographic areas where people can gather together to talk directly about public affairs. The proposed national government would be in the form of a *republic*, in which the interests of the mass of individual citizens would be represented by a few people. This is different from what he terms a *pure democracy*, in which citizens could participate directly if they wished. He portrays pure democracies of the past as turbulent and short-lived experiments in redistribution of wealth so that everyone is equal not only in their *procedural rights*, but also in their *substantive* position in society.

In number 17, Hamilton intends to counter the claim of those who think the new national government will be so powerful it will encroach on state governments, and by implication the ability of citizens to govern themselves. The opening sentence identifies a primary concern of Anti-Federalists that the proposed Constitution, unlike the Articles of Confederation, allows the national government to tax individuals directly instead of requesting revenue from the states. It was feared this would permit the national government to interfere in matters traditionally of state concern. In a passage especially interesting when compared with the situation today, Hamilton writes in the second paragraph: "It will always be far more easy for the State governments to encroach upon the national authorities than for the national government to encroach upon the State authorities."

Hamilton bases this prediction on the natural tendency of people to identify with what is closest to them: family first, then neighborhood, then community, and so on. The new national government would be led by "speculative men," by which he means thoughtful people somewhat removed from the passions and interests of the citizenry. Hamilton likens the proposed structure of nation and states to a medieval kingdom and the several principalities within the larger kingdom. In medieval times, people were loyal to the principality except when they were mistreated by the prince; then, they were likely to feel greater loyalty to the kingdom and its sovereign. This, according to Hamilton, demonstrates that if states behave appropriately, they will be the primary focus of the loyalty of the people and will not be threatened by the national government.

*Federalist Paper* 51 defends the proposed separation-of-powers model for the national government, the system of federalism that divides authority between the national and state governments, and the Federalist view of human nature that makes it necessary to split the authority of government to minimize abuse of power. Madison describes the proposed structure as a compound republic. The power "surrendered by the people" is first divided into the state and national levels ("vertical" splitting of power), then each government is separated into branches, which Madison calls departments ("horizontal" splitting of power).

The language used to describe human nature makes this *Paper* notable. In the fourth paragraph, Madison writes that each part of the government must be given the means to defend itself from attack by the others, with the government structured in the Constitution so that "ambition must be made to counteract ambition." Thus, people are portrayed as power-hungry and aggressive, a view for which Madison makes no apology. Instead, acknowledging the meaning of the proposed form of government for the way people behave, he writes that "If men were angels, no government would be necessary," and that once a government is created to control the governed, then it is necessary to "oblige it to control itself."

## THE ANTI-FEDERALIST RESPONSE AND TOCQUEVILLE'S CONCERNS ABOUT DEMOCRACY

In the reading from *The Essential Anti-Federalist*, W. B. Allen, Gordon Lloyd, and Margie Lloyd note that supporters of the new Constitution who became known as Federalists maneuvered to take the name from those to whom it originally referred, people who supported the Articles of Confederation. This put critics of the proposed Constitution in the role of being against change rather than in favor preserving something good. The Anti-Federalists thought of themselves as the people who wanted a truly *federal* system, one in which the several states agreed to perform some functions cooperatively; this was in contrast to the idea of creating a strong national government that could act effectively on its own.

This reading outlines several areas of disagreement between Anti-Federalists and Federalists, but the essential difference was belief in small, local, and decentralized government as a way to protect individual liberty, contrasted with belief in a stronger national presence to create a prosperous commercial nation. The closing paragraph of the reading makes the point that Anti-Federalist concerns about social inequality and democracy that is close to the people remain with us in today's public issues and policy debates. These concerns are echoed by Americans who express frustration that public-service agencies are slow, unresponsive, or intrude into private lives. The current public-sector trend answering these concerns is to "run government like a business," by equating the public sector to the market and emphasizing economic efficiency. However, Americans who echo sentiments of the Anti-Federalists in favor of limiting government and keeping it close to the people are addressing a larger issue than how to make existing public services cost less. They are thinking about the role of government and public-service practitioners in a democratic society, a society shaped in part by the constitutional debate between Anti-Federalists and Federalists.

In 1831–1832, French politician and political scientist Alexis de Tocqueville toured the United States, studying social conditions and government. Several years later, this tour resulted in the book *Democracy in America*, the reflections of a European on the possibility of achieving liberty and stability in newly emerging democratic forms of government. Tocqueville found American democracy fascinating; to him, the people were energetic participants in society and there was remarkable freedom to form and join organizations and express opinions. Yet he thought that Americans exhibited little "independence of mind" or freedom of discussion, because the pressure of majority opinion urged people to conform rather than stand out from the crowd.

The passage from Tocqueville's book included here is not about his observations of America. Instead, it is a scenario he constructed of a possible future for democracy if the tendencies Tocqueville found on his tour of America were to be carried forward many years. He is suggesting to us that some features of American democracy, if allowed to develop to their logical extreme, might result in a society many people would find undesirable. In this scenario, everyone is secure, there is relative social equality, people are concerned mostly with trivial matters of personal pleasure, and citizens no longer exercise critical thought about their situation. The government is gentle, protective, and all-powerful, carefully directing and supervising individual lives. Thanks to elections, the people think they have chosen this form of government themselves. However,

> After having thus successively taken each member of the community in its powerful grasp and fashioned him at will, the supreme power then extends its arm over the whole community. It covers the surface of society with a network of small complicated rules, minute and uniform, through which the most original minds and the most energetic characters cannot penetrate, to rise above the crowd. The will of man is not shattered, but softened, bent, and guided; men are seldom forced by it to act, but they are constantly restrained from acting. Such a power does not destroy, but it prevents existence; it does not tyrannize, but it compresses, enervates, extinguishes, and stupefies a people, till each nation is reduced to nothing better than a flock of timid and industrious animals, of which the government is the shepherd. (2003, 12)

This scenario of a possible democratic future may seem exaggerated, but in such exaggeration can be found parts of reality worth examining. Tocqueville wrote 170 years ago, but his words resemble the thoughts of people today who believe that government in modern society has become excessively large and intrusive. Americans have always feared government that could become something like Tocqueville's vision of the future. The question for public administration is how, in a complex urban society requiring large public organizations operating multiple public services for millions of people, we can recognize the traditional American concern about the influence of government in our lives, at the same time we work to deliver effective and efficient public services. This is the balance—sometimes a trade-off—between democracy and bureaucracy, liberty and stability, at the core of the American experiment.

## THE IDEAS OF THE FOUNDING GENERATION IN THE PRESENT

The final reading in this chapter, from Richard Stillman's *The American Bureaucracy*, applies the ideas of three members of the founding generation directly to public administration throughout the nation's history. One must take care in applying ideas and practices from over 200 years ago to today's social, political, and administrative worlds. However, Stillman adapts the work of Hamilton, Madison, and Jefferson to show how founding-era thought about the role of government in society can be applied to understanding the development of modern public administration.

The quite different models of public administration derived from these three historical figures can be summarized as follows: a large, active public administration (Hamilton), a politically responsive public administration (Madison), and a limited public administration closely monitored by the people (Jefferson). Readers may wish to consider how Stillman's ideas can be applied in their current workplace or in public organizations with which they are familiar from reading or from media accounts. These are powerful models, allowing us to integrate concepts that shaped the nation into better understanding of current and future public administration in America.

# THE FEDERALIST PAPERS, NOS. 10, 17, & 51

**FEDERALIST NO. 10**

<u>The Same Subject Continued</u>
<u>(The Union as a Safeguard Against Domestic Faction and Insurrection)</u>
From the *New York Packet*.
Friday, November 23, 1787.

MADISON

To the People of the State of New York:

AMONG the numerous advantages promised by a well constructed Union, none deserves to be more accurately developed than its tendency to break and control the violence of faction. The friend of popular governments never finds himself so much alarmed for their character and fate, as when he contemplates their propensity to this dangerous vice. He will not fail, therefore, to set a due value on any plan which, without violating the principles to which he is attached, provides a proper cure for it. The instability, injustice, and confusion introduced into the public councils, have, in truth, been the mortal diseases under which popular governments have everywhere perished; as they continue to be the favorite and fruitful topics from which the adversaries to liberty derive their most specious declamations. The valuable improvements made by the American constitutions on the popular models, both ancient and modern, cannot certainly be too much admired; but it would be an unwarrantable partiality, to contend that they have as effectually obviated the danger on this side, as was wished and expected. Complaints are everywhere heard from our most considerate and virtuous citizens, equally the friends of public and private faith, and of public and personal liberty, that our governments are too unstable, that the public good is disregarded in the conflicts of rival parties, and that measures are too often decided, not according to the rules of justice and the rights of the minor party, but by the superior force of an interested and overbearing majority. However anxiously we may wish that these complaints had no foundation, the evidence, of known facts will not permit us to deny that they are in some degree true. It will be found, indeed, on a candid review of our situation, that some of the distresses under which we labor have been erroneously charged on the operation of our governments; but it will be found, at the same time, that other causes will not alone account for many of our heaviest misfortunes; and, particularly, for that prevailing and increasing distrust of public engagements, and alarm for private rights, which are echoed from one end of the continent to the other. These must be chiefly, if not wholly, effects of the unsteadiness and injustice with which a factious spirit has tainted our public administrations.

By a faction, I understand a number of citizens, whether amounting to a majority or a minority of the whole, who are united and actuated by some common impulse of passion, or of interest, adverse to the rights of other citizens, or to the permanent and aggregate interests of the community.

There are two methods of curing the mischiefs of faction: the one, by removing its causes; the other; by controlling its effects.

There are again two methods of removing the causes of faction: the one, by destroying the liberty which is essential to its existence; the other, by giving to every citizen the same opinions, the same passions, and the same interests.

It could never be more truly said than of the first remedy, that it was worse than the disease. Liberty is to faction what air is to fire, an aliment without which it instantly expires. But it could not be less folly to abolish liberty, which is essential to political life, because it nourishes faction, than it would be to wish the annihilation of air, which is essential to animal life, because it imparts to fire its destructive agency.

The second expedient is as impracticable as the first would be unwise. As long as the reason of man continues fallible, and he is at liberty to exercise it, different opinions will be formed. As long as the connection subsists between his reason and his self-love, his opinions and his passions will have a reciprocal influence on each other; and the former will be objects to which the latter will attach themselves. The diversity in the faculties of men, from which the rights of property originate, is not less an insuperable obstacle to a uniformity of interests. The protection of these faculties is the first object of government. From the protection of different and unequal faculties of acquiring property, the possession of different degrees and kinds of property immediately results; and from the influence of these on the sentiments and views of the respective proprietors, ensues a division of the society into different interests and parties.

The latent causes of faction are thus sown in the nature of man; and we see them everywhere brought into different degrees of activity, according to the different circumstances of civil society. A zeal for different opinions concerning religion, concerning government, and many other points, as well of speculation as of practice; an attachment to different leaders ambitiously contending for pre-eminence and power; or to persons of other descriptions whose fortunes have been interesting to the human passions, have, in turn, divided mankind into parties, inflamed them with mutual animosity, and rendered them much more disposed to vex and oppress each other than to co-operate for their common good. So strong is this propensity of mankind to fall into mutual animosities, that where no substantial occasion presents itself, the most frivolous and fanciful distinctions have been sufficient to kindle their unfriendly passions and excite their most violent conflicts. But the most common and durable source of factions has been the various and unequal distribution of property. Those who hold and those who are without property have ever formed distinct interests in society. Those who are creditors, and those who are debtors, fall under a like discrimination. A landed interest, a manufacturing interest, a mercantile interest, a moneyed interest, with many lesser interests, grow up of necessity in civilized nations, and divide them into different classes, actuated by different sentiments and views. The regulation of these various and interfering interests forms the principal task of modern legislation, and involves the spirit of party and faction in the necessary and ordinary operations of the government.

No man is allowed to be a judge in his own cause, because his interest would certainly bias his judgment, and, not improbably, corrupt his integrity. With equal, nay with greater reason, a body of men are unfit to be both judges and parties at the same time; yet what are many of the most important acts of legislation, but so many judicial determinations, not indeed concerning the rights of single persons, but concerning the rights of large bodies of citizens? And what are the different classes of legislators but advocates and parties to the causes which they determine? Is a law proposed concerning private debts? It is a question to which the creditors are parties on one side and the debtors on the other. Justice ought to hold the balance between them. Yet the parties are, and must be, themselves the judges; and the most numerous party, or, in other words, the most powerful faction, must be expected to prevail. Shall domestic manufactures be encouraged, and in what degree, by restrictions on foreign manufactures ... are questions which would be differently decided by the landed and the manufacturing classes, and probably by neither with a sole regard to justice and the public good. The apportionment of taxes on the various descriptions of property is an act which seems to require the most exact impartiality; yet there is, perhaps, no legislative act in which greater opportunity and temptation are given to a predominant party to trample on the rules of justice. Every shilling with which they overburden the inferior number, is a shilling saved to their own pockets.

It is in vain to say that enlightened statesmen will be able to adjust these clashing interests, and render them all subservient to the public good. Enlightened statesmen will not always be at the helm. Nor, in many cases, can such an adjustment be made at all without taking into view indirect and remote considerations, which will rarely prevail over the immediate interest which one party may find in disregarding the rights of another or the good of the whole.

The inference to which we are brought is, that the CAUSES of faction cannot be removed, and that relief is only to be sought in the means of controlling its EFFECTS.

If a faction consists of less than a majority, relief is supplied by the republican principle, which enables the majority to defeat its sinister views by regular vote. It may clog the administration, it may convulse the society; but it will be unable to execute and mask its violence under the forms of the Constitution. When a majority is included in a faction, the form of popular government, on the other hand, enables it to sacrifice to its ruling passion or interest both the public good and the rights of other citizens. To secure the public good and private rights against the danger of such a faction, and at the same time to preserve the spirit and the form of popular government, is then the great object to which our inquiries are directed. Let me add that it is the great desideratum by which this form of government can be rescued from the opprobrium under which it has so long labored, and be recommended to the esteem and adoption of mankind.

By what means is this object attainable? Evidently by one of two only. Either the existence of the same passion or interest in a majority at the same time must be prevented, or the majority, having such coexistent passion or interest, must be rendered, by their number and local situation, unable to concert and carry into effect schemes of oppression. If the impulse and the opportunity be suffered to coincide, we well know that neither moral nor religious motives can be relied on as an adequate control. They are not found to be such on the injustice and violence of individuals, and lose their efficacy in proportion to the number combined together, that is, in proportion as their efficacy becomes needful.

From this view of the subject it may be concluded that a pure democracy, by which I mean a society consisting of a small number of citizens, who assemble and administer the government in person, can admit of no cure for the mischiefs of faction. A common passion or interest will, in almost every case, be felt by a majority of the whole; a communication and concert result from the form of government itself; and there is nothing to check the inducements to sacrifice the weaker party or an obnoxious individual. Hence it is that such democracies have ever been spectacles of turbulence and contention; have ever been found incompatible with personal security or the rights of property; and have in general been as short in their lives as they have been violent in their deaths. Theoretic politicians, who have patronized this species of government, have erroneously supposed that by reducing mankind to a perfect equality in their political rights, they would, at the same time, be perfectly equalized and assimilated in their possessions, their opinions, and their passions.

A republic, by which I mean a government in which the scheme of representation takes place, opens a different prospect, and promises the cure for which we are seeking. Let us examine the points in which it varies from pure democracy, and we shall comprehend both the nature of the cure and the efficacy which it must derive from the Union.

The two great points of difference between a democracy and a republic are: first, the delegation of the government, in the latter, to a small number of citizens elected by the rest; secondly, the greater number of citizens, and greater sphere of country, over which the latter may be extended.

The effect of the first difference is, on the one hand, to refine and enlarge the public views, by passing them through the medium of a chosen body of citizens, whose wisdom may best discern the true interest of their country, and whose patriotism and love of justice will be least likely to sacrifice it to temporary or partial considerations. Under such a regulation, it may well happen that the public voice, pronounced by the representatives of the people, will be more consonant to the public good than if pronounced by the people themselves, convened for the purpose. On the other hand, the effect may be inverted. Men of factious tempers, of local prejudices, or of sinister designs, may, by intrigue, by corruption, or by other means, first obtain the suffrages, and then betray the interests, of the people. The question resulting is, whether small or extensive republics are more favorable to the election of proper guardians of the public weal; and it is clearly decided in favor of the latter by two obvious considerations:

In the first place, it is to be remarked that, however small the republic may be, the representatives must be raised to a certain number, in order to guard against the cabals of a few; and that, however large it may be, they must be limited to a certain number, in order to guard against the confusion of a multitude. Hence, the number of representatives in the two cases not being in proportion to that of the two constituents, and being proportionally greater in the small republic, it follows that, if the proportion of fit characters be not less in the large than in the small republic, the former will present a greater option, and consequently a greater probability of a fit choice.

In the next place, as each representative will be chosen by a greater number of citizens in the large than in the small republic, it will be more difficult for unworthy candidates to practice with success the vicious arts by which elections are too often carried; and the suffrages of the people being more free, will be more likely to centre in men who possess the most attractive merit and the most diffusive and established characters.

It must be confessed that in this, as in most other cases, there is a mean, on both sides of which inconveniences will be found to lie. By enlarging too much the number of electors, you render the representatives too little acquainted with all their local circumstances and lesser interests; as by reducing it too much, you render him unduly attached to these, and too little fit to comprehend and pursue great and national objects. The federal Constitution forms a happy combination in this respect; the great and aggregate interests being referred to the national, the local and particular to the State legislatures.

The other point of difference is, the greater number of citizens and extent of territory which may be brought within the compass of republican than of democratic government; and it is this circumstance principally which renders factious combinations less to be dreaded in the former than in the latter. The smaller the society, the fewer probably will be the distinct parties and interests composing it; the fewer the distinct parties and interests, the more frequently will a majority be found of the same party; and the smaller the number of individuals composing a majority, and the smaller the compass within which they are placed, the more easily will they concert and execute their plans of oppression. Extend the sphere, and you take in a greater variety of parties and interests; you make it less probable that a majority of the whole will have a common motive to invade the rights of other citizens; or if such a common motive exists, it will be more difficult for all who feel it to discover their own strength, and to act in unison with each other. Besides other impediments, it may be remarked that, where there is a consciousness of unjust or dishonorable purposes, communication is always checked by distrust in proportion to the number whose concurrence is necessary.

Hence, it clearly appears, that the same advantage which a republic has over a democracy, in controlling the effects of faction, is enjoyed by a large over a small republic—is enjoyed by the Union over the States composing it. Does the advantage consist in the substitution of representatives whose enlightened views and virtuous sentiments render them superior to local prejudices and schemes of injustice? It will not be denied that the representation of the Union will be most likely to possess these requisite endowments. Does it consist in the greater security afforded by a greater variety of parties, against the event of any one party being able to outnumber and oppress the rest? In an equal degree does the increased variety of parties comprised within the Union, increase this security? Does it, in fine, consist in the greater obstacles opposed to the concert and accomplishment of the secret wishes of an unjust and interested majority? Here, again, the extent of the Union gives it the most palpable advantage.

The influence of factious leaders may kindle a flame within their particular States, but will be unable to spread a general conflagration through the other States. A religious sect may degenerate into a political faction in a part of the Confederacy; but the variety of sects dispersed over the entire face of it must secure the national councils against any danger from that source. A rage for paper money, for an abolition of debts, for an equal division of property, or for any other improper or wicked project, will be less apt to pervade the whole body of the Union than a particular member of it; in the same proportion as such a malady is more likely to taint a particular county or district, than an entire State.

In the extent and proper structure of the Union, therefore, we behold a republican remedy for the diseases most incident to republican government. And according to the degree of pleasure and pride we feel in being republicans, ought to be our zeal in cherishing the spirit and supporting the character of Federalists.
PUBLIUS.

## FEDERALIST NO. 17

The Subject Continued
(The Insufficiency of the Present Confederation to Preserve the Union)

For the *Independent Journal.*

HAMILTON

To the People of the State of New York:

AN OBJECTION, of a nature different from that which has been stated and answered, in my last address, may perhaps be likewise urged against the principle of legislation for the individual citizens of America. It may be said that it would tend to render the government of the Union too powerful, and to enable it to absorb those residuary authorities, which it might be judged proper to leave with the States for local purposes. Allowing the utmost latitude to the love of power which any reasonable man can require, I confess I am at a loss to discover what temptation the persons intrusted with the administration of the general government could ever feel to divest the States of the authorities of that description. The regulation of the mere domestic police of a State appears to me to hold out slender allurements to ambition. Commerce, finance, negotiation, and war seem to comprehend all the objects which have charms for minds governed by that passion; and all the powers necessary to those objects ought, in the first instance, to be lodged in the national depository. The administration of private justice between the citizens of the same State, the supervision of agriculture and of other concerns of a similar nature, all those things, in short, which are proper to be provided for by local legislation, can never be desirable cares of a general jurisdiction. It is therefore improbable that there should exist a disposition in the federal councils to usurp the powers with which they are connected; because the attempt to exercise those powers would be as troublesome as it would be nugatory; and the possession of them, for that reason, would contribute nothing to the dignity, to the importance, or to the splendor of the national government.

But let it be admitted, for argument's sake, that mere wantonness and lust of domination would be sufficient to beget that disposition; still it may be safely affirmed, that the sense of the constituent body of the national representatives, or, in other words, the people of the several States, would control the indulgence of so extravagant an appetite. It will always be far more easy for the State governments to encroach upon the national authorities than for the national government to encroach upon the State authorities. The proof of this proposition turns upon the greater degree of influence which the State governments if they administer their affairs with uprightness and prudence, will generally possess over the people; a circumstance which at the same time teaches us that there is an inherent and intrinsic weakness in all federal constitutions; and that too much pains cannot be taken in their organization, to give them all the force which is compatible with the principles of liberty.

The superiority of influence in favor of the particular governments would result partly from the diffusive construction of the national government, but chiefly from the nature of the objects to which the attention of the State administrations would be directed.

It is a known fact in human nature, that its affections are commonly weak in proportion to the distance or diffusiveness of the object. Upon the same principle that a man is more attached to his family than to his neighborhood, to his neighborhood than to the community at large, the people of each State would be apt to feel a stronger bias towards their local governments than towards the government of the Union; unless the force of that principle should be destroyed by a much better administration of the latter.

This strong propensity of the human heart would find powerful auxiliaries in the objects of State regulation.

The variety of more minute interests, which will necessarily fall under the superintendence of the local administrations, and which will form so many rivulets of influence, running through every part of the society, cannot be particularized, without involving a detail too tedious and uninteresting to compensate for the instruction it might afford.

There is one transcendent advantage belonging to the province of the State governments, which alone suffices to place the matter in a clear and satisfactory light—I mean the ordinary administration of criminal and civil justice. This, of all others, is the most powerful, most universal, and most attractive source of popular obedience and attachment. It is that which, being the immediate and visible guardian of life and property, having its benefits and its terrors in constant activity before the public eye, regulating all those personal interests and familiar concerns to which the sensibility of individuals is more immediately awake, contributes, more than any other circumstance, to impressing upon the minds of the people, affection, esteem, and reverence towards the government. This great cement of society, which will diffuse itself almost wholly through the channels of the particular governments, independent of all other causes of influence, would insure them so decided an empire over their respective citizens as to render them at all times a complete counterpoise, and, not unfrequently, dangerous rivals to the power of the Union.

The operations of the national government, on the other hand, falling less immediately under the observation of the mass of the citizens, the benefits derived from it will chiefly be perceived and attended to by speculative men. Relating to more general interests, they will be less apt to come home to the feelings of the people; and, in proportion, less likely to inspire an habitual sense of obligation, and an active sentiment of attachment.

The reasoning on this head has been abundantly exemplified by the experience of all federal constitutions with which we are acquainted, and of all others which have borne the least analogy to them.

Though the ancient feudal systems were not, strictly speaking, confederacies, yet they partook of the nature of that species of association. There was a common head, chieftain, or sovereign, whose authority extended over the whole nation; and a number of subordinate vassals, or feudatories, who had large portions of land allotted to them, and numerous trains of INFERIOR vassals or retainers, who occupied and cultivated that land upon the tenure of fealty or obedience, to the persons of whom they held it. Each principal vassal was a kind of sovereign, within his particular demesnes. The consequences of this situation were a continual opposition to authority of the sovereign, and frequent wars between the great barons or chief feudatories themselves. The power of the head of the nation was commonly too weak, either to preserve the public peace, or to protect the people against the oppressions of their immediate lords. This period of European affairs is emphatically styled by historians, the times of feudal anarchy.

When the sovereign happened to be a man of vigorous and warlike temper and of superior abilities, he would acquire a personal weight and influence, which answered, for the time, the purpose of a more regular authority. But in general, the power of the barons triumphed over that of the prince; and in many instances his dominion was entirely thrown off, and the great fiefs were erected into independent principalities or States. In those instances in which the monarch finally prevailed over his vassals, his success was chiefly owing to the tyranny of those vassals over their dependents. The barons, or nobles, equally the enemies of the sovereign and the oppressors of the common people, were dreaded and detested by both; till mutual danger and mutual interest effected a union between them fatal to the power of the aristocracy. Had the nobles, by a conduct of clemency and justice, preserved the fidelity and devotion of their retainers and followers, the contests between them and the prince must almost always have ended in their favor, and in the abridgment or subversion of the royal authority.

This is not an assertion founded merely in speculation or conjecture. Among other illustrations of its truth which might be cited, Scotland will furnish a cogent example. The spirit of clanship which was, at an early day, introduced into that kingdom, uniting the nobles and their dependants by ties equivalent to those of kindred, rendered the aristocracy a constant overmatch for the power of the monarch, till the incorporation with England subdued its fierce and ungovernable spirit, and reduced it within those rules of subordination which a more rational and more energetic system of civil polity had previously established in the latter kingdom.

The separate governments in a confederacy may aptly be compared with the feudal baronies; with this advantage in their favor, that from the reasons already explained, they will generally possess the confidence and good will of the people, and with so important a support, will be able effectually to oppose all encroachments of the national government. It will be well if they are not able to counteract its legitimate and necessary authority. The points of similitude consist in the rivalship of power, applicable to both, and in the CONCENTRATION of large portions of the strength of the community into particular DEPOSITS, in one case at the disposal of individuals, in the other case at the disposal of political bodies.

A concise review of the events that have attended confederate governments will further illustrate this important doctrine; an inattention to which has been the great source of our political mistakes, and has given our jealousy a direction to the wrong side. This review shall form the subject of some ensuing papers.
PUBLIUS.

**FEDERALIST No. 51**

The Structure of the Government Must Furnish the Proper Checks and Balances Between the Different Departments

From the *New York Packet*.
Friday, February 8, 1788.

HAMILTON OR MADISON

To the People of the State of New York:

TO WHAT expedient, then, shall we finally resort, for maintaining in practice the necessary partition of power among the several departments, as laid down in the Constitution? The only answer that can be given is, that as all these exterior provisions are found to be inadequate, the defect must be supplied, by so contriving the interior structure of the government as that its several constituent parts may, by their mutual relations, be the means of keeping each other in their proper places. Without presuming to undertake a full development of this important idea, I will hazard a few general observations, which may perhaps place it in a clearer light, and enable us to form a more correct judgment of the principles and structure of the government planned by the convention.

In order to lay a due foundation for that separate and distinct exercise of the different powers of government, which to a certain extent is admitted on all hands to be essential to the preservation of liberty, it is evident that each department should have a will of its own; and consequently should be so constituted that the members of each should have as little agency as possible in the appointment of the members of the others. Were this principle rigorously adhered to, it would require that all the appointments for the supreme executive, legislative, and judiciary magistracies should be drawn from the same fountain of authority, the people, through channels having no communication whatever with one another. Perhaps such a plan of constructing the several departments would be less difficult in practice than it may in contemplation appear. Some difficulties, however, and some additional expense would attend the execution of it. Some deviations, therefore, from the principle must be admitted. In the constitution of the judiciary department in particular, it might be inexpedient to insist rigorously on the principle: first, because peculiar qualifications being essential in the members, the primary consideration ought to be to select that mode of choice which best secures these qualifications; secondly, because the permanent tenure by which the appointments are held in that department, must soon destroy all sense of dependence on the authority conferring them.

It is equally evident, that the members of each department should be as little dependent as possible on those of the others, for the emoluments annexed to their offices. Were the executive magistrate, or the judges, not independent of the legislature in this particular, their independence in every other would be merely nominal.

But the great security against a gradual concentration of the several powers in the same department, consists in giving to those who administer each department the necessary constitutional means and personal motives to resist encroachments of the others. The provision for defense must in this, as in all other cases, be made commensurate to the danger of attack. Ambition must be made to counteract ambition. The interest of the man must be connected with the constitutional rights of the place. It may be a reflection on human nature, that such devices should be necessary to control the abuses of government. But what is government itself, but the greatest of all reflections on human nature? If men were angels, no government would be necessary. If angels were to govern men, neither external nor internal controls on government would be necessary. In framing a government which is to be administered by men over men, the great difficulty lies in this: you must first enable the government to control the governed; and in the next place oblige it to control itself. A dependence on the people is, no doubt, the primary control on the government; but experience has taught mankind the necessity of auxiliary precautions.

This policy of supplying, by opposite and rival interests, the defect of better motives, might be traced through the whole system of human affairs, private as well as public. We see it particularly displayed in all the subordinate distributions of power, where the constant aim is to divide and arrange the several offices in such a manner as that each may be a check on the other that the private interest of every individual may be a sentinel over the public rights. These inventions of prudence cannot be less requisite in the distribution of the supreme powers of the State.

But it is not possible to give to each department an equal power of self-defense. In republican government, the legislative authority necessarily predominates. The remedy for this inconveniency is to divide the legislature into different branches; and to render them, by different modes of election and different principles of action, as little connected with each other as the nature of their common functions and their common dependence on the society will admit. It may even be necessary to guard against dangerous encroachments by still further precautions. As the weight of the legislative authority requires that it should be thus divided, the weakness of the executive may require, on the other hand, that it should be fortified. An absolute negative on the legislature appears, at first view, to be the natural defense with which the executive magistrate should be armed. But perhaps it would be neither altogether safe nor alone sufficient. On ordinary occasions it might not be exerted with the requisite firmness, and on extraordinary occasions it might be perfidiously abused. May not this defect of an absolute negative be supplied by some qualified connection between this weaker department and the weaker branch of the stronger department, by which the latter may be led to support the constitutional rights of the former, without being too much detached from the rights of its own department?

If the principles on which these observations are founded be just, as I persuade myself they are, and they be applied as a criterion to the several State constitutions, and to the federal Constitution it will be found that if the latter does not perfectly correspond with them, the former are infinitely less able to bear such a test.

There are, moreover, two considerations particularly applicable to the federal system of America, which place that system in a very interesting point of view.

First. In a single republic, all the power surrendered by the people is submitted to the administration of a single government; and the usurpations are guarded against by a division of the government into distinct and separate departments. In the compound republic of America, the power surrendered by the people is first divided between two distinct governments, and then the portion allotted to each subdivided among distinct and separate departments. Hence a double security arises to the rights of the people. The different governments will control each other, at the same time that each will be controlled by itself.

Second. It is of great importance in a republic not only to guard the society against the oppression of its rulers, but to guard one part of the society against the injustice of the other part. Different interests necessarily exist in different classes of citizens. If a majority be united by a common

interest, the rights of the minority will be insecure. There are but two methods of providing against this evil: the one by creating a will in the community independent of the majority that is, of the society itself; the other, by comprehending in the society so many separate descriptions of citizens as will render an unjust combination of a majority of the whole very improbable, if not impracticable. The first method prevails in all governments possessing an hereditary or self-appointed authority. This, at best, is but a precarious security; because a power independent of the society may as well espouse the unjust views of the major, as the rightful interests of the minor party, and may possibly be turned against both parties. The second method will be exemplified in the federal republic of the United States. Whilst all authority in it will be derived from and dependent on the society, the society itself will be broken into so many parts, interests, and classes of citizens, that the rights of individuals, or of the minority, will be in little danger from interested combinations of the majority. In a free government the security for civil rights must be the same as that for religious rights. It consists in the one case in the multiplicity of interests, and in the other in the multiplicity of sects. The degree of security in both cases will depend on the number of interests and sects; and this may be presumed to depend on the extent of country and number of people comprehended under the same government. This view of the subject must particularly recommend a proper federal system to all the sincere and considerate friends of republican government, since it shows that in exact proportion as the territory of the Union may be formed into more circumscribed Confederacies, or States oppressive combinations of a majority will be facilitated: the best security, under the republican forms, for the rights of every class of citizens, will be diminished: and consequently the stability and independence of some member of the government, the only other security, must be proportionately increased. Justice is the end of government. It is the end of civil society. It ever has been and ever will be pursued until it be obtained, or until liberty be lost in the pursuit. In a society under the forms of which the stronger faction can readily unite and oppress the weaker, anarchy may as truly be said to reign as in a state of nature, where the weaker individual is not secured against the violence of the stronger; and as, in the latter state, even the stronger individuals are prompted, by the uncertainty of their condition, to submit to a government which may protect the weak as well as themselves; so, in the former state, will the more powerful factions or parties be gradually induced, by a like motive, to wish for a government which will protect all parties, the weaker as well as the more powerful? It can be little doubted that if the State of Rhode Island was separated from the Confederacy and left to itself, the insecurity of rights under the popular form of government within such narrow limits would be displayed by such reiterated oppressions of factious majorities that some power altogether independent of the people would soon be called for by the voice of the very factions whose misrule had proved the necessity of it. In the extended republic of the United States, and among the great variety of interests, parties, and sects which it embraces, a coalition of a majority of the whole society could seldom take place on any other principles than those of justice and the general good; whilst there being thus less danger to a minor from the will of a major party, there must be less pretext, also, to provide for the security of the former, by introducing into the government a will not dependent on the latter, or, in other words, a will independent of the society itself. It is no less certain than it is important, notwithstanding the contrary opinions which have been entertained, that the larger the society, provided it lie within a practical sphere, the more duly capable it will be of self-government. And happily for the REPUBLICAN CAUSE, the practicable sphere may be carried to a very great extent, by a judicious modification and mixture of the FEDERAL PRINCIPLE.
PUBLIUS.

# *"INTRODUCTION"*

## FREDERICK QUINN

## INTRODUCTION

The long coastline was fair prey for foreign invaders. Roads were few, muddy when it rained, dusty otherwise. Transportation was slow and irregular, most dependable by water. The potentially prosperous, primarily agrarian economy was stagnant, owing to the recent eight-year war, and entrepreneurial people were not sure how it would improve. Scattered insurrections flared, and the prospect of angry mobs or unschooled peasants taking the law into their hands threatened whatever form of government the newly independent states selected. The central government was powerless, lacking authority to raise funds or an army, or to administer justice. Politicians debated at length whether the existing government should be patched up, or if there should be a strong president, a president and council with shared powers, or a legislature with most powers vested in it; but the discussions went nowhere.

The confederation's thirteen isolated states were in infrequent contact with one another, except for commerce along the main maritime arteries. Spanish, French, English, and other metallic coins still circulated long after the war; the Continental Congress's money was valueless. "Not worth a continental" was a popular expression. The wartime military leader, George Washington, wrote state governors in 1783 that he feared "the union cannot be of long duration, and everything must very rapidly tend to anarchy and confusion." Thomas Jefferson, then Minister to France, said, "We are the lowest and most obscure of the whole diplomatic tribe." A British cleric said Americans were "a disunited people till the end of time, suspicious and distrustful to each other, they will be divided and subdivided into little commonwealths, or principalities."

These conditions, which America faced two centuries ago, are applicable to many modern nations. *The Federalist Papers,* first published in 1787–88 in the middle of intense debates over what form the new government should take, explain how the authors of the U.S. Constitution arrived at that document. There is a congruence between basic issues of governance raised then in Philadelphia and now in Warsaw, Conakry, Brasilia, and Moscow. The questions are not rhetorical or theoretical, but are fundamental to the formation of a national government to which all citizens can subscribe, and that will endure.

If the American Revolution was a time of political upheaval, the writing and ratification of the U.S. Constitution was no less revolutionary. The Constitution's framers boldly exceeded their mandate to suggest ways of patching up the Articles of Confederation. The ratification debates required the people to decide whether they would adopt an untried form of government or hold on to an ineffectual one that was sure to result in the balkanization of the new nation. The temptation to maintain individual state sovereignty was strong, even though many conceded the necessity of regional defense treaties. If the Anti-Federalists had prevailed, the sketch in *Federalist* No. 2 of the potentially prosperous nation would have remained an exercise in mapmaking. Far from defending the status quo, *The Federalist Papers,* in measured argument, seek support for a revolutionary form of government, unknown in world history to that date.

Fortunately, the *Federalist* authors—Alexander Hamilton, John Jay, and James Madison— wrote their work at a propitious moment in American history. A few years earlier, in the shadow of the war and the British crown, a proposal for a strong central government would not have found acceptance.[1] Nor could such a concept have succeeded during the Age of Jackson, just around the corner. Its republican features, which kept the people somewhat distant from the reins of power, would have been voted down. *The Federalist Papers* thus explain a revolutionary document that faced a hotly contested ratification battle for the political soul of a nation at a critical turning point in its history.[2]

*The Federalist Papers* reflect the end of an era in America, a chapter that began with the Mayflower Compact of 1620 and the various covenants, declarations, and state constitutions that followed, and culminated in the Declaration of Independence and Constitution. During that period of more than a century and a half, American political thought was formulated and tried, and arguments were rehearsed and refined in press, pulpit, and legislative chamber, often to express opposition to the British crown, but also to give an expanding country a workable government. It was against such a background that *The Federalist Papers* emerged, combining the traits Robert A. Ferguson ascribes to the Constitution: "generic strength, manipulative brilliance, cunning restraint, and practical eloquence."[3]

Despite their length, the papers are remarkably concise—long enough to establish their argument and answer opponents, but free of invective, extraneous commentary, or florid embellishment. *The Federalist Papers'* grounding in eighteenth-century philosophy and economic theory is only tantalizingly suggested in brief sections seeded throughout the essays. We wish for an additional hour of tavern talk with Madison or Hamilton, or a public television interview program tying up loose ends on the origins of their ideas; but the information is not forthcoming. The authors were primarily practitioners rather than theorists, and *The Federalist Papers* were written for a specific purpose: to convince delegates to New York's ratification convention of the value of a particular course of action.

As such, the essays are a radical, revolutionary statement of well-reasoned political thought, carefully moving beyond the central ideas raised by theorists like Hume, Locke, or Montesquieu. Instead of dramatically overthrowing the old order of theory and practice, the Constitution writers, with careful study, took its best features and gave new meaning to them. As works of theory and guides for practice, the essays are more lasting than anything written by Marx, Lenin, Mao, Castro, or Metternich.

*The Federalist Papers* represent the most long-lived contributions of a golden age of pamphlet literature. It was a time when public service, most leaders believed, was a responsibility mandated by the Deity, and public documents often reflected a literary quality comparable to contemporary sermons or works of science, history, or political or moral thought.[4] Simultaneously there were improvements in the technology and availability of printing presses; the growth of a relatively affluent, lettered audience; and the emergence of urgent and revolutionary issues, like the coming of age of republican political thought and the question of assembling a machinery of government for the polities that had just defeated the British forces and now must govern themselves.

## THE ARTICLES OF CONFEDERATION

Had the Articles of Confederation not failed, there would have been no Constitution and no *Federalist Papers.* Two centuries later, it is difficult to imagine the chaotic state of America in the postrevolutionary period. A war h a d  been won, but the eastern seaboard lay vulnerable to potential invaders. The economy was plagued by multiple currencies and tariffs; state governments were bankrupt and ineffectual; and the central government was central in name only. From 1776 to 1787 America was a loose alliance of states governed by the Articles, whose fatal flaw was that power remained with individual states. The central government could neither raise revenues nor enact legislation binding on individual states. The votes of nine of the thirteen states were required to pass laws, and a unanimous vote was necessary to effect any fundamental change in the Articles.

The central government's weakness was intentional; the American settlers had bitterly resented the British crown's power to control commerce and collect taxes. The legislative body created under the Articles was powerless, and there was no executive or judicial branch. Moreover, the thirteen states each had separate political and commercial interests, and the temporary unity forged from a decade of active hostility toward Great Britain failed to produce a national identity. Nine states had navies; seven printed their own currency; most had tariff and customs laws. New York charged duties on ships moving firewood or farm produce to and from neighboring New Jersey and Connecticut. When soldiers remarked, "New Jersey is our country," they echoed the widespread sentiment of other states.

Also contributing to political chaos in the 1780s were the insolvent state governments. Hamilton, in a stinging attack on the Articles, remarked in *Federalist* No. 9 that they encouraged "little, jealous, clashing, tumultuous commonwealths, the wretched nurseries of unceasing discord." Madison had the bankrupt state governments in mind in *Federalist* No. 10 when he described the need to "secure the national councils against any danger from . . . a rage for paper money, for an abolition of debts, for an equal division of property, or for any other improper or wicked project." Madison wrote on October 24, 1787, to Jefferson in France that the unstable state legislatures "contributed more to that uneasiness which produced the convention, and prepared the public mind for a general reform, than those which accrued to our national character and interest from the inadequacy of the confederation to its immediate objects."[5]

## TOWARD PHILADELPHIA

Trade disputes festered among the states in this disruptive setting. A commercial quarrel between Maryland and Virginia over an oyster fishery and navigation rights triggered the meeting that produced plans for a constitutional convention. The Maryland and Virginia delegates invited representatives from other states to meet "to take into consideration the trade of the United States." The issue was larger than informal negotiators could solve, and, at a meeting in Annapolis, Maryland, in the fall of 1786 to resolve commercial disputes, Hamilton and Madison urged delegates from the five states present to convoke all thirteen "to meet at Philadelphia on the second Monday in May next, to take into consideration the situation of the United States, to devise such further provisions as shall appear to them necessary to render the Constitution of the Federal Government adequate to the exigencies of the union." Congress endorsed the convention but gave it the limited mandate of recommending revision of the Articles.

A galvanic event that occurred shortly before the convention met was an agrarian debtors' uprising in western Massachusetts, Shays's Rebellion. Daniel Shays, an officer who had served in the Revolution, was an impoverished debtor by the winter of 1786. In an attempt to gain relief from unpaid debts, Shays and over a thousand destitute followers, armed with pitchforks and staves, tried to prevent county courts from sitting. They wanted tax relief, paper money, and the state capital moved westward from Boston. Their attempt to seize a federal arsenal was thwarted by the local militia. Funded by merchants' subscriptions, the rebels then hastily recruited Boston college students. The rebellion's leaders were hunted down in a snowstorm, sentenced to death, then pardoned or given short prison terms.

The event, which was replicated in several western Massachusetts towns, sent shock waves across the states because of who the participants were, how the crisis was handled, and the divisive issues it raised. The uprising influenced the states' decision to support a central government capable of enforcing public order. Many believed that creating a popular democracy with power vested in local authorities invited anarchy, tyranny, or dismemberment of the body politic.

## THE CONSTITUTIONAL CONVENTION

The country's future form of government was shaped during four months' deliberations beginning in late May 1787. Twelve states named seventy-four delegates to the convention; only fifty-five came to Philadelphia, and some arrived well after a quorum assembled on May 25. The average age of the delegates was forty-two; five were less than thirty years of age. Madison was thirty-six, Hamilton thirty. Thirty-four representatives were trained as lawyers, many others were merchants and planters. About 60 percent had attended college; Princeton, Yale, Harvard, and Columbia were well represented. The delegates came mostly from their states' tidewater regions. They were professionals and property owners, although some ended their lives in penury. A stabilizing force during the deliberations was the presence of George Washington, convention president, war hero, and likely choice for first president of the United States. Another respected figure was eighty-one-year-old Benjamin Franklin, who contributed infrequently, but importantly, to discussions, and who entertained delegates often at his nearby lodgings during the hot Philadelphia summer.

In modern times, it would be unthinkable for more than fifty leaders to spend nearly four months in intense political deliberations without press coverage, media leaks, or detailed copies of speeches and votes made public. Yet the convention agreed to meet secretly and keep no record of votes, allowing issues to evolve and delegates to change positions. Madison's notes, which were not published until 1840, remain the principal source of subsequent information about the proceedings.

The four-month convention produced several compromises that made the Constitution possible. Shortly after the convention opened, the Virginia delegation tabled a proposal creating three distinct branches of government and a national legislature with power to negate and supersede state laws. A dispute flared over Virginia's proposal that membership in both branches of the legislature should be proportionate to a state's population. Fearing a loss of power, smaller states objected, and New Jersey's William Patterson offered a counterproposal, suggesting that states have an equal number of delegates regardless of size—one state, one vote. Alexander Hamilton observed correctly, "It is a contest for power, not for liberty." Roger Sherman of Connecticut then proposed a compromise, later adopted; membership in the Senate would be limited to two senators per state, and membership in the House of Representatives would be relative to population. Additionally, tax bills and revenue measures would originate in the House, where citizens would be represented by a greater number of members.

Once these compromises were reached, the remaining issues—many as potentially significant as the questions already decided—were soon sorted out. Powers to collect taxes, regulate commerce, and support a national army, which had been denied to the Continental Congress in the Articles of Confederation, were included in the Constitution with little opposition.

Although delegates differed on the new government's structural details, they accepted awarding the national government a more powerful role than anything previously contemplated in America. Basically, the government's powers were expanded, the new Constitution became the "supreme law of the land," and the national government was no longer subservient to state governments. It could reach the people directly through laws, courts, administrative agents, and the previously denied power to raise revenues and armies. "The federal and state governments are in fact but different agents and trustees of the people, with different powers and designed for different purposes," Madison wrote in No. 46. There were thus two distinct governments, state and national, derived from the people and responsible to them. The constitutional government was a democratic republic—democratic because the people were represented, republican because there were restraints on what people and government could do.

Madison and his colleagues engineered a solution to a problem plaguing American government at that time. Instead of depending on the states for its lifeblood of power and funds, the new national government reached down and around states directly to the citizenry, establishing both the central government's authority and creating possibilities for perennial tension as well. Individual rights, states' rights, and the intrusions of "big government" would later become substantive issues in American law and politics, especially with the exponential rise in population and the federal government's influence in the late twentieth century.

When mid-September arrived, some tempers were short, but the convention's principal work was done, and several delegates departed. Thirty-nine of the original fifty-five delegates signed the document. Only three delegates declined to endorse it, and four others who opposed it were absent.

Madison's final notes of the historic event state:

Whilst the last members were signing it, Doctor Franklin, looking towards the President's chair [Washington was the presiding officer], at the back of which a rising sun happened to be painted, observed to a few members near him, that painters had found it difficult to distinguish in their art a rising from a setting sun. I have, said he, often and often in the course of the session, and the vicissitudes of my hopes and fears as to its issue, looked at that behind the President without being able to tell whether it was rising or setting: but now at length I have the happiness to know that it is a rising and not a setting sun.[6]

With a deft turn of phrase, Gouverneur Morris made it appear consent was unanimous in the document's concluding lines, since at least one representative from each state in attendance signed: "Done in Convention, by unanimous consent of the States present the 17 September." Then, at 4 P.M. that same day, "The Members adjourned to the City Tavern, dined together, and took cordial leave of each other."

Although the prescient Franklin compared the proceedings to a rising sun, the Constitution's fate was uncertain. The slavery issue had been dealt with to the satisfaction of neither northerners nor southerners; inland farmers and pioneers feared the political and commercial power of tidewater planters; and landed interests, like New York's Governor George Clinton, were apprehensive of Congress's taxation powers. Creditable leaders of the revolutionary generation, like Patrick Henry and Samuel Adams, cast a cold eye on the new government's centralized powers, and there was growing sentiment throughout the states for a bill of rights.

In spite of the fact that their deliberations focused on immediate problems, the Constitution writers, in several instances, both showed a keen awareness of historical precedent and realized they were writing for posterity. Madison records Gouverneur Morris: "He came here as a representative of America; he flattered himself he came here in some degree as a Representative of the whole human race; for the whole human race will be affected by the proceedings of this Convention. He wished gentlemen to extend their views beyond the present moment of time; beyond the narrow limits of place."[7]

Hamilton reflected this larger perspective in *Federalist* No. 34:

We must bear in mind that we are not to confine our view to the present period, but to look forward to remote futurity. Constitutions of civil governments are not to be framed upon a calculation of existing exigencies, but upon a combination of these with the probable exigencies of the ages, according to the natural and tried course of human affairs.

Still, in September 1787 the delegates were not certain their results would last more than a few years. Someone said all they had was a piece of paper and George Washington. They believed the summer's deliberations and compromises had resulted in a practical, workable frame of government, but many were pessimistic. Alexander Hamilton brooded that the Constitution represented "a weak and worthless fabric"; another delegate called it "the Continental Congress in two volumes instead of one."

The delegates' harmonious evening interlude of September 17, 1787, was a brief calm before a yearlong storm of argument over ratification.

## AN UNCERTAIN FATE

The Constitution was forwarded to the Continental Congress on September 20, three days after being signed. Reaction was mixed; some members of Congress wanted to censure convention delegates for exceeding their mandate. Nevertheless, Congress moved on September 28 to ratify the document. A majority of nine states was required for the Constitution to become law. Convention delegates—skeptical of popular democratic power and of entrenched powerbrokers, who would oppose any centralization of power—had written into the new Constitution's Article 7 that special conventions, not state legislatures, were required for ratification. The Constitution's supporters eschewed popular referendums, believing that special conventions, whose members were thus "refined" or "filtered," in Madison's language, would produce enlightened delegates and assure the Constitution's passage.

But victory was far from certain. Considerable literature was directed against the Constitution, much of it of a high order.[8] Distinguished patriots opposed the document, including Virginia's George Mason, landowner and author of the Virginia Constitution and Bill of Rights, who called George Washington an "upstart surveyor"; and Thomas Paine, who echoed a widespread sentiment: "That government is best which governs least." The Constitution's opponents included well-known patriots, backcountry farmers, small landholders, artisans, and laborers. They resisted a strong presidency, which "squints toward monarchy," and a Congress with powers of taxation, which could "clutch the purse with one hand and wave the sword with the other," as Henry put it. Madison's opponents believed the ideas incorporated in the Constitution were elitist (*aristocratic* was the word used in the 1780s), favoring the ruling elements, owners of businesses and property, and ignoring small farmers, workers, and those at the margins of society.

The lack of a bill of rights was a major issue with opponents and the undecided, as it was with many of the new Constitution's supporters. Historians of a later generation called it "an *Iliad,* or Parthenon, or Fifth Symphony of statesmanship"; but New York's John Lansing said the Constitution was "a triple headed monster, as deep and wicked a conspiracy as ever was invented in the darkest ages against the liberties of a free people." It was against this backdrop that *The Federalist Papers* appeared.

## THE NAME *FEDERALIST*

Proponents of the American Constitution gained a tactical advantage over those who opposed it or had reservations by claiming the name *Federalist* for themselves and by calling opponents *Anti-Federalists*. In the 1780s, partisans of a strong national government were called *nationalists*. Federalists, in contrast, supported state sovereignty and opposed a dominant national government. By preempting the title *Federalist,* Hamilton and his coauthors gained an advantage for their position and avoided an all-out confrontation over the issue of state versus national power. They appeared as supporters of states' rights, a theme elucidated in *The Federalist Papers,* yet were clearly advocates as well of a strong national government.

## PUBLIUS

The essays were signed "Publius." Classical pseudonyms and allusions to Greek and Roman history were popular with eighteenth-century American authors, and towns across the country were named Athens, Sparta, Rome, and Ithaca. Statues of heroes, like Washington or Jefferson, showed them clad in togas, with Roman features. Hamilton, originator of *The Federalist Papers,* made a shrewd choice in Publius. The name referred to Publius Valerius, the state builder who restored the Roman republic following the overthrow of Tarquin, Rome's last king. Plutarch compared Publius favorably to Solon, Greece's law giver. Now a modern Publius would help build the new American republic.

In selecting a name like Publius, the *Federalist* authors followed a practice common among eighteenth-century public document writers. They published a collaborative work under a pseudonym rather than a byline. If Hamilton, Madison, and Jay had publicized their authorship of *The Federalist Papers*, they would have been identified as advocates of particular positions, and they, rather than their arguments, would have become part of the debate over the Constitution. Likewise, if Americans in 1787 had known it was Pennsylvania's aristocratic, conservative Gouverneur Morris who had taken the Constitution's penultimate drafts and composed the final document, it could have diminished chances for ratification. But Morris's anonymity was preserved until the 1830s, as was authorship of *The Federalist Papers* until the 1840s. Hamilton, Madison, and Jay thus upheld a practice common in that era, creating a veil of anonymity that forced readers to focus on arguments rather than authors. This allowed politicians to develop ideas free from public pressures, change their minds during deliberations, and explore differences until conclusions were reached.

Choosing anonymity was also a function of the rivalry between Hamilton and New York Governor George Clinton. Hamilton was the only New York delegate who signed the Constitution. Clinton was an Anti- Federalist leader in a state where those opposing the new Constitution held a commanding majority. Choosing Publius was, in part, an attempt to move the debate away from the strong personal animosity between Hamilton and Clinton.

## THE FEDERALIST PAPERS PLAN

A modern reader, interested in the papers' core arguments, will be drawn to approximately twenty essays that retain lasting appeal. Their broad categories were enumerated by Hamilton in *Federalist* No. 1, in which he proposes that his plea for union and republicanism be divided clearly into six "branches of inquiry": (1) the "utility of union," (2) "the insufficiency of the present Confederation," (3) the necessity of energetic government, (4) the Constitution's republican nature, (5) its compatibility with state constitutions, and (6) the security it provides both liberty and property.

The series has a careful unity, not always apparent to later readers. The first thirty-seven papers detail shortcomings in the Articles of Confederation. Numbers 2–14 advocate a federal union as opposed to independent states and argue that a large country supports democracy better than a small country, a much-debated issue in the 1780s.

Madison's No. 10, which in modern times has become the best-known single essay, contains his famous definition of faction and his observations on human nature. After an enumeration of the Articles' flaws in Nos. 15–22, Nos. 23–36 make the case for the new government.

The argument for the Constitution is then stated positively in Nos. 37–51. The House of Representatives is described in Nos. 52–61, the Senate in Nos. 62–66, the presidency in Nos. 67–77, and the judiciary, including the principle of judicial review, in Nos. 78–83. Number 84 contends a Bill of Rights is not necessary because rights are provided throughout the Constitution; in No. 85 Hamilton presents an upbeat finale, closing the series on the optimistic note with which it began.

Although the series has a clearly enunciated argument line and organization, certain papers and sections of papers—for example, Nos. 10, 37, 39, 47, 51, 78, and 84—move beyond skilled polemics to a place among the most profound commentaries ever published on human behavior in political society. These include some carefully shaped darker passages about the failure of the Articles of Confederation and humanity's propensity to greed and self-interest, and passages of controlled enthusiasm on future hopes for the young republic.

Even though the papers were turned out quickly and the principal authors represented different viewpoints, they were still the product of careful planning and execution. It was Hamilton's tactical brilliance that saw the need for this comprehensive project. He conceived the idea of writing *The Federalist Papers,* produced the outline, and recruited the writers, deciding against other potential contributors whose attempts at essays he regarded as mediocre. Hamilton and Madison were in close contact for over a year, lived under the failed Articles, and spent the summer of 1787 debating constitutional issues in Philadelphia. The careful reasoning and construction of some papers and sections of others, like Nos. 10 and 51, suggest their authors prepared some arguments in advance and reshaped them for use in the series. Madison could draw on several years' worth of carefully refined notes on earlier experiments in government, the results of long study and reflection.

The writers did not lack subject matter; the challenge was to order arguments convincingly. This took place against a barrage of pamphlets and newspaper commentaries from opponents like "Cato" and "Brutus" and the support of bombastic friends like "Caesar." Still, *The Federalist* authors seized the high ground with a thoughtful format and carefully crafted arguments, avoiding personal attacks and the ambushes and broadsides of traditional political journalism.

## PUBLICATION: OCTOBER 27, 1787–AUGUST 15, 1788

*The Federalist Papers* were originally published to win New York's support for ratification of the Constitution. Ratification was to be decided at a special convention; thus, the essays aimed to influence the delegate selection process by building support for the Federalist candidates for election to the June 1788 convention.

The essays, part of a deluge of pamphlets and newspaper articles for and against the Constitution, first ran in New York's *Independent Journal* on October 27, 1787, and appeared on Wednesdays and Saturdays; on Tuesdays they were printed in another paper, and on Wednesdays and Thursdays in still another publication. When the *New York Journal* began circulating in 1788, it carried them as well. This outpouring extended until the following April, with the publication of No. 77, Hamilton's concluding essay on the presidency. The series resumed on June 17, with the important essays on the judiciary, and continued until August 15, when the last of the 85 works was published.

*The Federalist Papers* appeared in book form before the newspaper series ended; Hamilton dispatched them to Madison in Virginia to distribute to that state's delegation prior to its vote on the Constitution. Otherwise, their circulation was confined largely to New York, with scattered printings in Pennsylvania and some New England states. A French edition appeared in 1792, followed in the next two centuries by over a hundred editions or reprints in English and at least twenty foreign language editions. A Portuguese edition was published in Rio de Janeiro in 1840, a condensed German version in Bremen in 1864, and a Spanish edition in Buenos Aires in 1868.

Ironically, a French diplomat, writing in New York in 1788, found *The Federalist* not worth commenting on. It was "of no use to the well-informed, and . . . too learned and too long for the ignorant." A contemporary New York newspaper lamented "the dry trash of Publius in 150 numbers," and "Twenty-Seven Subscribers" protested the *Journal's* "cramming us with the voluminous Publius," which "has become nauseous, having been served up to us no less than in two other papers on the same day."[9]

Alexander Hamilton arranged for the publication of *The Federalist Papers*. According to popular lore, he wrote *Federalist* No. 1 on board a ship bringing him down the Hudson River from Albany to New York City in early October 1787. He invited Madison to coauthor the series only after other choices either had declined or written unusable essays. Hamilton and Madison were both in New York in the fall of 1787 as congressional delegates and could confer with one another. Hamilton probably wrote Nos. 1, 6–9,11–13, 15–17, 21–36, 59–61, and 65–85. Madison wrote the important Nos. 10 and 51, and 14, 37–58, and most likely 62–63. Madison and Hamilton may have jointly authored Nos. 18–20. Madison's contributions ended in March 1788 when he returned to Virginia for that state's ratification debates. John Jay, who was originally expected to have a more significant role in the project, became ill during the winter and wrote only five essays, Nos. 2–5 and 64. The authors, although drawn from the political and economic leadership of their time, represented both similarities and differences in viewpoints.

## THE AUTHORS

Alexander Hamilton (1757–1804) was born on the island of Nevis in the West Indies of a Scottish merchant father and a mother of Huguenot descent. His family origins gave rise to romantic speculation; John Adams, his avowed opponent, called him the "bastard brat of a Scotch peddler." Hamilton entered Columbia University, then called King's College, at age sixteen. During the American Revolution he rose to officer rank and became George Washington's aide-de-camp and private secretary for four years. After the war, Hamilton studied law and practiced successfully in New York, entering Congress in 1782. In addition to originating *The Federalist Papers,* Hamilton was an energetic author on other subjects, having written in support of the Boston patriots and later founding a newspaper in New York. Handsome, intense, aggressive, and self-assured to the point of arrogance, Hamilton married the socially prominent daughter of a rich New York merchant. Hamilton was not an original thinker, but possessed a well-disciplined legal mind, skills in public debate, and an ability to lay issues before the public in a compelling manner. In Washington's cabinet, Hamilton served as secretary of the treasury until 1795, when he resigned to return to the practice of law, remaining a close advisor to Washington until the latter's death. Hamilton was a leader in the Federalist Party and in later years was often in conflict with his coauthor Madison. A proponent of nationalism but not direct democracy, Hamilton once said, "Men are reasoning rather than reasonable animals." This dashing figure, filled with promise, was killed in a duel with Aaron Burr in Weehawken, New Jersey, in 1804.

Hamilton and Madison (1751–1836) were a study in contrasts. Scion of an established Virginia family, Madison was a deliberate, rather than a dramatic, public figure, who counted on his careful preparation, an instinct for politics, and meticulously crafted arguments to carry the day. Ralph Ketchum, biographer of Madison, calls his subject "an ardent revolutionist, resourceful framer of government, clever political strategist, cautious, sometimes ineffectual leader."[10] Madison was raised on a four-thousand-acre tidewater plantation. He later studied at Princeton, where he stayed to tutor in political thought with John Witherspoon, the Scottish pastor, intellectual, and the university's president. Madison was well-read in classical and modern writers on politics and history, had thought long and carefully about the relationship of Protestant Christianity to the state, and knew Latin, Greek, Hebrew, and French. He served in the Continental Congress from

1779 to 1783 and in the House of Representatives from 1789 to 1797. He was Jefferson's secretary of state from 1801 to 1808, and president from 1808 to 1816, after which he retired to his Orange County, Virginia, estate, living there until his death in 1836. He is most remembered in history as principal drafter of the Constitution and the Bill of Rights, although he had argued originally a Bill of Rights was not needed, reasoning that the state and federal constitutions guaranteed individual rights sufficiently.

John Jay (1745–1829) was born into an established New York merchant family. His contribution to *The Federalist Papers* was minimal. Severe rheumatism limited him to writing essays Nos. 2 through 5, and No. 64 on the Senate. Like Hamilton, Jay was a successful lawyer and graduate of King's College. An author of the New York State Constitution, he served as president of the Continental Congress in 1778, as ambassador to Spain, and as secretary for foreign affairs from 1784 to 1789. In 1781 he participated in negotiating the treaty that ended hostilities with Great Britain. Jay became the first chief justice of the United States in 1789, and in 1795 he began the first of two terms as governor of New York. At age fifty-six, he retired from active political life to his Westchester County, New York, estate. Jay was a landowner who believed "the people who own the country ought to govern it."

## NATIONAL SECURITY: THE PREEMINENT ISSUE

There were several issues in the "great national discussion" of 1787 and 1788 to which *The Federalist Papers* spoke. But the authors began with the threat of external and internal danger, the "safety" of the young republic. With memories of the recent war with Britain fresh and the weakness of the Continental Congress apparent, no issue was more important to the Constitution writers than national security. The Federalists believed only a strong central government could defend the country's borders and promote commerce. Hamilton wrote in No. 34, "Let us recollect that peace or war will not always be left to our option; that however moderate or unambitious we may be, we cannot count upon the moderation, or hope to extinguish the ambition of others." Hamilton spoke in No. 34 of the "fiery and destructive passions of war," which are more prevalent than "the mild and beneficent sentiments of peace." He urged a strong national government to provide defenses the republic lacked, and observed, "To model our political systems upon speculations of lasting tranquility would be to calculate on the weaker springs of the human character."

## DEMOCRATIC VERSUS REPUBLICAN GOVERNMENT

Certain key words recur in *The Federalist Papers*. Their use is deceptively simple. At first glance, they appear to be common adjectives and nouns; in reality, they carefully move republican political thought of the time decisively ahead, from episodic theoretical insights to a bold but yet untried plan for governing a new nation. Madison recognized the challenge. His explanation of the inadequacies of political language in No. 37 is more than a philosophical aside:

> Besides the obscurity arising from the complexity of objects, and the imperfection of the human faculties, the medium through which the conceptions of men are conveyed to each other adds a fresh embarrassment. The use of words is to express ideas. When the Almighty himself condescends to address mankind in their own language, his meaning, luminous as it may be, is rendered dim and doubtful by the cloudy medium through which it is communicated.

Here are the essential words. The authors wanted a *robust, energetic,* and *vigorous* government; they regarded *faction* as a great enemy of constitutional government; the dangers of uncontrolled popular government had *to be filtered* and *refined* through republicanism. This was done through "framing a government," Madison wrote in *Federalist* No. 51. *Framing* meant not only defining government's outer limits or parameters, but giving government internal form and cohesion as well.

In the worldview of *The Federalist Papers'* authors, the domains of politics, science, and religion were interwoven, and a graduate of one of the handful of eastern universities would be as conversant about the ideas of reformed Protestantism in politics as about developments in Newtonian physics. *Sphere, body,* and *orbit* are words lifted from eighteenth-century natural science; *The Federalist Papers'* writers move them directly into political literature, suggesting the order the new Constitution will provide.

There is a carefully planned use of political space in *The Federalist Papers.* The compact land described by Jay in No. 2, reminiscent of scenes depicted by early American landscape artists, extends gradually as the Confederation's limited confines are pushed back. By the time a defense of the new Constitution is introduced by Hamilton in No. 23, geographic and conceptual horizons are expanded. Hamilton, less the philosophe and more the power broker than Madison, wants *ample* authority, *ample* power, the *extension* of authority, and resists the idea that "we ought to contract our views." Amplitude as an idea in science, and with it the broadening of conceptual horizons, fit Hamilton's political goal of fashioning a political system to govern "so large an empire."

For the task of constructing a system of government, the Founders drew on Newton's understanding of a universe "moving according to mathematical laws in space and time, under the influence of definite and dependable forces." This concept was illustrated by David Rittenhouse, a Philadelphia scientist-politician and Pennsylvania's treasurer, whose orrery displays the motion of solar bodies through the rotation of metal balls moved by wheelworks.[11]

How can there be effective government that is truly representative of the people and that works in a "robust," "vigorous," "energetic" way? The focal point of the question was the clear division over republican government, with access to power separated and checked at various points in the political system, versus a broadly based popular democracy. Instead of votes under the village tree or in town meetings, with larger councils setting national policy, the Constitution writers were architects of an intricate machine whose structural components included such concepts as separation of powers, checks and balances, federalism, and an independent judiciary with the power of judicial review over the acts of legislative and executive bodies.

There were further barriers to a quick or sustained seizure of power: a bicameral legislature, indirect elections, the presidential veto, legislative control of the budget, and limitations on who was eligible to vote. It was almost impossible for a zealous movement to sweep like wildfire through the structures of government and seize control. Likewise, because the safeguards engineered into the system were so elaborate, almost like mechanical safety devices, it was unlikely a tyrant could seize and hold the government for long. Madison used the words *refine* and *filter* to explain how the process differed from direct democracy. In *Federalist* No. 10 he said republican government would "refine and enlarge the public views by passing them through the medium of a chosen body of citizens whose wisdom may best discern the true interest of their country, and whose patriotism and love of justice will be least likely to sacrifice it to temporary or partial considerations." Here Madison deftly appropriated the word *republican* for a specific use, as had been the case with *Federalist*. Madison's republic was not a popular democracy; in it power was not left directly in the hands of the people but with elected officials, thus providing a protective barrier from impulsive or unwise mob governance.

Jay believed the filtering process produced more enlightened, able candidates for national than for state office. In *Federalist* No. 3 he argued that once an efficient national government was established, "the best men in the country . . . will generally be appointed to manage it." The national government "will have the widest field for choice, and never experience that want of proper persons which is not uncommon in some of the States." *Wisdom, regularity, coolness, temperate, reasonable,* and *deliberate* were words the three authors used to describe the leadership the national government would attract through its filtered and refined selection process. This protected the country against impulsive decisions by uninformed mobs who would put self-interest first, the sort of persons Pennsylvania's Gouverneur Morris described in 1774:

> I stood on the balcony and on my right hand were ranged all the people of property, with some few poor dependents, and on the other the tradesmen, etc., who thought it worth their while to leave daily labour for the good of the country. . . . The mob began to think and reason. Poor reptiles! It is with them a *vernal* morning: they are struggling to cast off their winter's slough. They bask in the sunshine, and ere noon they will bite, depend on it.[12]

Opponents like Patrick Henry and Richard Henry Lee rejected such views as elitist republican rhetoric. Lee wrote, "Every man of reflection must see that the change now proposed is a transfer of power from the many to the few."

The Anti-Federalists favored town meetings, public assemblies, frequent elections, and large legislative bodies—the larger the body, the more representative it was of the general will, an idea borrowed from Rousseau. In such a view, government mirrors, rather than filters, popular interest. Hamilton's opponent, Melancton Smith, articulated this position at the New York ratification convention. Smith believed officials were elected to defend the interests of their constituents; he pleaded for "a sameness . . . between the representative and his constituents." He feared "the middling class of life" would be barred from political participation in the system Madison, Hamilton, and Jay proposed. Madison was no supporter of frequent elections. He used the words *energy* and *stability* to describe government's ideal characteristics; and such government required wise, dispassionate leaders having both distance from constituencies and duration of appointment to represent a national, rather than a local, interest.

The *Federalist* writers, in short, were explicit about the difference between a pure democracy, in which liberty prevails and the people decide all questions, and a republican government, in which powers are carefully delineated and divided among the government's different parts. The shift from liberty to order reflected a transformation from ideas prevalent in America in 1776 to those current in 1787. The *Pennsylvania Packet* in September 1787 wrote, "The year 1776 is celebrated for a revolution in favor of liberty. The year 1787 it is expected will be celebrated with equal joy for a revolution in favor of government."[13] It reflected Alexander Hamilton's argument that in 1776 "zeal for liberty became predominant and excessive," and in 1787 the issue was "strength and stability in the organization of our government, and vigor in its operations."[14]

Hamilton, Madison, and Jay knew the national and state governments' weaknesses. States were debtors, so were individuals. Moreover, the revolutionary period's small circle of educated, purposeful national leaders had been replaced in state legislatures by less able figures. Madison in 1788 said the state governing bodies were filled with "men without reading, experience, or principle." Jay worried about states being governed by people whom "wisdom would have left in obscurity."

Although the Federalists won and the Constitution was accepted, the debate never completely ended; the issues remain two centuries later in appeals to populism or republicanism, state and local rights versus national responsibility.

## WHO PARTICIPATES IN THE POLITICAL PROCESS?

The analysis of political society Hamilton sketched favored "landholders, merchants, and men of the learned professions." In No. 35 he argued, "we must therefore consider merchants as the natural representatives of all these classes of the community." Mechanics and manufacturers "will always be inclined . . . to give their votes to merchants in preference to persons of their own professions or trades" because "they know that the merchant is their natural patron and friend." Learned professions "truly form no distinct interest in society." Hamilton acknowledged that his portrait of society was limited to a small circle of land-owning leaders. He deftly sidestepped the issue of popular democracy. "If it should be objected that we have seen other descriptions of men in the local legislatures," he wrote in No. 36, "I answer that it is admitted there are exceptions to the rule, but not in sufficient number to influence the general complexion or character of the government."

Still, the door to upward political, economic, and social mobility was not closed. Hamilton's words were autobiographical: "There are strong minds in every walk of life that will rise superior to the disadvantages of situation and will command the tribute due their merit, not only from the classes to which they particularly belong, but from the society in general."

He concluded, "for the credit of human nature . . . we should see examples of such vigorous plants flourishing in the soil of federal as well as of state legislation," but these will be exceptions.

American constitutional history can be charted by the continuing expansion of the voting franchise. The elimination of property requirements, the Fifteenth, Nineteenth, Twenty-third, Twenty-fourth, and Twenty-sixth amendments, and the Voting Rights Act are all aspects of the growth of suffrage rights.

Parenthetically, the Constitution was not ratified by plebiscite; property requirements for voting eliminated many small farmers and artisans who opposed the document. If the Constitution had been submitted directly to the people for a vote, it probably would not have passed. State constitutional convention delegates were elected on the same basis as delegates to state legislatures, which favored established tidewater interests. Nevertheless, in 1788 the voting franchise was broader than it had been when either the Declaration of Independence or the Articles of Confederation was adopted, and New York expanded its electoral rolls and recognized universal manhood suffrage for the election of delegates to its state ratification convention.

## FACTION

No question of governance received more of Madison's attention than how to have a vigorous, energetic, effective government without allowing a single majority or minority faction, or combination of interests, to seize control of it. Madison weighed both the aftermath of Shays's Rebellion in the north and the trouble hundreds of southern landowners, farmers, artisans, merchants, debtors, and failed property owners would make if allowed into the political arena as equals. He described the problem in *Federalist* No. 10:

> The most common and durable source of factions has been the various and unequal distribution of property. Those who hold and those who are without property have ever formed distinct interests in society. . . . creditors . . . debtors. . . . A landed interest, a manufacturing interest, a mercantile interest, a moneyed interest. . . . The regulation of these various and interfering interests forms the principal task of modern legislation.

Madison believed "all civilized societies" were "divided into different sects, fashions, and interests, as they happened to consist of rich and poor, debtors and creditors, the landed, the manufacturing, the commercial interests, the inhabitants of this district or that district." Enlarge the circle of political participants, he argued, while dividing the community into numerous interests and parties, and it will be increasingly difficult for a special interest group to consolidate power and dominate the country or ignore a minority within the nation.

In discussing faction, Madison foresaw not only a vociferous, intransigent minority, but the dangers a majority, bent on working its will, could wreak on society. It was the great mass of restless, propertyless people and small farmers that the Constitution writers both sought to include in a democracy and control in a republic.

Although *Federalist* No. 10 provides an encompassing statement of Madison's idea of faction, he elaborated on the concept elsewhere. In an October 24, 1787, letter to Jefferson he wrote, "*Divide et imperia,* the reprobated axiom of tyranny, is, under certain qualifications the only policy by which a republic can be administered on just principles."[15] Four months earlier, in a speech to the Constitutional Convention, he described his ideas in greater detail. The problem: to have a working republican government yet protect minority interests. This can only be done if government is

> to enlarge the sphere and thereby divide the community into so great a number of interests and parties, that in the first place a majority will not be likely at the same moment to have a common interest separate from that of the whole or of the minority; and in the second place, that in case they should have such an interest, they may not be apt to unite in pursuit of it. It was incumbent on us then to try this remedy, and with that view to frame a republican system on such a scale and in such a form as will control all the evils which have been experienced.[16]

Madison, in short, faced a balancing act; and a misformulation could tilt the new government, so full of hope and promise, into the hands of an authoritarian president, or worse, a tyrant, or an equally oppressive legislative body. In Madison's view, government was a framework, a mechanical structure to keep political currents within acceptable limits, as a carefully engineered watercourse contains raging streams. Madison was much like Locke in this regard and saw government as a neutral agent brokering competing interests, an umpire among contending forces, an agent to protect property rights, on which the well-being of the fragile new nation rested.

## SEPARATION OF POWERS

After the Revolution, Americans understandably opposed conferring political power on a strong ruler. The memory of George III was fresh, and a much more attractive prospect was a strong legislature. The Constitution failed to award such concentrated power to the legislature. Instead, it created a strong presidency, but power was shared among the executive, legislative, and judicial branches; within the legislative branch, it was further partitioned between two houses. Madison believed the new political system could be wrecked easily by an imbalance in the distribution of power or its concentration in one place, especially in the legislature. In *Federalist* No. 47 he wrote, "The accumulation of all powers, legislative, executive, and judiciary, in the same hands . . . may justly be pronounced the very definition of tyranny."

The only reason such a powerful presidency was approved was because everyone knew George Washington would be the first president and would set a clear precedent for how the office should be conducted. Congress, too, would be a strong institution, every bit as capable of despotic rule as the presidency. Madison wrote of the legislature's tendency to draw everything into its vortex; Jefferson earlier had said 173 legislators could be as dictatorial as 1. A strong counterweight in the presidency was important for that reason as well.

Thus the raw confrontation of power against power, ambition against ambition, was counteracted, not through any assumption of goodwill on the participants' part, but through a clear process of separation of powers, distinct checks and balances, and an independent judiciary with the power of judicial review (the right to initiate review of the constitutionality of any act undertaken by the legislative or executive branches, as well as state laws). Judges could face impeachment proceedings in Congress and, while appointed by the president, would be subject to confirmation hearings and sometimes rejection by the Congress.

Madison wrote, in one of the most often-quoted passages from the eighty-five essays, "What is government itself, but the greatest of all reflections on human nature? If men were angels no government would be necessary." Thus, "ambition must be made to counteract ambition"; the government must establish "a policy of supplying by opposite and rival interests, the defect of better motives." Madison's intent was clear: to create a governmental structure in which interests would vigorously contend but not obliterate one another. Elsewhere in No. 51 he stated, "Comprehending in the society so many separate distinctions of citizens . . . will render an unjust combination of a majority of the whole very improbable, if not impracticable."

## THE PRESIDENCY

It was in the presidency that the "energy" and "vigor" of the new republic fused. Hamilton devoted Nos. 67–77 to the presidency. In *Federalist* No. 70 he wrote, "Energy in the executive is a leading character in the definition of good government. . . . A feeble executive implies a feeble execution of the government. A feeble execution is but another phrase for a bad execution: and a government ill executed, whatever it may be in theory, must be, in practice, a bad government."

The president was given powers to veto laws made by Congress; a two-thirds vote of both houses was required to override the veto. The president was, likewise, commander in chief of the armed forces, but Congress declared war and financed the military. The chief executive could conduct foreign affairs, make treaties, and appoint federal judges with the "advice and consent" of the Senate, and pardon those who commit crimes against the nation. What emerged from the convention was a strong presidency, which opponents believed had "powers exceeding those of the most despotic monarch we know of in modern times."[17] Still, presidential power was both separate from legislative and judicial power, and checked and balanced in numerous ways carefully structured into the basic law by Madison and his contemporaries. For example, a president could be impeached for "treason, bribery, or other high crimes and misdemeanors."

## CONGRESS

The *Federalist* authors were careful to delineate the powers of Congress, bearing in mind the legislature was the principal governing body in the states. This would not be the case in the new national government. Still, many of the powers given Congress by the Articles were transferred wholesale to the Constitution, including the right to borrow money, declare war, maintain an army and navy, and establish a post office and post roads. In addition, Congress could "lay and collect taxes," regulate commerce with foreign nations and among the states, and invoke the so-called "elastic clause," expanding congressional powers "to make all laws which shall be necessary and proper for carrying into execution the foregoing powers."

## THE JUDICIARY

The Constitution and *The Federalist Papers* presented a radically new concept of an independent judiciary with, implicitly, the right to rule on the constitutionality of actions originated by the executive and legislative branches and by state governments. An independent judiciary would only work if judges were given long-term appointments "during good behavior." The judiciary was seen, in No. 78, as having "neither force nor will but merely judgment." Hamilton argued that an independent judiciary was essential to a creditable government because "no man can be sure that he may not be tomorrow the victim of a spirit of injustice, by which he may be a gainer today."

The judiciary's authority to nullify unconstitutional state laws did not come until this century. The Framers did not anticipate this, and the First Congress rejected a Madison amendment that would have applied certain fundamental rights to state governments as part of the Bill of Rights. Once this power was vested in the judiciary, Justice Oliver Wendell Holmes was the first to note that it was probably more critical to the preservation of liberty than the authority to declare federal actions unconstitutional.

## RATIFICATION

When the first *Federalist* paper appeared in print in New York on October 27, 1787, the outcome of the ratification debates was still uncertain. There was widespread opposition, and the Constitution passed by only a narrow margin in several states. John Adams believed "the Constitution was extorted from a reluctant people by a grinding necessity."

Delaware was the first state to accept the Constitution by a unanimous vote on December 7, 1787, and Pennsylvania followed a week later by a margin of 46 to 23. In late December, New Jersey's convention ratified the document unanimously. Georgia ratified the document on January 2, 1788, as did Connecticut by a wide margin seven days later.

In early February the Constitution passed by a vote of 187 to 168 in Massachusetts, following a month of acrimonious debate. Twenty-nine of the 355 delegates meeting in Boston had fought with Captain Shays; many of them urged that the Constitution be sent to the towns for a vote. One of the Massachusetts delegates expressed opponents' fears of a strong central government:

> These lawyers and men of learning, and moneyed men, that talk so finely, and gloss over matters so smoothly, to make us poor illiterate people swallow down the pill, expect to get into Congress themselves; they expect to be managers of this Constitution, and get all the power and all the money into their own hands, and then they will swallow up all us little folks like the great *Leviathan;* yes, just as the whale swallowed up Jonah.[18]

Massachusetts ratified the Constitution, but proposed adding a bill of rights. The proposals were not binding, but they removed the Constitution's supporters from a nettlesome dilemma. The proposals secured votes needed for passage and preserved the flexibility to deal with rights issues reflectively in the drafting room rather than as an up-or-down vote in a public assembly. Many undecided or moderately Anti-Federalist voters were thus willing to give the document a chance.

Three states followed Massachusetts in quick succession with clear majority votes: Maryland in April, South Carolina in May, and New Hampshire in June. They gave the Constitution the needed votes for passage and authorized the formation of a new government.

Meanwhile, several important states, representing about 40 percent of the population, were not heard from, including Virginia, New York, North Carolina, and mercurial Rhode Island. The Virginia and New York votes were crucial if the Constitution was to gain national acceptance. In late June 1788, Virginia voted 89 to 79 to ratify the Constitution, ending a lengthy debate. Madison's presence at the Virginia deliberations was important in gaining the votes needed to support the Constitution. The Richmond contest pitted Madison against Patrick Henry, the colorful orator and patriot who spoke against the document for seven hours one day. Henry's effort, and that of the other Anti-Federalists, eventually resulted in a bill of rights—the first ten amendments to the Constitution—being added to the document. Henry, articulate in debate, was gracious in defeat. "I will be a peaceable citizen," he said. "My head, and my heart, shall be at liberty to retrieve the loss of liberty, and remove the defects of the system in a constitutional way."[19]

"One shudders to think what would have happened had Patrick Henry prevailed in Richmond," Warren E. Burger has written. "Earlier, there had been close votes in Massachusetts and New Hampshire in favor of ratification; Rhode Island had emphatically rejected it by popular referendum. With the Anti-Federalist views of Governor Clinton leading the opposition, sentiment in New York was sharply divided."[20]

New York was pivotal. The national capital and an important commercial center, it was also a geographic link between the nation's two halves. Governor Clinton, like other important New York landowners, opposed the Constitution, fearing both increased taxes and the loss of the state's profitable customs revenues to the national government. In a courthouse in Poughkeepsie during June and July 1788, delegates debated the issues raised in *The Federalist Papers* and at other ratification conventions. As had been the case with Madison in Virginia, Hamilton's spirited participation in the New York debate was crucial. News of the New Hampshire and Virginia votes endorsing the Constitution finally left New York's Anti-Federalists in disarray. On July 26, the New York convention approved the Constitution by a vote of 30 to 27.

That same month North Carolina rejected the proposed law but overturned the vote a year later. Rhode Island, which had not bothered to send a delegation to the Constitutional Convention, finally approved the document 34 to 32 in late May 1790, giving the United States a Constitution ratified by all the states.

It was time now to celebrate. There were parades and civic dinners in major cities; federal punch was a favorite drink, and federal hats were popular. Some parades included a horse-drawn replica of a ship; "The sloop of Anarchy has gone ashore on the rock of Union," read one banner.

Congress accepted the newly ratified Constitution. States sent presidential electors, senators, and representatives to New York, the temporary capital. The new House of Representatives and Senate both organized in March 1789, and George Washington was elected first president; but it took an additional week for the news from New York to reach Washington at his home in Virginia. After a triumphal carriage ride north, Washington took the oath of office on April 30 in New York, reciting the words spelled out in Article II, clause 7, that presidents have used for two centuries: "I do solemnly swear that I will faithfully execute the office of President of the United States and will, to the best of my ability, preserve, protect, and defend the Constitution of the United States."

The Bill of Rights was introduced by Madison in the House of Representatives on June 8, 1789, and approved by Congress on September 25. As approved, the Bill of Rights was part of a series of constitutional amendments Madison introduced. Not all were accepted by the House, and those that were passed were reformulated as an appendix to the Constitution rather than interlarded into the text. The Senate did not approve all of them either, sending twelve of the proposals to the states. Ratification took until December 15, 1791, when Virginia's favorable vote made ten of the twelve proposed amendments part of the Constitution.

## THE COMMERCIAL REPUBLIC

The new American experiment in government worked both because the Constitution was a practical, workable document, and because it was launched in an economically viable country. Hamilton, Madison, and Jay realized the importance of a strong commercial republic, although *The Federalist Papers* contain few expanded references to this subject. One of the most detailed was written by Hamilton *in Federalist* No. 12:

> The prosperity of commerce is now perceived and acknowledged by all enlightened statesmen to be the most useful as well as the most productive source of national wealth, and has accordingly become a primary object of their political cares. By multiplying the means of gratification, by promoting the introduction and circulation of the precious metals, those darling objects of human avarice and enterprise, it serves to vivify and invigorate all the channels of industry and to make them flow with greater activity and copiousness.

"If we mean to be a commercial people," Hamilton argued in *Federalist* No. 24, the nation must have an army and navy. In No. 6, he described the darker side of commercial life in language not unlike that which Madison employed: "Are there not aversions, predilections, rivalships, and desires of unjust acquisitions that affect nations as well as kings? Are not popular assemblies frequently subject to the impulses of rage, resentment, jealousy, avarice, and of other irregular and violent propensities?"

In *Federalist* No. 11, Hamilton was lyrical about "what this country can become." Led by a "vigorous national government, the natural strength and the resources of the country, directed to a common interest, would baffle all the combinations of European jealousy to restrain our growth." Europe would cease being "mistress of the world," America would be the dominant political-economic presence. Hamilton disputed those who argued "that even dogs cease to bark after having breathed awhile in our atmosphere" and described "the adventurous spirit, which distinguishes the commercial character of America." He asserted in No. 12 that in such a republic "the assiduous merchant, the laborious husband-man, the active mechanic, and the industrious manufacturer . . . look forward with eager expectation and growing alacrity to this pleasing reward of their toils."

## THE INFORMING VISION

The Founders were in the direct tradition of David Hume and other figures of the Scottish Enlightenment as well as English republican theorists who opposed arbitrary rule and supported popular sovereignty.[21] The great question was one of balance—how to create a strong, acceptable, workable government while avoiding the pitfalls of mob rule or despotism. The Founders' dilemma was expressed by a New England clergyman, Jeremy Belknap: "Let it stand as a principle that government originates from the people: but let the people be taught . . . that they are not able to govern themselves."[22]

The Constitution writers knew Jean Calvin's views on the easy corruptibility of human nature. A seaport town trader or general practicing attorney was rarely a starry-eyed idealist, but was often a person with a clear idea of what was required for honest, workable government, even if such hopes were not always realized. Richard Hofstadter described the Founders' outlook: "Having seen human nature on display in the market place, the courtroom, the legislative chamber, and in every secret path and alleyway where wealth and power are courted, they felt they knew it in all its frailty."[23]

There is a distinctly moral, but not sectarian, cast to *The Federalist Papers*. The authors described "malignant passions" and the "disease," "defect," and "evil propensity" of human behavior in political society. Madison's moralism was evident in the somber analysis of human nature in *Federalist* No. 37. He depicted both "the obscurity arising from the complexity of objects" and "the imperfection of the human faculties." The world is a place with "dark and degraded pictures which display infirmities and depravities of the human character." "Discordant opinions" clash, as do "mutual jealousies . . . factions, contentions, and disappointments." On the positive side, he expressed "wonder" and "astonishment" at the constitutional convention's achievement and suggested only "a finger of that Almighty hand" could give mortals adequate understanding to produce a Constitution governing an unruly citizenry.

No influence on the Constitution writers was more important than the Scottish Common Sense school of philosophy. Authors like David Hume, James Harrington, and John Locke acknowledged both the theology of Calvin and the realities of human nature, especially as displayed in the ferment of mid-eighteenth-century Scottish religious, political, and economic councils. John Dickinson, representing Delaware, echoed this viewpoint: "Experience must be our only guide; reason may mislead us." By "reason" he meant the political theory of the Enlightenment. Madison, author of much of the Constitution, was taken with Montesquieu's idea of the separation of powers expressed in the French writer's *Esprit des Lois*; but what made the concept of checks and balances acceptable to the delegates was probably less political theory than economic practice, the desire, expressed in a modern idiom, "to play the game on a level playing field." The Founders' view of human nature was optimistic, but vigilant. "You trust your mother, but you cut the cards," in more recent language.

Madison was an Anglican who had studied with a Presbyterian tutor, John Witherspoon. The problem of morality in public life was central to the Virginian's thought, more so than personal piety. Ketchum calls Madison's beliefs "eclectic, sensible, and reasonable, if not always wholly consistent," containing "realism about human nature, a comprehensive concept of political obligation, and an instinctive admiration for . . . moderation. From the Christian tradition he inherited a sense of the prime importance of conscience, a strict personal morality, an understanding of human dignity as well as depravity, and a conviction that vital religion could contribute importantly to the general welfare."[24]

## THE SUMMING UP

The contrast between the Articles of Confederation and the Constitution and the two ideas of government they represented is evident in the preambles to the two documents. The first was rambling: "To all to whom these presents shall come, we, the undersigned, delegates of the states affixed to our names, send greetings." The second was focused: "We the People of the United States, in Order to form a more perfect Union, establish justice, insure domestic Tranquility, provide for the common defence, promote the general Welfare, and secure the Blessings of Liberty to ourselves and our Posterity, do ordain and establish this Constitution for the United States of America." In short, the difference was between a bankrupt, ineffectual effort at governance, built on some useful ideas but held together only with the dry sticks of rhetoric, and a bold new plan that balanced popular participation with numerous constraints on what forms that participation might take.

Instead of choosing to locate all political power and ideology in one place, the new idea of government balanced passion against passion, power against power, and harnessed them in an active political process. This is part of what was behind *The Federalist Papers'* studied description of "vigorous," "robust," "energetic" government.

Not often do political essays endure beyond their appointed moment; even more rarely does a lengthy collection of such works, fired off in the heat of polemical politics, claim any lasting place in literary tradition. *The Federalist Papers* endure because the debate is of an unusually high order, about fundamental questions of how society should be governed, by participants who were major actors in shaping the new government.

George Washington signaled the lasting quality of *The Federalist Papers* in a letter to Hamilton on August 28, 1788:

> When the transient circumstances and fugitive performances which attended this Crisis shall have disappeared, That Work will merit the Notice of Posterity; because in it are candidly and ably discussed the principles of freedom and the topics of government, which will be always interesting to mankind so long as they shall be connected in Civil Society.[25]

## NOTES

1. Page Smith describes this transition as the demise of Classical-Christian and the rise of Secular-Democratic thought in American life, see Page Smith, *The Shaping of America: A People's History of the Young Republic*, vol. III (New York: McGraw-Hill, 1980), xxiii.

2. The extensive debates, ample political literature, and intellectual ferment of this period are chronicled in Gordon S. Wood, *The Creation of the American Republic, 1776–1787* (New York: W.W. Norton, 1972).

3. Robert A. Ferguson, "'We Do Ordain and Establish': The Constitution as Literary Text" (unpublished manuscript, Williamsburg, VA, 1987), 26.

4. For a discussion of pamphlet literature and intellectual ferment in this period, see Bernard Bailyn, *The Ideological Origins of the American Revolution* (Cambridge, MA: The Belknap Press of Harvard University Press, 1992).

5. Julien P. Boyd, ed., *The Papers of Thomas Jefferson,* vol. 12 (Princeton, NJ: Princeton University Press, 1955), 276.

6. Quoted in Catherine Drinker Bowen, *Miracle at Philadelphia: The Story of the Constitutional Convention May to September 1787* (Boston: Little, Brown, 1986), 263.

7. Quoted in Samuel Eliot Morison, *The Oxford History of the American People* (New York: Oxford University Press, 1965), 308.

8. The literature on the Anti-Federalists is voluminous; for example, see Jackson Turner Main, *The AntiFederalists, Critics of the Constitution, 1781–1788* (New York: W.W. Norton, 1974).

9. Quoted in James Madison, Alexander Hamilton, and John Jay, *The Federalist Papers,* ed. Isaac Kramnick (London: Penguin Books, 1987), 75.

10. Ralph Ketchum, *James Madison, A Biography* (Charlottesville: University of Virginia Press, 1990), ix.

11. I am grateful to Prof. William F. Harris of the University of Pennsylvania's political science department for calling my attention to this image.

Ketchum, in *James Madison,* also writes: "Madison saw at Princeton David Rittenhouse's intricate orrery, demonstrating the clock-like precision of the heavenly bodies as they moved in their perfectly predictable orbits. Its patterned motion was always the metaphor for Madison's concept of the way his world operated" (50).

12. Morris to J. Penn, May 20, 1774, in U.S. Congress, *American Archives,* comp. Peter Force, vol. I (Washington, DC: 1837–1853), 342–343.

13. Bowen, *Miracle at Philadelphia,* 232.

14. *The Debate of the State Conventions on the Adoption of the Federal Constitution, as Recommended by the General Convention at Philadelphia in 1787,* ed. J. Elliot, vol. II (Philadelphia: 1886), 301.

15. Boyd, *The Papers of Thomas Jefferson,* 278.

16. Max Farrand, ed., *The Records of the Federal Convention of 1787,* vol. I (New Haven, CT: Yale University Press, 1966), 136.

17. "Philadelphiansis: A Critic of the Constitution," in *The Declaration of Independence and the Constitution,* ed. E. Latham (Lexington, MA: D.C Heath, 1976), 176–177.

18. Quoted in William Lee Miller, *The Business of May Next, James Madison and the Founding* (Charlottesville: University of Virginia Press, 1992), 228.

19. Henry quoted in Morison, *Oxford History,* 315.

20. Warren E. Burger, foreword to Bowen, *Miracle at Philadelphia*, ix.

21. Forrest McDonald, *Novus Ordo Seclorum: The Intellectual Origins o f the Constitution* (Lawrence: University Press of Kansas, 1985).

22. Richard Hofstadter, *The American Political Tradition and the Men Who Made It* (New York: Vintage Books, 1955), 7.

23. Ibid., 3.

24. Ketchum, *James Madison,* 50.

25. John C. Fitzpatrick, ed., *The Writings of George Washington,* vol. 30, June 20, 1788 January 21, 1790 (Washington, DC: U.S. Government Printing Office, 1939), 66.

# "INTERPRETATIVE ESSAY"

## W.B. ALLEN, GORDON LLOYD, AND MARGIE LLOYD

"The Antifederalists" is the name commonly used to designate those people who opposed the ratification of the Constitution. This negative nomenclature was not consciously chosen by the opposition; they would have preferred the positive image of Republicans. Our surest guide to understanding how the Antifederalists received their name is to recognize in the term a political concept immediately applicable in the context of American politics. From the adoption of the Articles of Confederation in 1781, and coming to a head in 1783–85, those who aimed to "cement," "support," or "protect" the Confederation (in which they identified the Union) were perceived as federal-minded as opposed to state-minded. By 1787, those who opposed the movement to strengthen the Confederation were being widely described as "antifederal men" with an "unfederal disposition."

Federal-minded men finally persuaded the Confederation Congress in February 1787 to call a Constitutional Convention to revise the Articles of Confederation. They proposed a plan for a strengthened union which radically departed from the Articles. The antifederal men countered the nationalism of the new Constitution by emphasizing the extent to which it departed from the federal principles of the Articles. This position is manifest above all in the pre-eminent Antifederalist, Richard Henry Lee, whose "Federal Farmer" essays sought to arrogate to himself a title which for four years past he energetically resisted. A common Antifederalist argument during the ratification struggle of 1787–88 was that they were the true federalists because they took their bearings from the principles of federalism laid down in the Articles. They accused the Federalists of abandoning the principles of federalism and instead substituting a consolidated system.

Each side in the debate over the ratification of the Constitution was known by the reputation which it had earned in the struggles of the 1780's over strengthening the Union, rather than in terms of the principles which they articulated in the post-convention struggle. The initial argument for strengthening the Confederation was reinforced by Alexander Hamilton, John Jay, and James Madison in the eighty-five essays which they wrote urging ratification of the Constitution.

These essays, known as *The Federalist,* presented the choice in terms of the survival or dismemberment of the union. The opposition was accused of disunion sentiment, being self-serving, uninformed, and inconsistent. Despite the best efforts of opposition leaders to shed the "anti" image, the nomenclature "Antifederalist" was preserved for posterity.

Despite recent efforts to retrieve Antifederalist thought from oblivion, the Antifederalists are still widely viewed as irrelevant and disorganized. They even receive a negative evaluation from those scholars who give the Antifederalists a comprehensive and sympathetic treatment. Their opposition is viewed as "piecemeal," their attitude one "of little faith," and their perspective "reactionary." Their legacy to the American political tradition is viewed as minimal and inconsequential. We believe, however, that Antifederalist Melancton Smith was correct when he argued that "there is a remarkable uniformity in the objections made to the constitution, *on the most important points.*"

When the leading Antifederalists are permitted to speak for themselves on these important points, a hitherto unappreciated positive critique and coherent alternative emerges.

The task of evaluating Antifederalist thought would be much easier had three people sat down and written *The Antifederalist.* Instead, the Antifederalist literature (as is the case with most of the Federalist literature other than *The Federalist*) is immense and heterogeneous, encompassing speeches, pamphlets, essays, and letters. Tensions, inconsistencies, and disagreements among the Antifederalists over a particular issue can easily be discovered. The same is true for the "other" Federalists. There is a sense in which *The Federalist* has made our understanding of the founding too easy. Its coherence and coverage have lulled us into the belief that these easily accessible essays contain all we need to know.

But *The Federalist* represents only one side of the debate, presenting the opposition in the worst possible light. Unfortunately, much subsequent scholarship has accepted *The Federalist*'s view of the Antifederalists at face value. Although the Federalists won the debate, they did not get everything that they desired. Antifederalist political theory entered the very nature of the American system, for example, by virtue of its adherents' successful appeal for a bill of rights. This victory was less than they would have liked, but more than can justify rejection as insignificant by their inheritors. Precisely because American politics is so often a debate over the possibilities and limitations of the separation of powers, an independent judiciary, federalism, and representative government, it is vital that the potentcy of Antifederalist political analysis be restored. A healthy appreciation for "the other side" completes the picture of our founding, a drama which has yet to be equalled in the political world.

The Antifederalists did not sense a need to fundamentally change the government under the Articles of Confederation. They liked the union as it was: well-constructed and protective of its citizens' liberties. What and where was the danger in letting the people enjoy the fruits of their labors, protected by state laws against private violence? Faction, majority or minority, was troublesome, and many Antifederalists thought that it could have been better contained under the Articles; but wasn't that the reason the Philadelphia convention was called?—to preserve or enhance the present condition of liberty with "the least burdensome system which shall defend those rights."[1]

The Antifederalists admitted that the Articles were weak but they believed that the political and economic conditions did not approach the crisis proportions portrayed by those who wanted a radical overhauling of the Articles. The Antifederalist position on the eve of the Constitutional

Convention in May 1787 was to grant more power to Congress over domestic commerce and foreign affairs without abandoning the structure of the Articles (which guaranteed the traditional assumption that republicanism can only thrive in a small territorial orbit). With the emergence of the new plan of government from Philadelphia in September 1787, the Antifederalists fleshed out this argument in order to warn their fellow citizens that the new plan departed radically from traditional republicanism.

Two principles formed the basis of the Antifederalist idea for the American republic: the people are the sovereigns of the government; and an express reservation of powers qualified the government by affirming the people's sovereignty. Once the people proclaimed their power, they appointed a select few to administer that power, which filtered back to the people. They declared their rules, taking into account the chief negative feature of civil society—minority faction—and turned out a positive yet minimal framework which enabled them to govern themselves by honesty and virtue. Minority faction could not possibly exist in a government where all interests of the people were represented:

> The perfection of government depends on the equality of its operation, as far as human affairs admit, upon all the parts of the empire, and upon all the citizens. Some inequalities indeed will take place. One man will be obliged to travel a few miles further than another man to procure justice. But when he has traveled, the poor man ought to have the same measure of justice as the rich one. Small inequalities may he compensated.[2]

The publication of the Constitution signalled an onslaught of Antifederalist literature throughout the nation for the next nine months. They agreed that four basic questions were inadequately addressed in the document; the Antifederalists wanted the answers built into the Constitution, or they would encourage the delegates to their respective state ratifying conventions to oppose the Constitution. The first problem they had was with the notion of a strong central government. The Constitution provided neither recall elections nor frequent rotation in office. The Antifederalists feared that once elected and comfortable in his job, the representative would not relinquish his power. Being some distance away from his district would alienate him from his constituents' wishes; he would be inclined to favor his own affairs and those of the national city. Another shortcoming of the strong central government would be governing the regional differences between the states and their inhabitants. Making general laws that would apply to the needs of Maine or Georgia would not address the important local needs of either state. The new government would have to become more tolerant to the local issues (as under the Articles) for the people not to be suspicious of its huge power.

The Antifederalists understood the basic choice facing mankind to be either republicanism or despotism. The operating principle of republicanism was self government while that of despotism was force. They believed that republics fell into despotism if the principle of self government became corrupted. Thus, special care must be taken regarding those things which constitute the pillars of self government. They believed that if the pillar of small size was to be removed then it became imperative to erect an alternative system of support. They believed that special attention should be given to the habits and customs of the people on the one hand and to securing the dependency of the representatives on the other hand. The problem with the new plan was that it neglected the mores of the people, and it bestowed more power on the representatives without providing for the corresponding checks on that power.

A big complaint of the Antifederalists was the actual framework of the Constitution; it seemed to encourage the growth of minority faction. Remember, the Antifederalists were very suspicious of privilege, and the proposed Constitution appeared to offer plenty of public jobs for an aristocracy. There were no places in the government for the yeoman middle class, and no checks upon those civil servants who would not govern in the interests of the middle class. Wealthy office-holders would appoint their wealthy friends to public office. Instead of a governmental framework based on the middle class values of honesty and virtue, the proposed framework would be based on and encourage the love of wealth and greed. A lack of checks against this problem would allow a tyranny of minority faction over majority values. The Antifederalists wanted a middle class government, where

> . . . the people are sovereign and their sense or opinion is the criterion of every public measure . . . if you vest all the legislative power in one body of men (separating the executive and judicial) elected for a short period, and necessarily excluded by rotation from permanency, and guarded from precipitancy and surprise by delays imposed on its proceedings, you will create the most perfect responsibility.[3]

The third problem of the proposed Constitution was that Congress had the power to impose barriers against commerce. Antifederalists believed that commerce should remain free to pursue natural courses. Since commercial interests would not be supported by the government, trade would have to be allowed to develop new interests, thus benefitting more people. To this end, the Antifederalists pushed for either more checks on Congress or less power of Congress over the people and their affairs. Following Montesquieu, the Antifederalists believed that those governments which favored commerce seemed better suited to ameliorate the human condition, while those which *most* favored commerce served *best* of all. They feared that the power to regulate interstate commerce and foreign trade as well as the power over internal taxation granted to the Congress under the new plan would produce a restrictive and burdensome economy. They felt that Congress would use its authority to grant monopolies, thereby reducing economic opportunity even further. The Antifederalists were apprehensive about the restrictive economic message that America would be sending the world. More policy control over our markets would mean less productivity, less business, and less prosperity to reinvest into the businesses overseas. This unfriendliness to trade would be a signal to Europe that the new nation would not be working in congruence with its markets at home and abroad.

The key criticism the Antifederalists had with the Constitution was the representation issue. In order to truly represent the interests of the people and reduce the possibility of governmental corruption, the proposed government had to increase the number of representatives in proportion to the size of the nation (or conversely, decrease the size of the nation in proportion to the number of representatives now planned in the Constitution). For the nation to be flexible to the demands of representation and successful in maintaining good government, the scheme of representation had to expand in size along with the population. While it is true that aristocrats and demagogues would be elected over the middling class as a matter of course, expanding the representation meant that the majority opinion—middle class in this instance—would be guaranteed an ear in Congress. Antifederalists noted that under the proposed scheme of representation, the power of recall, frequent elections, and rotation of office were not built into the model. Their fear from this omission was that the local interests of the citizens would not be kept before Congress, thus causing their frustration with the new centralized government: the power of the representative would constantly rest over the peoples' heads, and not vice versa. If the proposed government had restored this power to the people, then minority tyranny, represented by aristocratic public servants, would be dissipated (if not completely done away with) by the majority opinion and/or representatives.

By widening the class of those with power we approach republican perfection.[4]

The stability of the country will be encouraged by the yeoman or small businessman as he has the nation's interests at heart; let him and the many others like him, thought the Antifederalists, represent the positive middling interests of this country.

The Antifederalists wanted a national government to reply to the peoples' local needs. Under the Articles, the people had a voice in the national government and they had basic freedoms to enjoy the results of their hard work. When the new government scrapped the Articles in the summer of 1787 for a more centralized framework, the Antifederalists reacted strongly, warning citizens of what freedoms and power they had before the new plan, and how quickly they would lose what had been fought for and cherished. The pamphlet war the opponents waged lasted until the adoption of the Bill of Rights, when the Antifederalists were placated with certain freedoms and powers built into the Constitution.

Therefore, respecting the main points, the Antifederalists were not only in agreement but their position was coherent. They believed that republican liberty was best preserved in small units where the people had an active and continuous part to play in government. They thought that the Articles secured this concept of republicanism. They argued that the Constitution placed republicanism in danger because it undermined the prop of small size. As a consequence, they argued that the system of representation under the new plan must be altered to secure what was formerly secured by small size. They believed that under the new plan the representatives would become independent from rather than dependent on the people. Lastly, they warned that unless restrictions were placed on the powers of the Congress, the Executive, and the Judiciary, the potentiality for the abuse of power would become a reality. These various approaches culminated in their insistence on a Bill of Rights which they saw satisfying the same function that small territory, representative dependency, and strict construction would perform for republicanism.

The Antifederalist perspective and recommendations are still alive in the American political tradition. When we hear the argument that the Founding Fathers feared power and thus separated government into three branches and tried to ensure that no one branch would dominate, we are in fact hearing what the Antifederalists feared and desired. Their perspective did not prevail in 1787; nevertheless, the contemporary argument that the Founding Fathers created a "deadlocked" system is in fact assuming that the Founding Fathers were Antifederalists. And when we hear the argument that the Founding Fathers were wealthy and ambitious men who designed an undemocratic government in secret for their own benefit, we are in fact hearing a vulgarization of the Antifederalist critique of the leading proponents of change. Moreover, when we hear the claim that our representatives are drawn from a minority of the population, that they operate in a manner which is independent of the people, and that the Congress does not represent the broad cross-section of interests which it is supposed to represent, we are in fact repeating the Antifederalist critique of the scheme of representation found in the Constitution. When we hear the argument that the federal government is out of control, that it interferes too much with the life of American citizens and that state and local officials understand the needs of the people far better than do representatives in Washington, we are in fact echoing the warnings of the Antifederalists. Finally, when Americans instinctively associate the Constitution and the meaning of democracy with the Bill of Rights, we are in fact honoring the essential contribution of the Antifederalists.

## NOTES

1. Agrippa: Essay III.
2. Agrippa: Essay VII [*Massachusetts Gazette* (Boston), December 1787 to January 1788].
3. Centinel: Letter I [*Independent Journal* (Philadelphia), 5 October, 1787].
4. Melancton Smith, 20 June 1788 [Speech given before the New York Ratifying Convention].

# "WHAT SORT OF DESPOTISM DEMOCRATIC NATIONS HAVE TO FEAR"

## Alexis de Tocqueville

*I HAD remarked during my stay in the United States that a democratic state of society, similar to that of the Americans, might offer singular facilities for the establishment of despotism; and I perceived, upon my return to Europe, how much use had already been made, by most of our rulers, of the notions, the sentiments, and the wants created by this same social condition, for the purpose of extending the circle of their power. This led me to think that the nations of Christendom would perhaps eventually undergo some oppression like that which hung over several of the nations of the ancient world.*

A more accurate examination of the subject, and five years of further meditation, have not diminished my fears, but have changed their object.

No sovereign ever lived in former ages so absolute or so powerful as to undertake to administer by his own agency, and without the assistance of intermediate powers, all the parts of a great empire; none ever attempted to subject all his subjects indiscriminately to strict uniformity of regulation and personally to tutor and direct every member of the community. The notion of such an undertaking never occurred to the human mind; and if any man had conceived it, the want of information, the imperfection of the administrative system, and, above all, the natural obstacles caused by the inequality of conditions would speedily have checked the execution of so vast a design.

When the Roman emperors were at the height of their power, the different nations of the empire still preserved usages and customs of great diversity, although they were subject to the same monarch, most of the provinces were separately administered; they abounded in powerful and active municipalities; and although the whole government of the empire was centered in the hands of the Emperor alone and he always remained, in case of need, the supreme arbiter in all matters, yet the details of social life and private occupations lay for the most part beyond his control. The emperors possessed, it is true, an immense and unchecked power, which allowed them to gratify all their whimsical tastes and to employ for that purpose the whole strength of the

"What Sort of Despotism Democratic Nations Have to Fear," Chapter VI, Section IV, of Volume II, *Democracy in America*, by Alexis de Tocqueville. Retrieved April 11, 2003 from http://xroads.virginia.edu/~HYPER/DETOC/ch4_06.htm.

state. They frequently abused that power arbitrarily to deprive their subjects of property or of life; their tyranny was extremely onerous to the few, but it did not reach the many: it was confined to some few main objects and neglected the rest; it was violent, but its range was limited.

It would seem that if despotism were to be established among the democratic nations of our days, it might assume a different character; it would be more extensive and more mild; it would degrade men without tormenting them. I do not question that, in an age of instruction and equality like our own, sovereigns might more easily succeed in collecting all political power into their own hands and might interfere more habitually and decidedly with the circle of private interests than any sovereign of antiquity could ever do. But this same principle of equality which facilitates despotism tempers its rigor. We have seen how the customs of society become more humane and gentle in proportion as men become more equal and alike. When no member of the community has much power or much wealth, tyranny is, as it were, without opportunities and a field of action. As all fortunes are scanty, the passions of men are naturally circumscribed, their imagination limited, their pleasures simple. This universal moderation moderates the sovereign himself and checks within certain limits the inordinate stretch of his desires.

Independently of these reasons, drawn from the nature of the state of society itself, I might add many others arising from causes beyond my subject; but I shall keep within the limits I have laid down.

Democratic governments may become violent and even cruel at certain periods of extreme effervescence or of great danger, but these crises will be rare and brief. When I consider the petty passions of our contemporaries, the mildness of their manners, the extent of their education, the purity of their religion, the gentleness of their morality, their regular and industrious habits, and the restraint which they almost all observe in their vices no less than in their virtues, I have no fear that they will meet with tyrants in their rulers, but rather with guardians.

I think, then, that the species of oppression by which democratic nations are menaced is unlike anything that ever before existed in the world; our contemporaries will find no prototype of it in their memories. I seek in vain for an expression that will accurately convey the whole of the idea I have formed of it; the old words despotism and tyranny are inappropriate: the thing itself is new, and since I cannot name, I must attempt to define it.

I seek to trace the novel features under which despotism may appear in the world. The first thing that strikes the observation is an innumerable multitude of men, all equal and alike, incessantly endeavoring to procure the petty and paltry pleasures with which they glut their lives. Each of them, living apart, is as a stranger to the fate of all the rest; his children and his private friends constitute to him the whole of mankind. As for the rest of his fellow citizens, he is close to them, but he does not see them; he touches them, but he does not feel them; he exists only in himself and for himself alone; and if his kindred still remain to him, he may be said at any rate to have lost his country.

Above this race of men stands an immense and tutelary power, which takes upon itself alone to secure their gratifications and to watch over their fate. That power is absolute, minute, regular, provident, and mild. It would be like the authority of a parent if, like that authority, its object was to prepare men for manhood: but it seeks, on the contrary, to keep them in perpetual childhood: it is well content that the people should rejoice, provided they think of nothing but rejoicing. For their happiness such a government willingly labors, but it chooses to be the sole agent and the only arbiter of that happiness; it provides for their security, foresees and supplies their necessities, facilitates their pleasures, manages their principal concerns, directs their industry, regulates the descent of property, and subdivides their inheritances: what remains, but to spare them all the care of thinking and all the trouble of living?

Thus it every day renders the exercise of the free agency of man less useful and less frequent; it circumscribes the will within a narrower range and gradually robs a man of all the uses of himself. The principle of equality has prepared men for these things; it has predisposed men to endure them and often to look on them as benefits.

After having thus successively taken each member of the community in its powerful grasp and fashioned him at will, the supreme power then extends its arm over the whole community. It covers the surface of society with a network of small complicated rules, minute and uniform, through which the most original minds and the most energetic characters cannot penetrate, to rise above the crowd. The will of man is not shattered, but softened, bent, and guided; men are seldom forced by it to act, but they are constantly restrained from acting. Such a power does not destroy, but it prevents existence; it does not tyrannize, but it compresses, enervates, extinguishes, and stupefies a people, till each nation is reduced to nothing better than a flock of timid and industrious animals, of which the government is the shepherd.

I have always thought that servitude of the regular, quiet, and gentle kind which I have just described might be combined more easily than is commonly believed with some of the outward forms of freedom, and that it might even establish itself under the wing of the sovereignty of the people.

Our contemporaries are constantly excited by two conflicting passions: they want to be led, and they wish to remain free. As they cannot destroy either the one or the other of these contrary propensities, they strive to satisfy them both at once. They devise a sole, tutelary, and all-powerful form of government, but elected by the people. They combine the principle of centralization and that of popular sovereignty; this gives them a respite: they console themselves for being in tutelage by the reflection that they have chosen their own guardians. Every man allows himself to be put in leading-strings, because he sees that it is not a person or a class of persons, but the people at large who hold the end of his chain.

By this system the people shake off their state of dependence just long enough to select their master and then relapse into it again. A great many persons at the present day are quite contented with this sort of compromise between administrative despotism and the sovereignty of the people; and they think they have done enough for the protection of individual freedom when they have surrendered it to the power of the nation at large. This does not satisfy me: the nature of him I am to obey signifies less to me than the fact of extorted obedience. I do not deny, however, that a constitution of this kind appears to me to be infinitely preferable to one which, after having concentrated all the powers of government, should vest them in the hands of an irresponsible person or body of persons. Of all the forms that democratic despotism could assume, the latter would assuredly be the worst.

When the sovereign is elective, or narrowly watched by a legislature which is really elective and independent, the oppression that he exercises over individuals is sometimes greater, but it is always less degrading; because every man, when he is oppressed and disarmed, may still imagine that, while he yields obedience, it is to himself he yields it, and that it is to one of his own inclinations that all the rest give way. In like manner, I can understand that when the sovereign represents the nation and is dependent upon the people, the rights and the power of which every citizen is deprived serve not only the head of the state, but the state itself, and that private persons derive some return from the sacrifice of their independence which they have made to the public.

To create a representation of the people in every centralized country is, therefore, to diminish the evil that extreme centralization may produce, but not to get rid of it.

I admit that, by this means, room is left for the intervention of individuals in the more important affairs; but it is not the less suppressed in the smaller and more privates ones. It must not be forgotten that it is especially dangerous to enslave men in the minor details of life. For my own part, I should be inclined to think freedom less necessary in great things than in little ones, if it were possible to be secure of the one without possessing the other.

Subjection in minor affairs breaks out every day and is felt by the whole community indiscriminately. It does not drive men to resistance, but it crosses them at every turn, till they are led to surrender the exercise of their own will. Thus their spirit is gradually broken and their character enervated; whereas that obedience which is exacted on a few important but rare occasions only exhibits servitude at certain intervals and throws the burden of it upon a small number of men. It is in vain to summon a people who have been rendered so dependent on the central power to choose from time to time the representatives of that power; this rare and brief exercise of their free choice, however important it may be, will not prevent them from gradually losing the faculties of thinking, feeling, and acting for themselves, and thus gradually falling below the level of humanity.

I add that they will soon become incapable of exercising the great and only privilege which remains to them. The democratic nations that have introduced freedom into their political constitution at the very time when they were augmenting the despotism of their administrative constitution have been led into strange paradoxes. To manage those minor affairs in which good sense is all that is wanted, the people are held to be unequal to the task; but when the government of the country is at stake, the people are invested with immense powers; they are alternately made the play things of their ruler, and his masters, more than kings and less than men. After having exhausted all the different modes of election without finding one to suit their purpose, they are still amazed and still bent on seeking further; as if the evil they notice did not originate in the constitution of the country far more than in that of the electoral body.

It is indeed difficult to conceive how men who have entirely given up the habit of self-government should succeed in making a proper choice of those by whom they are to be governed; and no one will ever believe that a liberal, wise, and energetic government can spring from the suffrages of a subservient people.

A constitution republican in its head and ultra-monarchical in all its other parts has always appeared to me to be a short-lived monster. The vices of rulers and the ineptitude of the people would speedily bring about its ruin; and the nation, weary of its representatives and of itself, would create freer institutions or soon return to stretch itself at the feet of a single master.

# "THE FUTURE OF THE AMERICAN BUREAUCRATIC SYSTEM"

## Richard J. Stillman II

What is the future of the bureaucratic system in the United States? Will it become more effective and efficient in delivering public goods and services for society? More equitable and fairer in distributing those goods and services? Will it be more accountable to the public at large as well as more responsive to individual and group needs? Will it retain its significant core functions in U.S. society? Will it grow or decline in influence? In short, what is tomorrow's role for public bureaucracy within America's democracy?

This [reading] will attempt to answer these and other critical questions by arguing that the future of public bureaucracy in the U.S. is fundamentally a normative value problem, one rooted in a unique, changing triad of historic national values. In other words, the central thesis of this [reading] is that our peculiar past national values will shape our future bureaucratic institutions. The chapter will be devoted to understanding these values and how they have influenced U.S. bureaucracy. It will begin by outlining the essential nature and content of these historic norms that are labeled Hamiltonianism, Jeffersonianism, and Madisonianism and explain how these values decisively influenced the course of American bureaucracy over the last two centuries. The chapter will conclude by stressing that the future role of U.S. bureaucracy will ultimately be determined by " trade-offs" among these competing values.

## THREE FOUNDING FATHERS' NORMATIVE MODELS FOR BUREAUCRACY IN A DEMOCRACY

The U.S. Constitution largely ignored the existence of *or need for* a bureaucratic system. It was mostly silent on this subject. But three of the founding fathers did give some attention to the subject in their writings—Alexander Hamilton, Thomas Jefferson, and James Madison.

### Alexander Hamilton: Maximizing Administrative Efficacy

Of all the founding fathers, none displayed more interest in and enthusiasm for administration and organization than Alexander Hamilton. Hamilton was a man of action from the time he was

From *The American Bureaucracy: The Core of Modern Government*, 2d ed. (Chicago: Nelson-Hall, 1996): 360–396. Copyright © 1996 Thomson Learning. Reprinted with permission.

the brilliant twenty-three-year-old aide-de-camp to General Washington during the Revolution. Throughout his tenure as the first secretary of the treasury in President Washington's Cabinet, he demonstrated masterful planning, control, and organization of national finances. As Leonard White writes, "In the Federalist Papers, Hamilton set out the first systematic exposition of Public Administration, a contribution which stood alone for generations. In his public life, he displayed a capacity for organization, system and leadership which after a century and a half is hardly equalled."[1]

The role Hamilton saw for bureaucracy in government as well as in society was an expansive one. He was an ardent, enthusiastic nationalist who envisioned a big, bold, broad role for the American nation—politically, economically, and militarily. His writings are studded with glowing ideas for promoting "the public interest," "the public good," "the good of the general society," and "the national interest." Some say he valued the nation more than its people. At least there was nothing timid or modest about Hamilton's vision for the future of the United States, for he was a very early believer in positive government framed essentially to promote a strong nation and its interests. He argued for setting up a national bank, a national university, a professional army and navy, a public school system, and a variety of national public works projects, such as building roads, ships, canals, and dams and mining metals for industrial development. In short, he laid out a bold blueprint for the nation's future.

Hamilton favored a strong, energetic administration[2] based on maximizing the efficiency and effectiveness of public organizations to bring about his expansive vision of the future of the country. His normative model of the place of bureaucracy within a democracy thus contained these elements: (1) broad discretionary and activist roles for public agencies, characterized by strong, decisive leadership that evidenced "energy" and "tone" (words he repeatedly used); (2) unified public organizations with responsibility for administrative action undivided and preferably concentrated in one individual (as opposed to being divided among boards or committees); (3) administrative power allocated to individuals and governmental units commensurate with the responsibility for the tasks assigned; (4) adequate time in office to ensure administrative effectiveness, long-term planning, and operational stability for implementing public programs; (5) preference for paid, trained professionals (as opposed to part-time volunteers) in staffing governmental positions; (6) emphasis on national planning, sound fiscal management, and responsible exercise of creative public leadership; and (7) popular control of public organizations achieved by means of the election of responsible, capable chief executives with adequate political power and support in order to ensure that tasks are performed competently and well.

### Thomas Jefferson: Maximizing Administrative Accountability to the Public at Large

If the values of nationalism fired Hamilton's conception of administration, Jefferson's values were shaped primarily by concerns for THE PEOPLE. His commitment to individual liberty, freedom for humanity, and the pursuit of personal happiness was evident throughout his writings, but never more forcefully than in the Preamble to the Declaration of Independence, which dedicated the United States to popular values of "life, liberty and the pursuit of happiness." Unlike Hamilton, Jefferson exhibited an abiding faith in human nature and its unlimited potential for growth and development unfettered by governmental authority. And unlike Hamilton, who repeatedly spoke of "administrative discretion," "energy," and "tone," Jefferson stressed "limits on government," "individual rights," "freedom," and "liberty." As he said, "I am for a government that is frugal and simple." The New England town meeting perhaps came closest to his ideal of a polity that was "simple" and "frugal" and maximized citizens' participation in governing their own affairs.

A Jeffersonian normative model regarding the relationship of bureaucracy to democracy therefore, as Lynton Caldwell observed,[3] placed a heavy emphasis upon numerous devices to ensure bureaucracy's strict accountability to the general public and included such elements as: (1) extensive popular participation, especially voluntary mass involvement in administration (as opposed to staffing administration with paid, full-time professionals); (2) maximum decentralization of functions in order to limit activities and bring public activities under close and constant popular scrutiny; (3) operational simplicity and economy—simplicity so that administrative activities could be easily understood by the average citizen, and economy so it would not be economically burdensome to the public; (4) strict legal limitations that clearly spell out organizational purposes and restrict administrative discretion and authority in order to protect human rights; (5) a weak leadership role for public administrators through defining them as narrow functional specialists and technicians rather than as broad-ranging general managers or educated professionals exercising wide discretionary powers; (6) lack of concern for promulgating national planning, long-term operational stability, and effective program implementation but rather a focus on developing voluntary citizen efforts, private initiatives, and the free market alternatives to the performance of public tasks; and (7) administrative power in public organizations that flows from the bottom up, not from the top down, in order to sharply limit administrative outputs and ensure public oversight.

## James Madison: Balancing Administrative Interest Group Demands

James Madison shared many of Jefferson's concerns about protecting human liberty through limitations placed upon government, but he also held little enthusiasm for his fellow Virginians' idealization of the broad abstraction THE PEOPLE, which served as Jefferson's basic value premise and upon which his normative conceptions of bureaucracy-democracy relationships were founded. Madisonian analysis of U.S. government and the exercise of political authority rested instead upon the faction (or in modern terms, the interest group) which he saw as the chief fount of government's authority.

Federalist 10 indicates that while Madison saw factions as the prime movers of U.S. politics, he had little liking for any sort of faction. He defined a faction as "a number of citizens whether amounting to a majority or a minority of the whole, who are united and actuated by some common impulse or passion, or of interests adverse to the rights of other citizens, or the permanent and aggregated interests of the community."[4]

Madison "fathered" a Constitution that employs numerous "checks and balances" to mitigate the pernicious influence of factions upon governing institutions. As he underscores in Federalist 51, while the root causes of factions can never be eliminated, a framework of government can be designed so that their harmful influence over government institutions and public decisions is reduced. His fundamental advice in Federalist 51 on the framing of a stable, enduring government is well known: "Ambition must be made to counteract ambition," so that "you first enable the government to control the governed; in the next place, oblige it to control itself."[5]

Madisonian analysis basically was oriented toward structuring a process that would reduce but not eliminate factional influence to enable government *both* to govern adequately *and* to ensure its public accountability. Hence, he sought a "mixed government" that would balance competing interest group demands. His genuine contribution to political philosophy was his conception of "an extended republic" that would broaden geographic size and diversity in order to balance competing interests and thus protect human liberty and promote social stability. Indeed, throughout his writings, unlike either Hamilton or Jefferson, Madison, as historian Ralph Ketchum indicates,[6] places a special premium on achieving the values of organic social balance, political equilibrium, and the Aristotelian "Golden Mean."

Yet, unlike either Hamilton or Jefferson, Madison says little *explicitly* about the role of administration in the context of U.S. politics. But from what he said about the executive branch, the execution of public policies, and his conceptions of faction-based politics, we can conclude that Madison conceived of a very different role for bureaucracy in democracy. The normative elements of this Madisonian model include the following elements: (1) public organizations that are involved in a pluralistic political process rooted in the divergent, changing factional interests of society; thus, their administration and relationship to politics can be neither static nor clear-cut but rather are dynamic, complex, intertwined, and interconnected with a diversity of organic social interests; (2) bureaucracies are political in that they share in the processes of exercising political authority with other branches of government—courts, executive, and legislature—and thus they should, like the other branches, engage in balancing social interests to promote consensus, stability, and representation of divergent points of view; (3) public agencies, though they may formally be separate bodies because of functional differentiation, in practice share power with executive, legislative, and judicial branches in order to undertake effective action and to operate within a continuous, complex *vertical* system of checks and balances upon the other three branches; (4) public entities that operate in a continuous, complex set of *horizontal* power-sharing arrangements between federal, state, and local units and acquire power to take effective action as well as to operate checks and balances upon one another; (5) in this fragmented world of political authority, public administrators that exercise the "art of the possible" in dealing with these competing interests; their roles would principally entail political negotiation, compromise, and bargaining; (6) that social consensus and equilibrium between competing interest groups, not organizational efficacy nor accountability to an abstract will of THE PEOPLE, should be the primary aim of public officials; (7) that administrative power to drive actions come from neither the top down nor the bottom up but must be picked up piecemeal by public administrators from the top, bottom, and sides of government agencies within a politically fragmented, constantly fluid environment.

In sum, these three founding fathers produced three very different normative models of bureaucratic-political relationships. They are summarized in Table [2.5.1].

## HISTORIC PATTERNS OF BALANCING ADMINISTRATIVE EFFICACY, PUBLIC ACCOUNTABILITY, AND INTEREST GROUP DEMANDS

Americans have never made up their minds throughout their two-hundred-year history as to which of the three normative models they prefer. They have remained uncertain about finding a place for their bureaucracy within their democracy. From time to time, the stress has been placed on promoting the values of administrative efficacy over the other two values; at other times, accountability to the general public has predominated; and at still other times, responding to diverse interest

Table 2.5.1

**Three Founding Fathers' Normative Models**

| Topic | Hamilton | Jefferson | Madison |
| --- | --- | --- | --- |
| overall goal | strong, sustained; focused organizational efficacy to promote national interests | strict public accountability to maximize personal liberty | organic balancing of interest group demands to promote social stability (i.e., "factions") |
| key method | unified administrative processes | decentralized, participatory processes | horizontal/vertical checks and balances; an extended republic |
| degree of administrative discretion | broad | narrow | mixed and interdependent with other branches |
| degree of centralization | high | low | varied with the capacity to acquire power |
| organizational autonomy | high | low | interdependent with societal interests |
| ideal public offical | professional careerist—i.e., "a doer" | citizen volunteer—i.e., "a servant of the people" | negotiator and compromiser—i.e., interest brokers |
| sources of power | flows from top down | flows from bottom up | flows from all around—top down, bottom up, and side to side |
| agency outputs and capacity to shape future of the nature | strong | strong | mixed—driven by diverse conflicting needs of special interests |
| degree of seperation of politics from administration | sharp—to promote agency efficiency | sharp—to promote public control | complex, mixed and unclear, depending upon many processes |
| social status of bureaucrats | high-status "professionals" | low status "technicians" | mixed status as "interest brokers" |

group demands has been clearly an overriding priority. Yet within any single historic period, where one value has held sway over the other two, the others have never been entirely neglected or ignored. Calibrating the proper emphasis has never been easy nor have the results ever been permanent, though four distinct eras where one value has tended to predominate over the other two can be discerned.

### The Nineteenth-Century Dominance of Jeffersonian Values

... During most of its first century the U.S. government operated with little bureaucracy (with the exception of the Civil War era). Hence its Constitution, erected upon republican ideals, was largely compatible with its limited bureaucratic institutions. Limited functions were demanded from government. Accident of geography had much to do with creating these conditions. High agricultural productivity made the nation largely self-sufficient, and continental isolation made a large standing army unnecessary. Further, there was virtually no popular demand for extensive social services. Rural farmers and small communities that dotted the landscape were relatively independent from public institutions for key support services. Only briefly, in the 1790s, did an activist (Alexander Hamilton) favoring rapid national modernization seriously press the case for an expansionist positive government, complete with a large professional army and a trained civil service to perform a broad array of nation-building tasks. The rapid decline of the Federalists and the rise of Jeffersonian-Jacksonian Democrats committed to the political dogma that "government governs best that governs least" ensured the continuation of the Jeffersonian ideals supporting negative government throughout most of the nineteenth century.

Ideology and geographic accident that sharply restricted administrative functions were external constraints on bureaucracy. They, in combination with three other important internal constraints in this era, maximized the Jeffersonian values favoring tight controls and public accountability of bureaucracy. First, direct popular controls over administrative machinery waxed because of the rapid growth of a party system that awarded administrative jobs based upon party loyalty and activism. Particularly after the election of Andrew Jackson, the spoils system grew and became a deeply ingrained institutional process. The belief that any job in government could and should be done by the average person was accepted as a given. Thus party affiliation was stressed over professional expertise as a central requirement for holding public office. Most of the federal government jobs were with the post office and required the performance of menial and repetitious tasks that could indeed be performed by the lay citizen, making bureaucratic expertise unnecessary. Further, the lack of serious external threats, except for those from Native Americans, meant that the United States could rely upon untrained citizen-soldiers in state militias for its primary defense. Even the top military posts in this era were largely filled by political appointees; of the thirty-seven generals appointed between 1802 and 1861, not one was a West Pointer and twenty-three were without any military education or experience.[7] Political appointment became common practice in most civil administrative offices as well (with citizen-volunteers providing most public services at the local level). Probably these personnel trends peaked in the 1860s when Lincoln used patronage more effectively than any previous president to run a government and to fight the Civil War.

Money, or more precisely tight fiscal constraint, was the second critical instrument ensuring the primacy of public accountability. Despite Alexander Hamilton's early efforts to develop a comprehensive executive budget in order to strengthen executive branch autonomy and managerial planning, Congress, not the chief executive, firmly grasped the reins of budgetary controls over public agencies. Surprisingly, presidents gave up this authority over the purse without much of a fight.[8] In his first annual message to Congress, Thomas Jefferson recommended that appropriations be made as "specific sums to every purpose susceptible of definition"[9]; and each public agency operated through a complex voucher system expending funds incrementally authorized by Congress in the absence of centralized treasury oversight. This fiscal pattern of legislative control made the nation's financial system unique by comparison with that of every other nation (both then and today) by giving Congress, not the president, authority over policy formulation and internal administrative matters of agencies. The multiplicity of fiscal controls over public agencies by legislatures at every level of government was further extended by dividing up responsibility for fiscal oversight among several special committees and subcommittees. Not only was it customary for several legislative subcommittees to exercise financial oversight over the same public agency—thereby serving to fragment bureaucratic fiscal integrity—but the process of financial oversight was further divided into two elements: first, legislative authorization to approve the programs, and second, appropriations to fund the programs.

A third critical strategy for extending public accountability over public organizations throughout the nineteenth century involved the structuring of their organizational designs so as to prevent organizational autonomy and enhance their dependency upon other branches in order to function. From the earliest period onward, public agencies, their organizations, and their procedures were subject to intense congressional scrutiny. Nothing was considered beyond the bounds of legislative concern. As Don Price notes, "The term executive branch . . . is a misleading metaphor. Organization charts and television pundits to the contrary, there is no such thing as an executive branch of the U.S. government. The Constitution gives the President certain executive powers, but it does not mention the executive branch. Instead, it lets the Congress by legislation set up executive departments and control their organization and procedures to any degree it likes."[10]

Beginning with Jefferson, Congress developed the habit of setting up governmental organizations through highly detailed legislative statutes that exhibited minute technical controls over their designs, purposes, internal procedures, and external relationships. The general thrust of these legislative mandates was not only to make these agencies creatures of the legislature but also to foster functional specialization. The military, for example, was, throughout much of the nineteenth century, organized around strong specialized bureaus—cavalry, infantry, engineers, and ordnance—rather than around a unified command structure staffed with military professionals who were servicewide generalists, not technical specialists. Officers identified with bureaus in which they served since they were not general military professionals with broad training and experience. Not until the National Security Act of 1921 was this preference for bureau specialists over military generalists altered. The same was true for civilian agencies where specialized bureaus rather than broad departmental interests were of paramount influence and concern. Institutional fragmentation served not only to inhibit unity of purpose and generalized public professionalism but also to divide and conquer. By creating small "bureau governments," staffed with political appointees in various specialized fields, Congress easily controlled these numerous small bureaucratic entities through detailed subcommittee oversight.

Throughout most of the nineteenth century, U.S. public agencies were creatures of Congress, not the president. President James Garfield in 1882, for example, could enumerate *all* of his presidential duties without ever mentioning "administration of the executive branch" as a significant responsibility. And the young Woodrow Wilson in 1885 wrote his political science Ph.D.

dissertation, which became a best-selling book entitled *Congressional Government*, as a criticism of legislative control over most federal administrative machinery. The irresponsible actions that ensued from fragmented, haphazard political oversight by congressional subcommittees were Wilson's primary target. Indeed, most of the great leaders of federal departments during this period were lawyers and legislators—Albert Gallatin, Jefferson's secretary of the treasury; John C. Calhoun, Polk's secretary of war; and William Seward, Lincoln's secretary of state—who knew the workings of the law and Congress and gained most of their fame in legislative halls. Bureaucracy offered no opportunity for bolstering one's reputation in that century.

But were the other values—administrative efficacy and interest group demands—entirely neglected during this era? Hardly, but they were not predominant values. Flashes of concern for administrative efficacy appeared in various parts of government from time to time in the nineteenth century, such as during Amos Kendall's tenure as postmaster general in Jackson's presidency, described in Mathew Crensen's *Federal Machine*,[11] and during the Civil War, when professionals gained prominence in Lee's army, as described in Douglas Southall Freeman's *Lee's Lieutenants*.[12] The organization of special interests, such as farmers, who pressed their case for the first clientele department, the Department of Agriculture (1889), also began in the nineteenth century.[13] "However, political authority was exercised by fairly homogeneous political communities, not by organized special interests.[14]

There were, of course, sectional interests that loomed large over the entire century's politics especially prior to the Civil War, but the organization of government services *around or directed at* particular special interest group demands had to await the twentieth century and a fundamental shift in the underlying nature of U.S. political authority.

### The Dominance of Neo-Hamiltonian Values 1883–1945

The year 1883 saw the passage of the Civil Service Act. It also marks the beginning of a decisive shift in national values, away from the Jeffersonian ideal of limited government and its attendant emphasis upon strict administrative accountability to THE PEOPLE. Instead, stress upon neo-Hamiltonian values favoring administrative efficacy began to appear. Not that Hamiltonianism ever entirely eclipsed Jeffersonianism, only that new methods, outlooks, and perspectives tended to give priority during this period to improving overall efficacy of public institutions.

If Jeffersonianism favored limited functions, popular representation, fiscal constraints, and organizational dependency as the keys to "marrying" bureaucracy with democracy, Hamiltonianism sought to broaden the range of public action, enhance public professionalism, and strengthen executive management and organizational autonomy. This decisive shift in national values did not come all at once but grew gradually over time and was to a great extent caused by a rapidly changing sociopolitical and economic environment that required a new approach to national governance.

The nation itself was rapidly modernizing from an agrarian republic to a contemporary industrial society with significantly differentiated and expanded functions. A modernizing nation required a modernized government. Industry during this era replaced agriculture as the major employer, thus creating needs for new public regulatory agencies such as the Interstate Commerce Commission [ICC]. Growing international responsibilities required a standing military and diplomatic presence abroad; and a vast influx of immigrants to the United States turned towns into cities, creating heterogeneous urbanized communities requiring effective local administrative services of many kinds. These new socioeconomic and political realities of life at the turn of the century made Jeffersonian values less relevant to the growing responsibilities of a modernizing nation. Jeffersonian values, for most Americans at the dawn of the twentieth century, just did not make sense or contain much meaning in a rapidly changing society that suddenly demanded that new tasks be performed with efficiency and dispatch (though some, such as William Jennings Bryan, clung steadfastly to the old Jeffersonian values). Effective public organizations were now required to carry out the myriad and expanding responsibilities of a modernizing nation-state. On the eve of this transformation of values, in 1871, only 51,020 civilians worked for the federal government, of whom 36,696 (73%) were postal employees. The remaining 14,424 constituted the entire national government for 40 million Americans. By 1940, nearly 1 million federal workers were employed in a wide range of social, economic, regulatory, and public services, with a mere 5 percent employed in the post office.

This period 1883–1945 witnessed not only the rapid expansion and differentiation of governmental services but also the development of skilled, specialized public personnel. Any effective public bureaucracy must contain, at its heart, a career service, with dedicated employees who are offered opportunities for career development and have advanced education, specialized expertise, and at least some degree of freedom from politics in order to exercise bureaucratic responsibilities. The Civil Service Act of 1883 was a first important step in that direction. Drawn largely from the British experience but adapted to American circumstances, the act developed merit criteria as opposed to political criteria for appointment to public office. While it took nearly a half-century to extend merit protection to most elements of government (helped by the passage of other such important laws as the Classification Act of 1923), the growth in the size and scope of skilled expertise inside government agencies was perhaps the most significant factor enhancing and extending administrative efficacy during this era. Also critical to strengthened public agencies was the establishment of various specialized professional groups, such as the foreign service, which was established by the Rogers Act of 1924. The extension upward, downward, and outward of professional expertise came gradually and piecemeal during this era with the growth of new specialists in various fields, such as public health, personnel, teaching, and city planning. . . .

Along with growing expertise, professionalization, and specialization, modernization of key management institutions and techniques was also instrumental in furthering the goal of administrative efficacy. Among the important management reforms were the establishment of the general staff by Elihu Root in 1902 to improve management planning, coordination, and control of the military (later extended to most civilian agencies); and the development of an executive budget, first used by the New York Bureau of Municipal Research for the New York Public Health Department and later established in the federal government through the 1921 Budget and Accounting Act. Also, new staff offices such as the Bureau of the Budget (established by the Treasury Department in 1922) and the post of chief of naval operations, created in 1915, were important organizational devices for centralizing executive control. At the grass roots level, the reorganization of state government pursued by Governors Lowden in Illinois and Byrd in Virginia set new patterns for increasing executive management effectiveness and centralizing state-level functions. The council-manager government in cities and towns, spurred on especially through the 1916 Model City Charter of the National Municipal League, modernized, centralized, and rationalized local institutions by putting trained management expertise at the core of expanding municipal functions. City government with city managers in charge soon equipped cities with new financial management and with budgeting, planning, and civil service capacities.

Equally critical to the enhancement of overall administrative efficacy during this period was the increasing institutional autonomy of executive branch agencies at every level of government. Separation of politics and administration was advocated as good government practice by reformers and theorists. In practice, though, institutional autonomy from congressional subcommittee oversight was promoted by Congress itself. Independent regulatory agencies, beginning with the ICC in 1887, and the first government corporations, starting with Panama Railway Company in 1905, were designed as autonomous units that could do what Congress either could not do or did not want to do. . . . World War I particularly accelerated the trends toward organizational autonomy and centralization of authority in executive agencies. Wartime emergencies, as always, necessitated rapid troop mobilization, national economic planning, press censorship, and emergency nationalization, as well as regulation of various sectors of industry.

The Hatch Act (refer to Figure [2.5.1]) formalized the separation of "administration" from "politics," thereby serving to strengthen administrative autonomy and discretion at the federal as well as at state and local levels.

While peacetime in the 1920s saw the return of many of these administrative powers to private authority, the Great Depression and World War II saw another set of emergencies that created overnight new bureaucratic institutions with autonomous authority for dealing with these crises. Perhaps the strongest influences in achieving organization autonomy and administrative independence from congressional oversight were the Reorganization Act of 1939 and Reorganization Plan No. 1, which implemented several of the Brownlow Commission's recommendations.

Brownlow synthesized most forcefully the neo-Hamiltonian values in arguing for the creation of a strong, energetic presidency by means of (1) placing the president in charge of an independent executive branch; (2) establishing adequate staff assistance in the White House in order that the executive branch functions could be properly managed; (3) transferring authority for budgeting, personnel, and planning to the White House in order to strengthen the managerial capacity of the president; (4) professionalizing civil service personnel by extending upward and downward merit protection to cover all nonpolicy-determining posts; (5) reducing the president's control and the lines of authority by reorganizing the more than one hundred agencies reporting to him into twelve major departments; (6) giving the executive "complete responsibility for accounts and current financial transactions while providing a genuine independent post-audit of all fiscal transactions by an auditor general reporting to Congress.

Here, in short, was the "high energy" model for the federal government that had been in the making since 1883—professional personnel, executive budgets, rationalized organizational span of control, "pre-audit" authority, autonomous executive units, general management and planning capacity, and political authority concentrated in political executives while leaving administrative work to the "pros" to allow maximum administrative efficacy. Brownlow, as Barry Karl observed,[15] transferred upward to the federal level many ideas that twenty years earlier had become the model for effective local government practices, especially as exemplified by the council-manager plan. And certainly long after the Brownlow Report, these neo-Hamiltonian ideals favoring administrative efficacy echoed in many postwar recommendations for governmental reorganization, such as the two Hoover Commission Reports (referred to by Herman Finer as "Mr. Brownlow's children"), the continued growth of the council-manager plan, the Ash Commission Report (1970), and, significantly, National Academy of Public Administration's report, *A Presidency for the 1980s* (1981). But in the post–World War II United States, neo-Hamiltonian values were no longer in ascendancy, having given way to a very different amalgam of values for designing "ideal" political-bureaucratic relationships.

Figure 2.5.1   **The Hatch Act Erected an Influential Legal Barrier for Separating "Politics" from "Administration" Activities at Federal and State Levels (1939)**

*Be it enacted by the Senate and House* of *Representatives of the United States of America in Congress assembled,* That it shall be unlawful for any person to intimidate, threaten, or coerce, or to attempt to intimidate, threaten, or coerce, any other person for the purpose of interfering with the right of such other person to vote or not to vote as he may choose, or of causing such other person to vote for, or not to vote for, any candidate for the office of President, Vice President, Presidential elector, Member of the Senate, or Member of the House of Representatives, Delegates or Commissioners from the Territories and insular possessions.

Sec. 2. It shall be unlawful for (1) any person employed in any administrative position by the United States, or by any department, independent agency, or other agency of the United States (including any corporation controlled by the United States or any agency thereof, and any corporation all of the capital stock of which is owned by the United States or any agency thereof), or (2) any person employed in any administrative position by any State, by any political subdivision or municipality of any State, or by any agency of any State or any of its political subdivision or municipalities (including any corporation controlled by any State or by any such political subdivision, municipality, or agency and any corporation all of the capital stock of which is owned by any State or by any such political subdivision, municipality, or agency), in connection with any activity which is financed in whole or in part by loans or grants made by the United States, or by any such department, independent agency, or other agency of the United States, to use his official authority for the purpose of interfering with, or affecting, the election or the nomination of any candidate for the office of President, Vice President, Presidential elector, Member of the Senate, Member of the House of Representatives, or Delegate or Resident Commissioner from any Territory or insular possession.

Sec. 3. It shall be unlawful for any person, directly or indirectly, to promise any employment, position, work, compensation, or other benefit, provided for or made possible on whole or in part by any Act of Congress, to any person as consideration, favor, or reward for any political activity or for the support of or opposition to any candidate or any political party in any election.

Sec. 4. Except as may be required by the provisions of subsection (b), section 9 of this Act, it shall be unlawful for any person to deprive, attempt to deprive, or threaten to deprive, by any means, any person of any employment, position, work, compensation, or other benefit provided for or made possible by any Act of Congress appropriating funds for work relief or relief purposes, on account of race, creed, color, or any political activity, support of, or opposition to any candidate or any political party in any election.

Sec. 5. It shall be unlawful for any person to solicit to receive or by in any manner concerned in soliciting or receiving any assessment, subscription, or contribution for any political purpose whatever from any person known by him to be entitled to or receiving compensation, employment, or other benefit provided from or made possible by any Act of Congress appropriating funds for work relief or relief purposes.

Sec. 6. It shall be unlawful for any person for political purposes to furnish or to disclose, or to aid or assist in furnishing or disclosing, any list or names of persons receiving compensation, employment, or benefits provided for or made possible by any Act of Congress appropriating, or authorizing the appropriation of, funds for work relief or relief purposes, to a political candidate, committee, or campaign Manager, and It shall be unlawful for any person to receive any such list or names for political purposes. . . .

## Madisonian Value Patterns in the Postwar Era

World War II proved to be another turning point in political-bureaucratic relationships. The crisis of wartime not only centralized political authority in the United States to unprecedented degrees, but the new postwar global responsibilities of an economic-political superpower also required the maintenance of a complicated international and military apparatus in order to carry out tasks imposed by free world leadership. Further, a welfare state, largely created by the New Deal, required numerous administrative agencies in order to carry out growing social tasks that Americans deemed essential.

The lives of millions of Americans in this period were directly touched for the first time by the activities of an expanded bureaucracy at every level of government; for payment of social security checks, auto licensing, FHA/VA home mortgages, regulating most sectors of the economy, and furthering new scientific developments, such as the atomic bomb. As a result, bureaucratic institutions became more numerous and more complicated. Their interconnections and relationships with politics likewise became more complex. Hence, the underlying American values associated with these relationships shifted as well. Compare, for example, two bureaucracies—both considered highly successful in accomplishing their particular missions during World War II— the Manhattan Project and the Selective Service System. By means of quite different institutional processes, both organizations achieved their objectives.

The purpose of the Manhattan Project was to develop an atomic bomb quickly. The complexity of this task (no one really knew if the bomb would work or even if it could be built in the first place), the diversity of personnel and material resources required for it (spanning a continent), and the requirements of speed and secrecy (only a chosen few, those at the very highest policy level, could know about its existence), forged a public entity that was unprecedented. U.S. bureaucracy showed remarkable inventiveness in designing this new public organization. The project was organized in such a way as to ensure tight control at the top by secretary of war Henry Stimson, chief of staff General George C. Marshall, and General Leslie Groves as operational director. A wide range of expert personnel and highly specialized material resources were pulled together from across the United States in the production of the bomb. Dispersion of resources prevented many individuals from knowing the overall extent and purpose of the operations. A highly restricted group composed of top scientists of that day, such as Vannevar Bush from MIT and James B. Conant of Harvard, and military officers such as navy Admiral Parnell and army General Styer, served as a joint sciences-military advisory policy committee for the project. Requirements for competitive bidding for contracts were eased to permit sole-source suppliers to build specialized parts of the bomb. The project maximized administrative efficacy—speed, secrecy, efficiency of implementation—through limiting, though not entirely neglecting, accountability requirements.

By contrast, the complex system for drafting individuals developed in World War II (which operated until 1973), the Selective Service System, created by Congress and directed by General Hershey, proved an equally effective bureaucratic instrument for inducting 12 million soldiers during World War II. Whether one agreed with its purposes or not, the draft system functioned well in World War II by maximizing public involvement at the grass roots. While there was a national headquarters, the bulk of its work was delegated to state headquarters and 6,443 local boards composed of three or more volunteer community citizens. In contrast with the highly centralized, scientifically driven institutional processes involved with the Manhattan Project, voluntary grass roots participation, coupled with federal- and state-level procedural controls, mobilized men rapidly and implemented the draft laws successfully and economically—it cost only $22.50 to draft a soldier during World War II.

The complexification and differentiation of bureaucratic processes continued in the postwar era . . .,with the establishment of large, unique entities for achieving the varied tasks of governance. Americans invented superdepartments at the federal level, such as the Departments of Defense and of Health and Human Services, as well as small but critical coordinative units, such as the Advisory Commission on Intergovernmental Relations, and units to control and extend scientific knowledge, such as the Atomic Energy Commission, and the National Science Foundation—each highly complicated, differentiated institutional processes deemed vital for servicing complicated, diverse national interests. As Herbert Simon, Dwight Waldo, and other postwar theorists stressed, simple pre-World War II principles of economy and efficiency (à la Brownlow) no longer applied to organizational life. The "one-best-way" gave way to "multiple-best-ways" of organizing and formulating political-administrative relationships.[16] Thus institutional complexity became the hallmark of bureaucratic operational processes after 1945.

The development and growth of postwar U.S. bureaucratic institutions, as "realistic" political scientists point out, involved numerous special interest groups. As administrative agencies touched more Americans in this era, more groups and individuals, out of self-interest, came to influence the course of administrative processes. In *The Governmental Process* (1951), David Truman depicted the reality of government as *a process*, a seamless web of competition and compromise between competing social interests, each pressing its claims on existing public institutions and creating new ones to service its needs. Like Madison, Truman analyzed in realistic and behavioral terms U.S. society as based upon interest groups and their interaction. According to Truman, "The behavior that constitutes the processes of government cannot be adequately understood apart from groups, especially the organized and potential interest groups."[17] Truman, echoing Madison, postulated that government provides social equilibrium through establishing and maintaining a "measure of order between groups." Bureaucracy became but one of "a multiplicity of points of access" for groups seeking to influence public policies.

Like Madison, Truman and other political scientists of this period viewed the problem of governance essentially in terms of *structuring access* of groups to government so that none gains the upper hand and so that, overall, balanced points of view are heard. The public thus as a whole was well served. Implicit within their writings was a faith that the processes of bargaining and compromise between interests over administrative goods and services would work out for the good of the total society or of *most of those within society*. As Truman writes, "Government functions to establish and maintain a measure of order in the relationship between groups." We find this evidenced also in Robert Dahl's classic, *A Preface to Democratic Theory*, where he writes: "The vast apparatus that grew up to administer the affairs of the American welfare state is a decentralized bargaining bureaucracy. This is merely another way of saying that bureaucracy has become a part of . . . the 'normal American political process.'"[18] Most Americans probably concurred with this uncritical, benevolent view of the self-regulating pluralistic system of interest groups operating in, around, and through bureaucracy. Theodore Lowi coined the phrase "interest group liberalism"[19] to symbolize the postwar mood of general support of these arrangements.

Perhaps no one has captured the spirit, attitudes, and faith in interest group liberalism as well as Charles E. Lindblom did in his famous 1959 essay, "The Science of 'Muddling Through,'"[20] Here, the bureaucrat is not a "doer" governed by the Hamiltonian values of promoting efficacy or efficiency, but rather an official, one of many, who practices "the art of the possible" in a complex world of competing interests. Negotiation and compromise are the tools of his trade, with which he tries to produce "agreeable compromise to all parties concerned." He "muddles," rather than manages. Yet in this difficult, confused act of incremental decision making, argued Lindblom, he produced the best pragmatic results for society as a whole. He achieved social harmony and political equilibrium through compromises in a fluid, unstable sea of ever-changing interest groups. Here were Madisonian values stated in their clearest, most concise, and most persuasive manner in the postwar United States. In the relatively calm period of the 1950s, when the United States enjoyed both superpower status and industrial prosperity, these Madisonian values stressing interest-group balance fitted hand in glove with the social stability and conservativism of the times.

Hence, the debate over how best to control bureaucracy in a democratic society between Carl Friedrich and Herman Finer,[21] which dominated the thinking of political scientists and government specialists after 1940 (i.e., were internal or external controls the most effective instruments for controls?), was largely an irrelevant question. *Both* Friedrich and Finer's ideas were in practice used. Both at times were highly effective. Both at times were highly ineffective. In the postwar world of Madisonian institutional complexity, built upon a liberal faith in self-regulating interest groups, a wide variety of institutional controls were put into place and found to be effective *and* wanting at the same time. The prior examples of the Manhattan Project and Selective Service point out this diversity quite well. But there were others. The Veterans Preference Act of 1944 created the Veterans Administration as an independent agency built upon open access to single-interest veterans groups. The act catered to veterans' interests, often at the expense of the interests of others. Conversely, new postwar institutions were created, such as the Department of Defense [DoD] at the federal level, which merged the army, navy, and air force under one superdepartment, and the Council of Governments (CoGs), which administers 701 grants to grass roots organizations that foster planning on a regional basis. These two organizations, which are decidedly important postwar bureaucratic innovations, promote diverse points of view on defense and metropolitan policy matters. But DoD did not prevent the rise of—and controls by—the military-industrial complex. Nor did CoGs solve critical political problems of postwar center-city decay. No absolute "objective" controls for bureaucracy by democracy seemed to exist; some only promoted certain values over others. The diversity of means *and* ends was quite relative and mixed. In short, Madisonianism thrived.

Table [2.5.2] sums up elements of bureaucratic-political relations in each of the three historic eras discussed so far in this chapter.

## THE PERSISTING RIVALRY OF THE THREE VALUE TRADITIONS

In the words of Samuel P. Huntington, the late 1960s and early 1970s witnessed a "democratic surge"[22] that characterized past eras of Jeffersonian-Jacksonian democracy and progressive reform, in which there was a vital reassertion of democratic idealism in all phases of life in the United States. As Huntington argues, the era reflected

> a general challenge to the existing system of authority, public and private. In one form or another, this challenge manifested itself in the family, the university, business, public and private associations, politics, the governmental bureaucracy and the military service. People no longer felt the same compulsion to obey those whom they had previously considered

Table 2.5.2

**Political-Bureaucratic Relations in Three Eras**

| | 19th Century (Jefferson Era) | Late 19th/Early 20th Century (Hamiltonian Era) | Post–World War II (Madisonian Era) |
|---|---|---|---|
| *socioeconomic setting* | largely rural, stable, isolationist | predominantly nation-building | mature welfare state with global responsibilities |
| *political authority basis* | homogenous community life | rapid sociopolitical-economic change | interest group liberalism, scientific changes, and international responsibilities |
| *bureaucratic duties* | limited | expanding | diverse |
| *generalist/specialist personnel* | generalist political personnel | growth of professional/specialized staffs | mixed generalist/specialist personnel |
| *fiscal controls* | sharp limits imposed by legislatures | increased executive control/authority through executive budgets | mixed controls on finances depending upon agency and policy area |
| *organizational autonomy/dependency* | highly dependent upon legislature | increasing autonomy | mixed organization autonomy, and dependence |
| *use of external or internal controls* | primarily external controls utilized, based upon laws | increasing reliance upon internal controls, based upon public professions and their norms | highly complex mix of *both* external and internal controls |
| *ideological basis of political-bureaucratic relationship* | "best government governs the least" | "politics—administrative dichotomy" | "interest group liberalism" |

superior to themselves in age, rank, status, expertise, character or talents. Within most organizations, discipline eased and differences in status became blurred. Each group claimed its right to participate equally—and perhaps more equally in the decision making which affected itself. In American society, authority had been commonly based on organizational position, economic wealth, specialized expertise, legal competence, or electoral representation. Authority based on hierarchy, expertise and wealth all obviously ran counter to the democratic and equalitarian temper of the times.[23]

Was the reassertion of Jeffersonian idealism due to the Vietnam war? The rise of a "youth culture"? The demands for equal rights by women and minorities? The television age? The Great Society programs? The reactions to Watergate and the Nixon presidency? While the reasons for the sudden "democratic surge" were complex and are even now still unclear, Huntington points to a number of significant consequences of the "democratic surge":

1. Increase in the size and scope of governmental activity, though with a concomitant decline in governmental authority;
2. Increased public interest and concern about government, coupled with a sharp decline in public trust and confidence in government;
3. Increased public activism in politics, yet with a commensurate decay in the traditional two-party system;
4. A noticeable shift away from coalitions supporting government to those in opposition to it.[24]

Popular in this period was a philosophical treatise by John Rawls, *A Theory of Justice*,[25] which defined justice in egalitarian terms. In the fields of history and politics, Arthur Schlesinger's *The Imperial Presidency*[26] found a wide, enthusiastic post-Watergate audience for its criticism of the flagrant abuses of strong executive institutions in the United States. This book stood in sharp contrast to a popular text on the presidency a decade earlier by Richard E. Neustadt, *Presidential Power*,[27] which had praised the values of a strong chief executive. Egalitarian themes found their way into the literature of economics, particularly in E.F. Schumacher's *Small Is Beautiful*,[28] which proposed that the goods and services in society be distributed more equitably and, in the words of its subtitle, *As If People Mattered.* In public administration, books such as that of Frank Marini (ed.), *Toward a New Public Administration*,[29] and Vincent Ostrom's *The Intellectual Crisis in American Public Administration*,[30] though grounded in radically different methodological traditions, argued for the similar popular values of broader participation, decentralization of authority, and social equality.

Throughout the 1970s many of these Jeffersonian values were translated into institutional reality by means of new legislation directly affecting the internal operation of most public agencies.[31] First, there was a sharp increase in formal and informal controls placed upon bureaucratic operations to improve public accountability. The Legislative Reorganization Act of 1970 began a rapid expansion of congressional oversight staff and functions aimed at improving public accountability. New legislation, such as the War Powers Resolution (1973), the Freedom of Information Act as amended in 1974, the Congressional Budget and Impoundment Control Act of 1974, the Ethics in Government Act of 1978, and Codes of Ethics for Public Services (1983) sought to place important new external controls on executive branch activities. New internal procedural controls, such as the extension of the Office of Inspector General throughout the federal executive branch, were approved in 1978, as were the various ombudsman offices and "sunset" and "sunrise" laws instituted with state and local bureaucracies during this same period. Such legislation tended to reduce public organizational autonomy, flexibility, and discretion and to increase institutional fragmentation and legislative oversight. Furthermore, efforts were made by the late 1970s to limit sharply bureaucratic functions. At the local level, Proposition 13 in California (1978) and Proposition 2½ in Massachusetts served as "model legislation" for those who advocated reducing the role of the public sector through placing tight revenue-raising lids in state constitutions. On the federal level, the movement toward deregulation of various economic sectors begun in the Carter administration, combined with sharp cutbacks in federal expenditures inaugurated by the Reagan administration in 1981, reflected these popular Jeffersonian values favoring less government. Furthermore, the scope of popular representation *within* public organizations at every level of government was extended and enlarged during this era. Generally, the roles of professionals and professional groups such as city managers, foreign service officers, and public health officials were not enhanced or supported by the public but rather gave ground to expanded groups of political appointees, minority group representation, citizen volunteers, and contractual employees. Administration efficacy yielded to extended public accountability, on many fronts, throughout the 1970s and 1980s.

By the mid-1990s, however, the picture . . . has become mixed. Strong concerns about finding effective means of achieving broader public accountability over bureaucracy are still expressed by many. The dominant ideological environment, sharply hostile to public bureaucracy, exhibits values that favor squeezing, cutting, and reducing bureaucracy and expanding public controls. However, . . . the competing demands of Madisonian interest groups on governmental activities and its bureaucracy have hardly declined. Indeed, . . . there is strong indication that PACs, issue networks, and pressure groups have strengthened their influence throughout the executive branch in the 1990s. Not infrequently, voices favoring Hamiltonian values for increased administrative efficacy have been raised to promote health care reform and environmental protection, conduct international diplomacy in the Post–Cold War era, enhance national competitive capabilities in international trade and economics, revitalize U.S. industry, rebuild local infrastructure, employ the unemployed, stimulate hi-tech industries, and improve educational opportunities. These and other insistent public demands for action by liberals, moderates, and conservatives turn on the capacity of some public agency to ultimately deliver the goods. The 1990s . . . present a crazy-quilt pattern of contradictory assertions of all three values—administrative efficacy, public accountability, and interest group demands.

## PUBLIC BUREAUCRACY IN AMERICA TOMORROW: THE NECESSITY FOR RECOGNIZING THE VALUE OF TRADE-OFFS

Americans have never quite made up their minds about the place of bureaucracy in their democracy. Certainly this confusion over values is apparent today. As with numerous other unsettled constitutional issues, the genius of the founding fathers in writing the U.S. Constitution was precisely in allowing succeeding generations of Americans to determine the shape of their government. Throughout the nineteenth century, when Jeffersonian values predominated, institutional arrangements affecting bureaucratic-democratic relationships were fashioned to stress the value of public accountability. In the late nineteenth and first half of the twentieth century, when the United States was in a nation-building mode, administrative efficacy replaced accountability as the overriding priority. As a mature superpower and welfare state in the socially stable era after World War II, a benign Madisonian liberalism favored an interplay of interest groups stressing organic social balance and institutional equilibrium.

No single value for structuring political-bureaucratic relations is inherently right or wrong. Each of the three has worked successfully in its own time and place. Each of the three also contains inherent limitations and problems. The dilemma the United States faces today is that all three of these values have come to the forefront of national debate and attention. Strong ideological currents and political interests in society loudly press their cases for Jeffersonian, Hamiltonian, and Madisonian perspectives. The issue of which *one* historic value should predominate over the others remains unresolved in the 1990s. Americans run their government with a confusing amalgam of all three value-emphases, which creates a situation of enormous confusion and complexity—and frustration. Realistically, one cannot expect the United States to suddenly decide to adopt one value perspective and neglect the others entirely. The problem is a matter of emphasis and accent.

The U.S. system of governance throughout its history has given at least some attention and concern to all three values as a basis for fashioning bureaucratic-democratic relations. But throughout most of its two-hundred-year history, only one value approach at a time has tended to predominate over the other two. What is perhaps most important for the future of bureaucracy in U.S. society is a frank appreciation of the strengths and limits of each value and of what trade-offs or consequences ensue from the adoption of one value over the others. Understanding the trade-offs can help clarify the alternatives facing Americans today as they fashion tomorrow's bureaucraticdemocratic relationships.

### Hamiltonian Values Maximizing Administrative Efficacy

A modern system of public bureaucracy constructed upon Hamiltonian values is depicted as a generalized model in Figure [2.5.2]. The system focuses on producing high-energy outputs rather than on maximizing public accountability or interest group demands. A bureaucratic system reflecting these values (1) ensures the adequacy of economic and political inputs in order to undertake the required tasks; (2) permits clear political objectives and policy directions from the top; (3) contains a highly professionalized cadre of public officials to carry out the tasks; (4) possesses the essential tools of effective leadership, broad administrative discretion, institutional autonomy, and unified orderly procedures to deliver public goods with speed and dispatch; and (5) manages and restricts feedback, communications, and publicity in such a way as to enhance, not detract from, the efficient programmatic outputs. Table [2.5.3] depicts representative public agencies today that approximate this model.

The values that ultimately shape the overall design of this bureaucratic system, however, are not purchased without some costs:

1. Doing a particular job is given precedence over temporary popular concerns or individual group demands. This perspective assumes that there is a fixed task that can be understood and accomplished. It thus tends to be rigid and inflexible in its pursuit of these goals.

2. The emphasis upon achieving administrative efficacy—getting the job done effectively and quickly by focusing on achieving certain objectives, sometimes at any cost, tends to ignore other important, though possibly secondary, tasks. Adaptability to multiplicity of goals is thus sacrificed or reduced.

Figure 2.5.2    **Public Bureaucracy (Hamiltonian Model)**

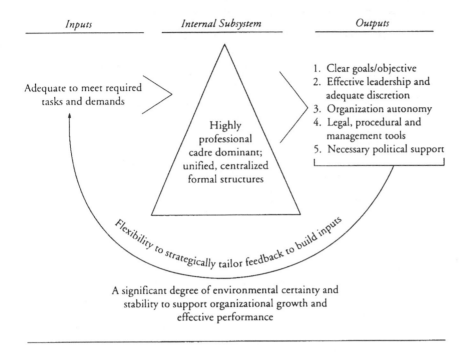

*Inputs*          *Internal Subsystem*          *Outputs*

Adequate to meet required tasks and demands

Highly professional cadre dominant; unified, centralized formal structures

1. Clear goals/objective
2. Effective leadership and adequate discretion
3. Organization autonomy
4. Legal, procedural and management tools
5. Necessary political support

*Flexibility to strategically tailor feedback to build inputs*

A significant degree of environmental certainty and stability to support organizational growth and effective performance

3.  Hamiltonian values demand the attainment of adequate inputs of economic and political power to perform the required tasks. In other words, authority must equate with responsibility. This ideal is considered sound management practice for fashioning any high-energy bureaucratic model. Yet more often than not, public bureaucracy today involves dealing with fragmented institutions, shifting interest groups, with radically opposing ideologies. Power to operate agencies usually must in real life be picked up piecemeal, and available input resources normally are never adequate to do any job.

4.  The internal dynamics of this system place a premium upon maximizing professional expertise and limiting political oversight. Increasing the autonomy of professionals assumes a degree of faith in their competency to make important judgment calls. Ultimately, professionalizing bureaucracy involves increasing discretionary authority of experts over lay citizens.

5.  A high-energy model achieves efficient output levels by building unified, stable administrative delivery systems with the necessary administrative discretion and staffing levels. The following elements are critical for inducing high-level outputs: limiting the span of control, clear lines of nonfragmented authority, authority for reorganization, internal systems for long-range planning, and tight hierarchical control. However, such centralized management tools for speeding service delivery reduce opportunities for political participation, decentralized community involvement, and intrusion of particularistic demands by interest groups.

6.  Hamiltonian values also stress the importance of bold, creative, and decisive leadership as central to making things happen in order to maximize administrative efficacy. Robert Moses,

Table 2.5.3

**Representative Public Agencies Approximating Hamiltonian Model**

|  | U.S. Marine Corps | New York Port Authority |
|---|---|---|
| *Clear purposes* | combat missions and defense preparedness | metropolitan transportation projects |
| *Organizational designs* | centralized military hierarchy—power flows from top down | independent special district with autonomous funding/personal authority |
| *Source of inputs* | strong popular support and congressional backing | autonomy for revenue raising and fiscal directions |
| *Internal subsystem control* | highly professionalized military cadre of officers | highly professional engineering and transit planners |
| *Outputs—the strengths* | swift combat force prepared for demanding, single missions of national need | effective regional-wide transit planning and building |
| *Outputs—the weaknesses* | complex missions, with multigoals or political goals offer problems in implementation | single-mindedness in transportation development; neglects other urban needs |

General George Patton, J. Edgar Hoover, and Alexander Hamilton exemplify this sort of high-energy public leader. But this high-octane leadership frequently is purchased at the price of individual rights, due process, and the accommodation of interest group demands.

7.   Feedback mechanisms necessary to achieve administrative efficacy require limits on communication, secrecy, and news restrictions to a high degree in order to gain agency publicity, popular support, future resources, and political approval. Openness, access to information, and public scrutiny are, therefore, sacrificed in adopting Hamiltonian values.

### Jeffersonian Values Maximizing Public Accountability

By contrast, a modern bureaucratic system constructed upon Jeffersonian values can be abstractly illustrated as in Figure [2.5.3]. Here the emphasis is upon maximizing accountability to the public as a whole, often at the expense of administrative efficacy or satisfying special interest group demands. Important elements of the model include (1) strict legal limitations placed upon economic and political inputs; (2) a high degree of political oversight and popular participation *within* the bureaucratic system; (3) emphasis on decentralizing delivery of outputs as far as possible; (4) weak overall executive leadership; and (5) constant public scrutiny over the entire system. Public organizations exhibiting these features are outlined in Table [2.5.4]. Some of the costs apparent in using this type of value-orientation to structure bureaucratic-democratic relations are as follows.

Figure 2.5.3    **Public Bureaucracy (Jeffersonian Model)**

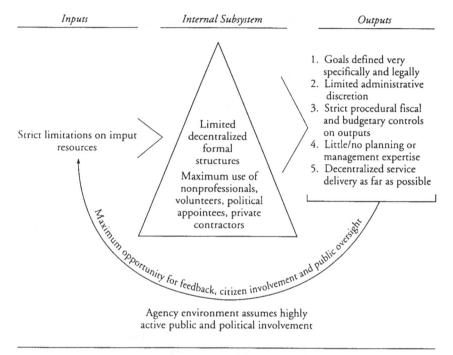

| Inputs | Internal Subsystem | Outputs |
|---|---|---|

Strict limitations on imput resources

Limited decentralized formal structures

Maximum use of nonprofessionals, volunteers, political appointees, private contractors

1. Goals defined very specifically and legally
2. Limited administrative discretion
3. Strict procedural fiscal and budgetary controls on outputs
4. Little/no planning or management expertise
5. Decentralized service delivery as far as possible

Maximum opportunity for feedback, citizen involvement and public oversight

Agency environment assumes highly active public and political involvement

1.  Extensive popular participation, expanded oversight, and political involvement serve to lengthen the time needed to achieve agreement upon organizational goals and create delay in the implementation of services, thereby reducing the overall administrative efficacy of the organization.

2.  This model assumes that there *is* a single, undifferentiated *public* to be served, not many groups, or that the PEOPLE, not groups, are paramount in making democracy work. It is thus "blind" to the essential requirements and problems of operating in a pluralistic society.

3.  Legal limitations on inputs, particularly economic resources, reduce the flexibility and ability of the governmental agency to meet changing and expanding demands, frequently leading to situations where responsibilities *exceed* the authority to act. This can lead to irresponsible, even corrupt, bureaucratic actions.

4.  The internal dynamics of such a bureaucratic system emphasize expanding the roles for political appointees, citizen volunteers, and nonprofessionals, thus frequently reducing knowledge, expertise, skilled planning, and management competency, as well as rationality and operational consistency.

5.  Decentralization of public services promotes flexibility and adaptability to different local and regional needs, but this also can create frequent opportunities for political intrusion by special interests as well as lack of uniform standards in services rendered and laws enforced.

6.  Strict legal internal procedures to ensure public accountability related to program implementation, staffing, budgeting, and organizing service delivery can reduce the likelihood of bold, creative, and innovative public leadership.

7.  Public scrutiny, ample information, and media publicity can enhance "honest" feedback to

Table 2.5.4

**Representative Public Agencies Approximating Jeffersonian Model**

|  | *Executive Offices of the Mayor, Governor, President* | *New England Town-Meeting Government* |
|---|---|---|
| Purposes | re-election of chief executive | representation of community interests |
| Source of inputs | voters selection | decisions of town meeting |
| Internal subsystem control | political appointees | citizen-volunteers |
| Outputs—the strengths | responsiveness to public opinion | mass citizen involvement in community life |
| Outputs—the weaknesses | little long-term planning/management capacity | little or no expertise in dealing with complex demands or problems |
| Organizational Design | fluid and changing according to chief executive needs | simple organizations with few employees |

citizens about agency actions, but not without possibly jeopardizing national security, personal privacy, law enforcement, security, and the confidentiality that is necessary for frank appraisals and thoughtful deliberations within the inner councils of government. Economic costs to government may increase due to delays from more public oversight.

## Madisonian Values Maximizing Balanced Interest Group Demands

A third alternative for modeling bureaucratic systems on Madisonian values is depicted in Figure [2.5.4]. Madisonian values stress attaining a social equilibrium through balancing interest group demands. The generalized elements of this system include (1) inputs principally based upon interest groups' demands; (2) bureaucratic outputs incrementally adjusted to meet special interest needs; (3) putting the principal focus of the internal bureaucratic subsystem upon "satisficing" the interest groups; (4) directing feedback toward promoting interest group satisfaction; and (5) devoting the bureaucracy not to efficiency, effectiveness, or to public service but to promotion of social equilibrium. Some of the typical units of government that operate in this mode are depicted in Table [2.5.5], but there are costs in adopting this bureaucratic model:

1.  The goal of maintaining social equilibrium through satisfying interest group demands is an essentially conservative doctrine favoring the status quo over change and innovation. The model implicitly assumes that the present arrangement of social interests is adequate and that their "caring and feeding" at present levels are acceptable, *and* that nonincremental change upsetting the existing status quo is unacceptable.

2.  The model thus favors the powerful and organized groups over the "voiceless," unorganized, weak, and underrepresented ones.

Figure 2.5.4    **Public Bureaucracy (Madisonian Model)**

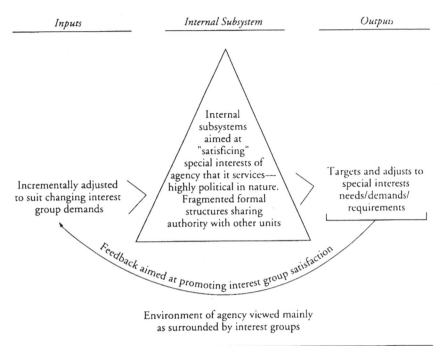

Inputs               Internal Subsystem               Outputs

Internal subsystems aimed at "satisficing" special interests of agency that it services— highly political in nature. Fragmented formal structures sharing authority with other units

Incrementally adjusted to suit changing interest group demands

Targets and adjusts to special interests needs/demands/ requirements

Feedback aimed at promoting interest group satisfaction

Environment of agency viewed mainly as surrounded by interest groups

3.  The Madisonian system, by its very nature, is a complex, fragmented, interconnected system of institutions operating with numerous checks and balances upon one another, but it also induces a high degree of institutional fragmentation and dispersal of authority, thereby making bureaucratic accountability and responsibility much harder to establish. Whom does the individual citizen turn to for help or justice when administration becomes so complicated? Which institution or public agency can be held accountable when power becomes so fragmented, interdependent, and shared? Public accountability tends to be reduced as fragmentation increases.

4.  Where inputs and outputs of the system turn largely upon "satisficing" special group demands, what ensures that *national* needs are considered, international economic or global security needs accomplished, and long-term future policy responsibilities for such imperatives as the fiscal deficit or environmental problems met? The narrow perspectives and short-term agendas of particular groups tend to overtake the long-term broad agendas. The big picture, in short, is sacrificed often for the short-term immediate needs of special interests.

5.  The internal dynamics of agencies that maximize representation of special interests inside agencies tend to open up agencies to political fun and games, not to doing work. Representation of interests, political diversity, and participation are given primary emphasis.

6.  Leadership of these public organizations, from top to bottom, is measured in terms of an individual's capacities to bargain, negotiate, and compromise effectively as opposed to his or her ability to make things happen and efficiently produce goods and services. Important institutional elements *necessary to lead* and deliver services are often overlooked. Gaining agreement, rather than achieving tangible goals, becomes an end in itself.

Table 2.5.5

**Representative Public Agencies Approximating Madisonian Model**

|  | Veterans Administration | State Road Dept. |
|---|---|---|
| Clear purpose | fulfill veterans needs | transportation functions |
| Organization design | permits limited executive control; maximizes veteran group oversight | permits limited executive control; maximizes highway interests control over department |
| Sources of inputs | large veterans organizations | powerful highway and transportation groups |
| Internal dynamics | highly political | highly political |
| Outputs—strengths | high amounts of social services for veterans | promotion of various state road-building and transit programs |
| Outputs—weaknesses | largely ignores other interest group needs and requirements | promotes road building and major capital projects at expense of alternatives |

7. Feedback, and indeed all agency activity, is judged by essentially political perspectives— will it strategically enhance the short-term interests of the agency? Gain bargaining chips? How will it "play" to the media? Build political support? Enhance reputations? The general public welfare or institutional efficacy for the long term tends to be disregarded.

## THE FUTURE OF U.S. PUBLIC BUREAUCRACY AS A VALUE PROBLEM: THE WORTH OF A SYSTEMS PERSPECTIVE

At the end of a recent book, Don Price recalls an incident that happened during his service as vice chairman of the Weapons Research and Development Board in the Pentagon.[32] Price was being flown in a navy plane out to Cape Hatteras (off the coast of North Carolina) to an aircraft carrier, where he was to watch the testing of new naval weapons. The weather was cloudy and the sea was rough. The navy lieutenant flying the plane had trouble finding the carrier. While cruising in search of the ship, he tuned in a radio station that announced the appointment of a new secretary of defense. While Price and others aboard expressed surprise at the appointee's selection, the navy lieutenant expressed little interest and then said, "By the way, who is the secretary of defense now?" Price tells how he felt a sense of outrage at a naval officer having so little comprehension of the broad governmental picture that he did not even know the name of his civilian superior. "But then," explains Price, "came a break in the clouds and far below—it seemed miles below—we could see the carrier on which we were to land. It looked about as big as a teacup, bouncing on the waves. And all of a sudden I did not want that navy pilot to have the slightest concern with the policies or the identity of his political superiors. I only wanted him to know how to land that plane."

In real life, especially when *our* lives depend upon it, we, like Don Price, want public bureaucrats to do what they are supposed to do, when they are supposed to, as effectively as possible. Knowingly or not, we depend upon numerous public agencies to do such important jobs as inspecting the food we eat, the roadways we use, and the drinking water we consume, securing our public safety, teaching the young, and protecting consumers in the marketplace as well as workers on the job. In real life, also, we want these numerous agencies to be responsible to the public they serve, protect the rights of citizens, obey the law, follow legally prescribed due processes, allow for popular oversight of their actions, and, above all else, be responsive to the general welfare and needs of the citizenry. In short, we prize public accountability as well. *And* in real life, too, we form associations to press our claims upon government as groups of farmers, veterans, steel workers, teachers, and many others who lay legitimate claim to benefits and services from one or several public agencies. The right to associate and secure *our group's interests* from government is regarded as important and essential by most Americans.

The point is that in reality Americans are not pure Jeffersonians, Hamiltonians, or Madisonians. Rather, we take public action based on bits and pieces of all three value systems, often contradictory and confused, regarding public bureaucracy. And such has been true throughout the course of U.S. history. While one value may have predominated over the other two for long periods, the others have never been entirely neglected or ignored. All three are vital to knitting together democracy with bureaucracy.

But, as the foregoing analysis suggests, there are always important trade-offs associated with the pursuit of one value over the other two. In maximizing one, sacrifices are required from the other two. All three cannot be obtained at the maximum levels simultaneously—nor would we want them if they could be. How ghastly is the prospect of living with (or under) the perfectly efficient bureaucracy! Or under one that is perfectly accountable to the public! Or perfectly responsive to all interest group demands!

And in real life, too, public agencies must operate with the constant dilemma of adjusting and juggling these three competing values in the course of carrying out their affairs. As long as they operate within the U.S. governmental system, they must constantly balance concerns for administrative efficacy, public accountability, and interest group demands in order to govern justly and well. No one value can entirely eclipse the other two. Certainly so far no one has discovered a magic tool with which to fine tune the relationship among the three values.

What is most important for citizens and public officials to appreciate is that important costs and benefits are associated with each value. And here, taking the broad systems perspective, as this book does, can give a better understanding of what happens to any public organization, to its inputs, outputs, and feedback, and to the general environment when a certain value is emphasized over others. In the imperfect world in which we live, there always will be trade-offs, costs, and benefits resulting from actions based upon particular value orientations. Being conscious of the results of our actions before they are taken makes not only good horse sense but wise public policies as well.

## SUMMARY OF KEY POINTS

The U.S. Constitution of 1787 was largely silent on the subject of bureaucracy. Now, more than two hundred years later, the United States operates with a large, equal public bureaucracy as its core system of governance. How to knit together our democratic ideals embodied in the written

Figure 2.5.5    **Significant Events in Shaping the Twentieth-Century Values of U.S. Bureaucracy**

| Neo-Hamiltonian Values<br>Strengthening Administrative Efficacy | Neo-Jeffersonian Values<br>Strengthening Administrative Accountability |
| --- | --- |
| Creation of General Staff (1902)<br>Taft Commission on Economy and Efficiency (1912)<br>Model City Charter (1916)<br>Budget and Accounting Act (1921)<br>Classification Act (1923)<br>Brownlow Report (1937)<br>Reorganization Act (1939)<br>Hatch Acts (1939 and 1940)<br>Government Corporation Act (1945)<br>Employment Act (1946)<br>First Hoover Report (1949)—Performance Budgeting<br>Organization for National Security Act (1947 and 1949)<br>Second Hoover Report (1955)<br>Kestnbaum Report (1955)<br>Creation of ACIR (1959)<br>PPBS Applied to DoD (1961)<br>BoB Circular A-95 (1969)<br>Reorganization Plan No. 2 (1970)<br>ZBB established (1977)<br>Civil Service Reform Act (1978)<br>Reagan's Cabinet Councils (1981)<br>REFORM 88 Management Improvement Initiative (1982)<br>President's Private Sector Survey on Cost Control (1983)<br>President's Council on Management Improvement (1984)<br>Federal Employees Retirement System Act (1986)<br>Federal Employee Pay Comparability Act (1990) | Freedom of Information Act (1967)—FOIA Amendments (1974)<br>Legislative Reorganization Act (1970)<br>War Powers Resolution (1973)<br>State "Sunset and Sunshine" Laws (mid-1970s)<br>Privacy Act (1974)<br>Budget Impoundment and Control Act (1974)<br>Civil Service Reform Act (1978)<br>Inspector General Act (1978)<br>Proposition 13 (California, 1978)<br>Ethics in Government Act (1978)<br>President's Council on Integrity and Efficiency (1984)<br>National Commission on the Public Service (1987)<br>Federal Quality Institute (1988)<br>Ethics Reform Act (1989)<br><br>**Neo-Madisonion Values Strengthening Special Interest Roles**<br><br>Veterans' Preference Act (1944)<br>Administrative Procedure Act (1946)<br>Federal Labor-Management Program established by E.O. 10987 and 10988 (1962)<br>Title VII of Civil Rights Act of 1964<br>Equal Employment Opportunity Act (1972)<br>Legislative Reorganization Act (1970)<br>Freedom of Information Act Amendments (1974)<br>Budget and Impoundment Control Act (1974)<br>Privacy Act (1974)<br>Creation of Expansion of PACs by Campaign Reform Act (1974)<br>National Partnership Council (1993) |

document and operational reality involving bureaucratic practices has been a recurring dilemma throughout U.S. history. No perfect "fit" between the two has yet been discovered. The issue involves a fundamental question of values: what is the place of bureaucracy in our modern democracy? Three value-approaches [See Figure 2.5.5] were outlined in this chapter—Jeffersonian, Hamiltonian, and Madisonian normative models—that have, in very different ways, served to answer this question. Throughout much of the nineteenth century, Jeffersonianism, emphasizing public accountability, dominated American bureaucratic-political relations. It fitted hand in glove with the largely rural, self-sufficient, isolationistic nation. The nation-building, expansionist forces of the late nineteenth and early twentieth centuries led to a strong assertion of Hamiltonian values reshaping fundamental democratic-bureaucratic relationships. The mid-century United States, with the creation of a mature welfare state and global international responsibilities, found Madisonian values, embodied in the interest group liberalism, formulating the design of democratic-bureaucratic relations. In the 1990s all three values find prominence and support. The chief theme of this chapter is that the selection of *any one* value over the other two involves certain costs and benefits. A frank recognition of the trade-offs involving adoption of any one value over the other two is essential. A systems perspective can be an invaluable tool for improving our knowledge about these potential trade-offs.

## KEY TERMS

| | |
|---|---|
| administrative efficacy | value trade-offs |
| public accountability | Hamiltonian values |
| administrative interest groups | Jeffersonian values |
| Brownlow Commission Report | Madisonian values |
| interest group liberalism | systems perspective |

## REVIEW QUESTIONS

1. What briefly are the nature and content of Jeffersonian, Hamiltonian, and Madisonian values?
2. Why is the problem of relating bureaucracy to democracy so complicated within the context of U.S. politics?
3. Can you describe the periods in U.S. history that were important turning points in political-administrative relations?
4. Why does the author argue that the future of U. S. bureaucracy involves recognizing trade-offs associated with different values involving political-bureaucratic relationships?
5. Briefly, how did three founding fathers—Hamilton, Jefferson, and Madison—conceive of connecting bureaucracy and democracy?

## NOTES

1. Leonard D. White, *The Federalists: A Study in Administrative History* (New York: Macmillan, 1948), 478.

2. Clinton Rossiter, *Alexander Hamilton and the Constitution* (New York: Harcourt Brace Jovanovich, 1964), 162.

3. Lynton K. Caldwell, *The Administrative Theories of Hamilton and Jefferson: Their Contribution to Thought on Public Administration* (Chicago: University of Chicago Press, 1944), 236–41.

4. James Madison, "The Federalist, No. 10," in *The Federalist*, ed. Edmund M. Earle (New York: Random House, 1937), 54.

5. Madison, "The Federalist, No. 51," ibid., 337.

6. Ralph Ketchum, *James Madison: A Biography* (New York: Macmillan, 1971), 301.

7. Samuel P. Huntington, *The Soldier and the State* (Cambridge, MA: Harvard University Press, 1957), 206.

8. Don K. Price, *America's Unwritten Constitution: Science, Religion, and Political Responsibility* (Baton Rouge: Louisiana State University Press, 1983), 83.

9. Ibid.

10. Ibid., 86.

11. Mathew Crenson, *The Federal Machine* (Baltimore: Johns Hopkins University Press, 1975).

12. Douglas Southall Freeman, *Lee's Lieutenants: A Study in Command*, 3 vols. (New York: Scribners, 1942–1944).

13. Refer to chapter 2 of [the original text where this chapter appeared] for an extended discussion of this subject.

14. Robert H. Wiebe, *The Search for Order, 1877–1920* (New York: Hill and Wang, 1967).

15. Barry Karl, *Executive Reorganization and Reform in the New Deal* (Cambridge, MA: Harvard University Press, 1963).

16. See especially Dwight Waldo, *The Administrative State* (New York: Ronald Press, 1948); and Herbert Simon, *Administrative Behavior* (New York: Macmillan, 1947).

17. David B. Truman, *The Government Process* (New York: Knopf, 1951), 501.

18. Robert A. Dahl, *A Preface to Democratic Theory* (Chicago: University of Chicago Press, 1956), 145.

19. Theodore J. Lowi, *The End of Liberalism* (New York: Norton, 1969), 37.

20. Charles E. Lindblom, "The Science of 'Muddling Through,'" *Public Administration Review* 19 (Summer 1959): 79–88.

21. Carl J. Friedrich, "Public Policy and the Nature of Administrative Responsibility," *Public Policy* (1940): 3–24; and Herman Finer, "Administrative Responsibility in Democratic Government," *Public Administration Review* 1 (Summer 1941): 335–50.

22. Samuel P. Huntington, "The United States," in *The Crisis of Democracy*, ed. Michael Crozier, Samuel P. Huntington, and Joji Watanuki (New York: New York University Press, 1975), 74–75.

23. Ibid.

24. Ibid.

25. John Rawls, *A Theory of Justice* (Cambridge, MA: Harvard University Press, 1971), 25.

26. Arthur Schlesinger, *The Imperial Presidency* (Boston: Houghton Mifflin, 1973).

27. Richard E. Neustadt, *Presidential Power* (New York: Wiley, 1960).

28. E. F. Schumacher, *Small Is Beautiful: Economics as If People Mattered* (New York: Harper and Row, 1973).

29. Frank Marini, ed., *Toward a New Public Administration: The Minnowbrook Perspective* (Scranton, PA: Chandler, 1971).

30. Vincent Ostrom, *The Intellectual Crisis in American Public Administration* (University: University of Alabama Press, 1973).

31. For a useful summary of these seminal pieces of legislation framing political-administrative relationships in the 1970s, refer to Part Four of Richard J. Stillman II, ed., *Basic Documents of American Public Administration Since 1950* (New York: Holmes and Meier, 1982).

32. Price, *America's Unwritten Constitution*, 177–78.

# PART III

# COMMUNITY
# AND THE INDIVIDUAL

The classical republican and classical liberal models of citizenship were compared in Chapter 3 of Part I. The republican model comes from ancient Athens in which citizenship—that is, active involvement in shaping and carrying out public decisions of the community—was considered the highest human good. It is reflected today in the *communitarian* movement. The liberal model was developed in the Age of Enlightenment in the eighteenth century, with an emphasis on freedom of the individual, who thinks and acts independently within society.

There has always been a tension in America between a vision of a community of people, relatively equal in influence and wealth, cooperating on public issues and drawing satisfaction from relationships with others, and a vision of a nation of separate, competitive individuals working hard to "get ahead" of others. Today, we live in a society that is largely urban and, in contrast with America up until the late 19th century, we are not a nation of farmers, shopkeepers, professionals, and craftspeople working on our own. Instead, we live and work within organizational and, at a larger level, institutional systems that shape and constrain our roles and behavior.

Despite this interrelatedness, we have a culture that portrays individualism as an especially valued ideal. Many books, films, television shows, and so on have strong, independent action figures who ignore or fight "the system," with its rules, procedures, and hierarchy. Children are absorbed for hours in video games emphasizing individual action, yet the world waiting for them when they enter the labor pool stresses skillful interpersonal communication and desire to be a "team player." It seems Americans use elements of an imagined past to express a desire to escape, if only in imagination, from the intricacies of obligation, responsibility, and behavioral expectations that surround and restrict us. This may be another way the tension between the community and the individual is apparent in daily life.

Among the many things in our surroundings that shape and constrain our knowledge and behavior are public-sector institutions, organizations, laws, programs, and policies. These things are not fixed or static but are created by people, and people constantly change them. They are not created, mostly, by disinterested, philosophically inclined wise persons (Hamilton's "speculative men") solely for the benefit of everyone in their jurisdiction, but by citizens and representatives who also have their own interests in mind, or those of a part of the public such as a state or a neighborhood. Public service practitioners are sometimes in a position to soften the impact on the general public of actions taken by those with wealth, power, knowledge of the policy-making process, and motivation to steer that process for the benefit of a specific few. Practitioners may sometimes find themselves in a position to protect individuals or minority groups from actions of the majority, and they may sometimes be able to propose or implement measures that improve life for those without significant influence or wealth.

Readings in Part II explore the nature of the idea of community in America and how it relates to the individual, linking conceptually with other concepts in the book, such as citizenship, liberty, and democracy. Perhaps most important for our work here, the readings offer an opportunity to question the role of the public-service practitioner in helping create a sense of community. We might ask:

- Is it within the scope of public practice to facilitate a sense of community, and if so, for what purposes?
- Should public-service practitioners work to protect private lives from the negative effects of governmental action?
- Does the role of public-service practitioners set them completely apart from others as neutral implementers of public policy, or do they share responsibility for conditions in society?

Americans are exposed frequently to the classical liberal model of the individual in society. The model focuses on protecting the individual from the larger collective body, but offers little about the feeling each person has toward community or society, or the duties and responsibilities citizens might be expected to fulfill. The readings in this section were chosen to give readers the opportunity to consider their relationship to the larger society and what it might mean for public administration.

## CONCERNS ABOUT THE NATURE OF COMMUNITIES

Not everything about the idea of community is constructive or desirable. The reading by Richard Box includes a vision of the use of power in local government that suggests citizens should be wary of appeals to community spirit or loyalty to specific administrations or leaders. Critics of the communitarian vision for society note that shared community values can be the basis for narrow, judgmental, even oppressive treatment or exclusion of people who are different in some way. The description, in Chapter 3 of Part I, of distinctions made between citizens and others in ancient Athens is an illustration of this phenomenon, as is the social environment of some small towns or places where one ethnic or religious group is dominant. Given this cautionary knowledge to balance naïve enthusiasm, exploration of the idea of shared memory, place, and sense of purpose can be inspirational and useful in public practice.

A good place to begin discussion of the relationship of the individual to the broader society is by exploring the nature of power and how it has influenced public institutions over time. The reading by Box from *Citizen Governance* focuses on the local community, the traditional locus of face-to-face public discourse. The intent is to go beyond description of typical policy issues, organizational structures, or power distribution within cities, to find underlying motivations for public action. Motivation is complex and cannot easily be reduced to a single factor, but patterns can be identified that help practitioners understand the human context of their work.

In the historical narrative of local government as an institution, we find a tension between citizen self-determination (democracy) and the influence of people and groups who wish to use government for their own purposes. The balance between these two polar ends of a continuum shifts over time; though it may be inaccurate to claim the balance is moving, long-term, toward one pole or the other, there is some indication currently of a movement on the local level toward expanding self-governance.

The reading includes description of four "community policy orientations," sets of characteristics drawn from the models of community power and policy-making presented in the reading. These orientations apply models of community power directly to the working environment of citizens, elected officials, and practitioners. The primary impact of this piece on the reader may be to challenge preconceptions about why power is exercised in the community and what would have to be done to promote democratic self-governance in such an environment.

## IN CELEBRATION OF COMMUNITY

In a wide-ranging introduction to an edited book on civil society, Don Eberly outlines what many would call a *communitarian* view of society. This does not mean big, socialized government, but citizen discussion and action based on shared values. Eberly's words counter possible misinterpretation of his intent regarding the role of government in society. He writes (xxxvii): "it is large omnipresent structures—such as the central state, ideological movements, or the national media culture—along with their totalizing effects on society, that may be the most destructive of community."

Eberly draws on Tocqueville and the work of Robert Bellah and associates (see the Kemmis reading, discussed below, for further description of Bellah's analysis of American society). He understands that strong communities can stifle liberty and individuality (by individuality he means the cultivation and expression of the person; this is different from individualism as being separate from society), but argues it is possible to have classical republicanism that also allows pluralism and democracy. The desired role of government in facilitating a sense of community, Eberly believes, cannot be the current trend toward "the science of public management," which "is inherently oriented toward constructing large central agencies from whence professional systems of command and control can design and implement solutions to problems by way of standardized rule-making" (xli–xlii).

The first question asked of those who would strengthen a sense of community in America is: how would this be done in a large, diverse society with a constantly changing mixture of cultures? One answer is not to seek consensus on many issues at once at larger geographic levels, such as states or the nation. At the level of the neighborhood or community, it may be possible to find agreement on several matters of importance. However, at larger levels of aggregated opinion, agreement (which at any level of government is often only temporary) may be reached in only a few key areas, such as protection of freedom and individual rights. This explains the emphasis on America as a *pluralist*, *liberal* nation, in the classical sense of emphasizing the individual rather than the collective whole, and allowing people to decide matters of values for themselves. At smaller levels of aggregated opinion such as a neighborhood, people have the opportunity to meet face-to-face to exchange ideas and agree on what to do. This allows people to find agreement on a wider range of issues with less likelihood their actions will inconvenience others or restrict individual freedom of choice.

Drawing from the work of Robert Bellah and colleagues, the distinction between a "first language of individualism" and a "second language of tradition and hope" is the conceptual framework for the reading from Daniel Kemmis's book, *Community and the Politics of Place*. Americans are accustomed to living in a "procedural republic," a place where the rights of citizens are protected and they regard themselves as having few if any obligations to those around them. Such "unencumbered selves" speak in the language of individualism, even when they are feeling a need to connect with others, with their shared history, and with shared aspirations for the future.

This is a predictable result of creating a society for the express purpose of avoiding aspects of European life in the seventeenth and eighteenth centuries that immigrants to America found oppressive. Kemmis's point is not that liberty and individuality are "bad," but that we can preserve them while simultaneously expressing our concerns about community and "common ground." Kemmis writes that his vision is not nostalgic, a yearning for an imagined communitarian past, but rather an argument for "the politics of cooperation." He uses the story of Albert and Lilly, neighbors on the high plains, as a metaphor illustrating how people can work together despite differences.

An additional point of importance in the reading is the attention given to physical place as a basis for "communities of memory." It has become common in the social sciences to downgrade place as a factor in human thought and action, in part because it is assumed that in the electronic age people have moved beyond mundane concern with physical surroundings. On the micro level of immediate surroundings, this seems untrue on its face, as we all have to deal with daily physical matters, beginning with our bodies and homes, and extending out into streets, neighborhoods, and places of work. On a macro level, resources, political movements, and culture are all related to place. The sensitivity to place exhibited by Kemmis is heightened by his identification with the American West, a place of wide-open spaces, where the connection between people and their physical environment is more direct and obvious than it is in places that are more urbanized or gentler.

READING 3.1

# "THE NATURE OF COMMUNITY GOVERNANCE"

## RICHARD C. BOX

### DESCRIBING THE INSTITUTION OF LOCAL GOVERNANCE

A good place to begin an examination of the nature of community governance is to describe how it serves as an institution in our lives. An *institution* is a large and enduring set of practices that we accept, take for granted, because these practices are so familiar and so much a part of daily life. Philip Selznick wrote that, "a social form becomes institutionalized as, through growth and adaptation, it takes on a distinctive character or function, becomes a receptacle of vested interests, or is charged with meaning as a vehicle of personal satisfaction or aspiration" (1992, 233). Such a social form, because of this "distinctive character," endures, lasts longer, than the more transitory organizations or practices not thought of as institutions. Thus, longevity and the weight of importance people attach to its values are key distinctions between an institution and other forms of human organization. As Selznick put it, "the more settled the practice, the more firmly vested the interests, the more values at stake, the more sense it makes to speak of 'an institution'" (1992, 233).

Anthony Giddens emphasized time and distance in the process of changing institutions. In his theory of "structuration," the practices of knowledgeable human actors create the structures of social systems. Practices with "the greatest time-space extension" (those that are found in the most places over the longest time) are those that "can be referred to as institutions" (1984, 17). For Giddens, institutions are not permanent and separate from human intent and action, they are a constantly changing result of the things that people do.

So it appears that institutions are not inflexible, concrete realities outside of human control and beyond the influence of changes in their environment. Instead, they are the result of choices made by many people over long periods of time. If this is true, we are not helpless as we deal with institutions. It makes no sense to observe an institution and say, "We must behave as all have behaved in the past in regard to this institution, because that's how it is." We have the option of changing our relationship to institutions and of working to change the institutions themselves. Recognizing the possibility of change does not diminish the importance of institutions to human activity. Nor does it mean that we in the public sphere should act as if we did not care about past institutional practice and for the accumulated experience of other people in other times and places that has created the "institution."

In applying the concept of the institution to American communities, we could examine either organizations or practices. For example, an organizational definition could include all local governmental jurisdictions or identify each organizational type as an institution (cities, counties, special districts). Practices could be defined as institutions as well. Certainly, local police service is an institution, as is provision of local streets. However, to the extent that specific organizational types or service-based practices are defined as institutions, there will be marginal calls and problems with how to treat organizations or services that do not meet the Selznick–Giddens tests of institutionalism. Are regional park and recreation districts institutions? Are local animal control or health inspection of restaurants practices that should be defined as institutions?

It makes sense to define all local organizations and practices of public governance as "the institution of U.S. local government." Within this definition are found a variety of institutionalized organizations and practices, as well as organizations and practices that are not often thought of as institutions in themselves, but are practices of the broader institution of local government. Though the ideas in this book are applicable across the great variety of local governmental organizations, it is common in such research to focus on cities rather than other forms of local government, and this book is no exception. In their development over several centuries, towns and cities reflect the desires of Americans as they come together to create a vision of how to live together as a community. Thus, the focus here is on "the institution of American community governance," which may be thought of as a subset of the larger institution of local government.

As you read about the way American local government developed and the nature of community power . . . you might think about the principles of Community Governance . . . (scale, democracy, accountability, and rationality). What is the impact of the history of communities and the characteristics of local political and economic power on these principles?

## THE SEPARATE PATH OF LOCAL GOVERNMENT

It is common to mix the history and structure of American communities with that of the national government as if they were the same, but the values, stages of development, and the resulting structures are very different. The . . . usefulness of the history of community governance in shaping the communities of the twenty-first century, points us toward events and issues little known by most people. Though elementary and high schools teach students about the history of our national government—albeit in a rather cursory manner—they teach little about the local level of government (Massialas 1990). This is unfortunate, because this is a level of government that has been and remains crucial in the lives of Americans. The events of the eras of community governance help explain where we are today, what services we expect from local government, why we have the governmental structures we do, who exerts the greatest power and control in communities, and what roles we expect citizens, elected representatives, and practitioners to play.

The history of the institution of community governance can be divided into sections in any number of ways. I have chosen to split it into four eras that correspond roughly with centuries, beginning in the colonial period.

## The Era of Elite Control

To understand the development of the institution of community governance, we may separate the institutional history into four broad eras that roughly coincide with the boundaries between centuries. The characteristics of these eras are cumulative—that is, new characteristics were added with time. So, though each era is given a name characterizing a primary feature of the institution of community governance during that time period, the name reflects an important addition to the nature of communities, not a complete shift or abandonment of earlier features.

The first era, roughly coincident with the seventeenth and eighteenth centuries, may be called the era of *elite control.* Early incorporations of many colonial American towns and cities were based on the English model of the borough, a corporate entity formed by prominent local landowners on grant of a charter from the crown. In the American colonies, charters were granted, and sometimes revoked, by colonial governors and assemblies. The English practice of the *close corporation,* in which the local notables who were granted the charter and governed the community also chose their replacements from among their ranks, was replicated in a portion of the new colonial incorporations, though this practice faded with time. In similar fashion, county government in many of the colonies was characterized by rule by appointed magistrates (Griffith 1938, 1: 191), though there were geographic variations. For example, Pennsylvania began its county form of government with commissioners chosen by the people (Martin 1993, 5).

Though most communities were not of the close corporation type, leaders who served on governing bodies were often wealthy citizens. Griffith noted the example of the borough of Bristol, Pennsylvania, in the mid-eighteenth century, in which five of the ten wealthiest men were members of the common council and were a majority of its nine members. Only one council member had an income below the community's 60th percentile (Griffith 1938, 1: 189–90). This situation was apparently typical of the "prevailing deference" granted those of the upper economic classes in colonial American communities, where "acceptance of a stratified society" was "the normal state of affairs" (Griffith 1938, 1: 189).

Circumstances were somewhat different in the New England "covenanted" communities in which people formed a town (likely to be, initially, a geographic area including several farms, rather than an urbanized center) based on a common commitment to religious standards of behavior. In contrast to "cumulative" communities that were formed of diverse individuals with a variety of interests, covenanted towns were "composed of individuals bound in a special compact with God and with each other. . . . This community, so covenanted, was the unique creation of New England Puritanism" (Smith 1966, 6). Such a community was a "Christian Utopian Closed Corporate Community" (Lockridge 1985, 16), allowing citizenship only to others dedicated to exactly the same principles as the town's founders. In Dedham, Massachusetts, these principles included everlasting love, exclusion of the "contrary minded," mediation of differences by members of the town, obedience to town policies, and obligation of successors to the town's principles, in perpetuity (Lockridge 1985, 4–7).

There were differences in founding circumstances between cumulative and covenanted communities, but the covenanted towns were initially, like the others, governed by an elite—in their case an elite of religious leadership. In Dedham, as in many similar towns, the pressures of population growth, religious differences, and attractions of economic opportunity in distant urban or frontier areas led to a gradual breakdown of elite control. The eighteenth century brought an increasing trend toward active policy debate and decision making by the collected members of the town meeting instead of the familiar deference to the wisdom of the selectmen chosen as town leaders. Over time, members of the town meeting took control by appointing ad hoc committees, investigating expenses of the selectmen, approving tax rates, and so on (Lockridge 1985, 119–80).

However, even as the town moved slowly away from its original "monolithic corporate quietism" (Lockridge 1985, 136) toward broader participation of citizens in town governance, the selectmen "still came for the most part from among the most wealthy quarter of the townsmen" and exercised considerable leadership in town affairs (Lockridge 1970/1985, 126). In fact, loosening of political consensus occurred simultaneously with, in the eighteenth century, increasing stratification by wealth in many New England communities. People of all economic classes served in leadership roles, but the relatively better-off predominated, partly because they had the ability to serve in the variety of largely unpaid and often time-consuming volunteer roles of town affairs (Cook 1976, 63–94). Also, consensus and religious conformity remained a factor. Zuckerman, who argued that eighteenth-century New England towns exhibited little or no control by elites, nevertheless wrote that

> a wide franchise could quite easily be ventured after a society that sought harmony had been made safe for such democracy. Most men could be allowed to vote precisely because so many men were never allowed entry to the town in the first place, because those who were there were of like minds. Indeed, within the compass of such conformity, extended participation ceased to be the danger it had been in England and became a source of strength. (1970, 187–8)

**The Era of Democracy**

The second era of the institution of community governance, roughly the nineteenth century, may be called the era of democracy, a time when "Jacksonian Democracy was the upsurge of a new generation of recently enfranchised voters" (Morison 1965, 423). This democratic tendency extended into localities, where Jackson's "populistic ideas had an important effect upon urban government" (Adrian and Griffith 1976, 2:178). Beginning in the late eighteenth century and setting the stage for the democratizing changes to come, the structure of city government often was modeled on the new national government, with separation of powers between the council and the mayor and sometimes a bicameral legislative body.

In the nineteenth century, with the increasing size of population centers and the complexity of operating public services, city councils came to use committees of their members to administer specific functions (police, streets, etc.). However, the administrative burden was often too great, resulting in the adoption of the *board* system in which the council appointed citizens with knowledge or interest in a particular area to serve on a supervisory body. By the late nineteenth century, there was a growing perception that large legislative bodies and dispersed responsibility for administrative functions was creating inefficiency, lack of coordination, and the opportunity for corruption in the form of contracting kickbacks, patronage jobs, and direct payments to decision makers to ensure policy outcomes (Griffith 1974, 3: 53–96).

In the search for administrative efficiency, there were several experiments with the idea of having a council member be individually responsible for a department; in the early twentieth century this came to be called the *commission plan,* but it was tried earlier by several cities, including

Nachitoches, Louisiana, in 1819, Sacramento in 1863, and New Orleans in 1870 (Adrian and Griffith 1976, 2: 161). From the 1880s to the end of the nineteenth century, the trend in the structure of city government was toward centralizing administrative power in the hands of a single elected official, the mayor, though not everyone thought this a wise idea. In his 1904 book, *City Government in the United States,* Frank Goodnow recognized the nationwide thrust toward concentrating executive power in a single person, whether in private industry, educational organizations, or government, but lamented the passing of the board system, which he found to exhibit the best blend of professional competence with public accountability (Goodnow 1991,189–200).

## The Era of Professionalism

In relation to the institution of community governance, much of the twentieth century has been spent creating and implementing structural reforms to limit the possibilities of patronage, spoils, and control of government by political machines, reforms such as at-large elections to dilute the political impact of neighborhood (often ethnic) groups, and the council-manager plan, which uses the private corporation model of a professional general manager or chief executive officer accountable to a board of directors (in the public sector, the city council, or county commissioners). Such reforms work well where there is relative goal consensus and the challenges facing a community are largely physical and technical, such as challenges of infrastructure and finance. However, where the challenges relate to mediating or arbitrating between contending groups in the community, or when there are significant socioeconomic problems, the corporate model can be less effective (Williams and Adrian 1963).

The twentieth-century rush to professionalism accomplished what was intended, bringing to bear efficiency and economy to solve the largely technical concerns of rapidly growing urban areas. Almost one half of American communities today use the council-manager structure, although the "pure" manager plan has in many places been modified to include a directly elected mayor and often council-member election by districts. The emphasis on professionalism and merit-based hiring and promotion practices has been felt in cities with strong-mayor structures and counties as well. However, there has been substantial resistance to professionalism (Box 1993), with resulting movement toward mixed structural accommodations that allow a more politically acceptable blend of administrative rationality and political responsiveness (Box 1995). The strong-mayor movement of the late nineteenth and early twentieth centuries is making a comeback, worrying the advocates of the professionalized city manager system and showing the strength of the citizen belief that government is out of the hands of the people who rightfully "own" it (Blodgett 1994; Gurwitt 1993).

## The Era of Citizen Governance

The name given here to a new era is speculative; to the extent that there is a new era, it is in the early stages and its outlines and future are as yet uncertain. However, it seems accurate to say that the era of professionalism is drawing to a close, and that, although the benefits of professionalism in community governance are clear, the reformist zeal generated in reaction to conditions in the nineteenth century has succeeded too well. A counter reaction to excessive bureaucratization and professionalization has set in, and this is a time of change, of movement back toward greater control by nonprofessionals, by citizens.

The contemporary challenge is not to achieve efficiency but to realize a community vision chosen and enacted by its residents, something that Lappe and Du Bois (1994) called "living democracy" Conceptually, this means a redefinition of the role of the citizen, from passive consumer of government services to active participant in governance (see Chapter 3). This redefinition requires that citizens take greater responsibility for determining the future of their communities.

The movement back toward local control and citizen self-governance has significant implications for the practice of public administration. If we are metaphorically coasting down the back side of the wave of reform that swept the early twentieth century—moving away from professionalism toward citizen control—then "traditional" public administration, based on administrative power, control, and a sense of positional "legitimacy," is a thing of the past. Naturally enough, there will be voices of protest with the passing of such a comforting model of what the public service practitioner should be doing on a daily basis, but it appears that it will indeed pass. This leaves the practitioner with serious problems to deal with, including how to make possible the desired citizen access in combination with an acceptable level of grounded rationality in policy making, how to avoid retribution from angry elites who see citizens and open dialogue as a threat to their interests, and how to prevent a return to spoils, patronage, and corruption, the evils professionalism was intended to eliminate.

## THE INSTITUTIONAL LEGACY

Much has been learned about the institution of community governance over these four eras. At the beginning of the twentieth century, the thrust was toward efficiency and rooting out corruption and patronage, as well as toward solving the increasingly complex problems of an urbanizing society. Professionalism and centralized administration replaced a free-wheeling combination of democracy and machine politics. At the end of the twentieth century we are in a period of transition as earlier values of localism and citizen involvement in governance reassert themselves.

Today, the emphasis is on shifting the balance from centralized, expert-based systems to decentralized, citizen-centered systems. This shift is only possible because of the success of the reform impulse. If the battles for efficient and effective community government remained to be fought, if streets were muddy dirt tracks and water and sewer systems in their crude infancy, if local action was hampered by overly complex governance structures and administration was in the hands of political machines and their patronage employees, the contemporary discussion about citizen self-determination would seem trivial and foolish. Instead, we would be worried about solving the basic problems of service delivery, as were the reformers a century ago.

The eras of community governance outlined in the previous sections are indeed cumulative. This does not mean that we are always moving toward a more perfect future, but that we are accumulating valuable experience that can be used to deal with the problems of today. The era of control by a wealthy elite was superseded by democratic structures and processes, yet local government still had hierarchies and leaders, whether based on wealth or the power of the political machine. In the twentieth century era of professionalism, many came to question the depoliticization of the community, the suppression of dissent and debate in favor of "unitary" goals of economic and physical development, yet the era also included broad opportunities for citizens to participate in local affairs.

As the twenty-first century approaches, the institution of community governance is a strong one, vibrant with value-laden controversies and containing a foundation of experience that allows for informed change. The lessons learned from centuries of practice are many and cannot be easily summarized, but I believe the institutional legacy includes two powerful, fundamental issues that are especially important today. The first is the question of the scope of local government, the question of what services local government should offer. Most people agree on services such as police and fire protection, streets, and water and sewer service (though even such basic services are sometimes offered by private residential associations in addition to governmental provision). However, it is more difficult to find agreement on a host of services that could be ·provided by the private sector but are often found in local government. These include items such as developing conference centers or housing projects, owning parking garages, hospitals, or golf courses, or providing ambulance, mass transit, or garbage collection services. In these areas, service provision by local government can compete with, or even preclude, provision by the private sector. Local provision of such services causes us to reflect on what we think is the proper model, in American local government, of the relationship of the public sector to the surrounding private market economy.

This question of the scope of local government also involves the degree to which government regulates private sector activities. Though we often think of regulation as the well-publicized actions of national organizations like the Environmental Protection Agency or the Food and Drug Administration, local government regulates a wide range of private activities, sometimes in concert with state or national agencies. These activities include the location and construction of buildings, streets, and utilities; who pays for provision of public streets and sewer and water systems; types of businesses that can locate in certain areas; the safety and cleanliness of public places like restaurants; the appearance of signs, billboards, landscaping, and building exteriors; provision of affordable housing; and many others. Different communities regulate such activities differently, and local debate about what to regulate and to what degree are often the source of heated controversy and changes in political and administrative leadership. The relationship of collective, governmental authority to private, individual behavior is a central theme in American history and in the history of the institution of community governance. People have strong feelings about it and are often willing to do battle in the arena of public discourse to have their way.

In this [reading], I spend little time on the substance of these matters, instead concentrating on the processes by which people reach decisions about them. This leads us to the second fundamental question growing out of the institutional legacy, the question of structural practices, or how to organize local government. In the history of community governance, people have expressed their values through the way they organize government. The desire for greater democracy led to the use of multiple boards and committees, the ward-based patronage system developed to allow for governance by political machines, and the desire for efficiency led to the centralized council-manager plan.

The structural practices question revolves around the opposing values of public responsiveness (fragmented systems, open citizen access to the policy-making process, organizational guidance by elected officials), and administrative rationality (centralized scientific-purposive systems, the citizen as outsider and consumer of services, professionalized decision making). The search for balance between these values is fraught with the hazards of a chaotic, inefficient, and possibly corrupt community on the one hand, and a coldly efficient, depoliticized, and bureaucratized community on the other. Decisions about where to locate on the continuum between these polar views must be made in a community political environment often dominated by an elite, sometimes well-organized and relatively hard to penetrate, sometimes loosely organized around issues and open to changing membership.

Structure is not everything, in the sense that organizations with the same structures can show very different operational characteristics depending on local combinations of personalities, economic circumstances, geographical setting, political history, demographics, and other factors. Even so, it cannot be denied that people believe structure and its associated practices are important. The four eras discussed previously each contain dominant ideas about the proper structure of local government. These ideas were a central, dynamic, living part of the goals community residents had for their attempts to create ideal communities. Within this area of belief in structure as an expression of community goals can be found commonalities of practice and points of contention, points at which the institution of community governance is changing to meet the new demands of the fourth era.

Commonalities in local public structures include the use of elected representatives of the people in place of the town meeting (though the town meeting is alive and well in some places; see Elder 1992); movement toward a directly elected mayor (though selection of the mayor by the rest of the governing body, a feature of the "pure" council-manager plan, is still in use); reintroduction of district elections or a mixture of at-large and district elections; professional administration of technical functions; and provision of basic services (though sometimes contracted or franchised) such as public safety, public health regulation, and the utilities/infrastructure functions of streets, water, sewer, waste removal, and drainage.

Points of structural contention include the appropriate role of the public professional, the strong-mayor versus council-manager debate, and efforts to increase citizen participation and control over decision making. This latter area is highlighted by criticism of overhead, or "loop" democracy, which is decision making by elected representatives instead of the people themselves (Fox and Miller 1995, 15–7). Concern about loop democracy leads to renewed interest in various forms of citizen participation, including use of committees, boards, and commissions to supervise governmental functions, thus injecting citizen influence directly into administration instead of routing it through elected officials.

In this [reading], it is assumed that the democracy principle is served if community residents are able to get information and make decisions about scope and structure if they wish to do so. This means that structure is very important, because it can hamper people who want to determine the future of their communities. For example, a local government with structures that block citizen self-determination might have some combination of these features: few opportunities for meaningful participation in decision making but instead meetings or hearings in which citizens are told what is happening and given a chance to give input; inconvenient public meeting times or long, confusing agendas and procedures; an overly professionalized internal decision-making process in which elected representatives appear to have few ideas of their own and almost always follow the lead of their full-time staff; and a complex bureaucratic system with lots of rules and little explanation. The scope of local government is an important issue, but in this [reading], the focus is on creating open and flexible structures and practices within the institution that will allow people to make their own decisions about scope and indeed about further structural change.

Such is the legacy of the institution of community governance at the end of the twentieth century. It includes concern about the following issues: localism (a preference for local as opposed to state or national decision making); citizen involvement and self-determination (making community governance open, accessible, and welcoming for those who wish to take part);

demystification of professionalized systems; a desire to avoid the excesses of political intrusion into routine administration; a political environment in which elite groups are an important feature; an impressive technical-professional capacity; and lively debates about the scope and structure of community government. Let us turn now to an examination of the economic and political forces in the community that shape the institution and the practices of governance.

## POWER IN THE COMMUNITY

It is remarkable how little we Americans know about our local government. We are surrounded by television and print media reports about national issues and events; people who pay attention to news and politics know about the president, the Congress, the Supreme Court, agencies of the national government, and issues being debated in Washington, DC News reports about state and local activities are given less attention and, even if people watch or read them, they often do not understand the political, structural, and economic contexts within which they occur.

Though most of us know something about how we came to have a national government with three branches, what each branch does, and how they "check and balance" one another—this knowledge may not be very broad or deep. Even graduate students in public administration may think that the act of creating the Constitution was a relatively placid event, a gathering of wise people who made a few technical changes to the existing government by consensus (it wasn't), or that there has always been a strong central government and general agreement that states are lower-level administrative subdivisions of secondary importance (not so).

However, this thin knowledge of the national government is more than what many people know about local government. Local government is the context of our everyday lives, the only level of government that has a constant impact on our physical and social environment. Despite this, even people studying for advanced degrees in fields related to public administration or working in the field frequently do not understand the basics of local government structure and function. They do not know why some cities have an elected person, the mayor, as their chief administrative officer and why some have an appointed official, the city manager (more than a few people do not know whether this person is elected or appointed), or why many counties have neither, but instead a long list of elected and appointed officials with complex interrelationships.

Knowledge of the structures and processes of local government is necessary for citizens and professionals to choose wisely among the ways they may govern their communities, but it is not sufficient. It is not sufficient because it ignores the community as a whole. How can we decide what governmental structure to have, how much to tax the citizenry, or what services to offer if we have not studied our community, the socioeconomic characteristics of its citizens, the condition of its streets, housing, sewer and water systems, or what the majority of the citizens want for the future and how they view government's role in achieving that vision?

The growing desire for local control over the fate of the community is evidenced by local resistance to mandates and regulations from "higher" levels of government, citizen demands for a more open and accessible decision-making process, and calls for a "communitarian" approach to governance that emphasizes citizen responsibility for local affairs. The tendency for citizens to protest actions of local government, to complain rather than participate, or to object to changes in land uses (the NIMBY, or not-in-my-back-yard, syndrome) is viewed by some as evidence of citizen alienation from government. It may also be regarded as the first step in citizen participation; many people do not become involved in local governance until an issue close to home causes them to think seriously about their community. As they participate they come to under-stand the structure of government and local laws, the politics of individual and group interests, and what is needed for them to make a difference in community governance. Some people stop participating after taking part in the debate over a local issue, but single-issue involvement is for others a turning point at which they are drawn into local affairs.

Local control means that people in communities decide their futures relatively independently, putting outside pressures in perspective and fighting against closed or elitist political systems to chart a course grounded on a broad and democratic political foundation. This sounds logical and easy enough. People want to determine the fate of their communities, and they do not want to be blocked by national or state governments or by a few powerful people or special-interest groups. But logical or not, it is not easy. There are strong political and economic forces working against such citizen self-determination.

It makes little sense to examine local public governance in a vacuum, as if professionals and involved citizens could sit quietly in meeting rooms and make decisions about public policy without taking account of the community around them. The variety of community political and economic circumstances is so great as to be difficult to describe or understand in a comprehensive way. In a book about the "ecology" of local government policy making, Robert Waste (1989, 7) listed ten community features that affect the policy process: age of the community, locale, the growth process of the governing body, policy types, policy conflict levels, reform activity, regu-latory activity, external factors, personality factors, and local political culture.

That is a lot to consider in trying to understand the workings of a local government. Despite this complexity, I present next some powerful and relatively straightforward ideas I have found to be especially useful among the many concepts available about community life, drawing them together into four community policy orientations that shape the environment of citizen and pro-fessional policy involvement. This narrative concentrates on models that broadly describe the political and economic nature of the community.

Running through the literature of community power and politics is the common theme of competition between communities for limited resources, competition that focuses on land specu-lation and development. In his book *City Limits*, Paul Peterson (1981) argued that the local community has limited control over the forces that affect its growth and development. This is because economic conditions are the result of regional, national, or global forces, national and state governments have cornered the policy market in matters of social welfare, and what re-mains to be dealt with locally is the use of land and buildings. Peterson wrote that, "Urban politics is above all the politics of land use, and it is easy to see why. Land is the factor of production over which cities exercise the greatest control" (1981, 25). Each community is in competition with others for whatever economic growth is available at a given point in time, so local leaders feel a sense of responsibility to promote development as a means to defend prop-erty values, jobs, and the condition of the physical infrastructure and the educational system, essential components of economic expansion.

Within communities, groups and individuals compete, attempting to use local government as a tool to gain advantage over each other. There are places and times when community life is characterized by consensus, cooperation, and gentle transitions, but this is the exception rather than the rule. More often, this is an environment of conflict, competition, and unsettling change—and this is not necessarily to be viewed as negative or abnormal but as the sign of a healthy democracy. As political scientist E.E. Schattschneider put it:

Nothing attracts a crowd so quickly as a fight. Nothing is so contagious. Parliamentary debates, jury trials, town meetings, political campaigns, strikes, hearings, all have about them some of the exciting qualities of a fight; all produce dramatic spectacles that are almost irresistibly fascinating to people. At the root of all politics is the universal language of conflict.

The central political fact in a free society is the tremendous contagiousness of conflict. ⁻ (1960, 1–2)

How has awareness of the political nature of communities developed and what does it mean for local governance? Can citizens meaningfully influence the fate of their communities, or are they caught in the grip of forces beyond their control?

## THE DEBATE ABOUT COMMUNITY LEADERSHIP

For decades, two opposing views of community leadership battled one another. The literature of urban power structures contains several notable milestones such as Floyd Hunter's (1953) work in Atlanta in the early 1950s and Robert Dahl's (1961) study of New Haven in the late 1950s. In the post–World War II period and into the 1970s, a debate raged between the *elite* theorists and the *pluralist* theorists. Elite theorists tended to use assumptions and research methods that yielded a portrait of communities as controlled by a relatively cohesive and closed socioeconomic class, whereas the pluralists used assumptions and research techniques that produced a view of community governance characterized by changing and accessible groups of people involved in specific issue areas such as education or redevelopment (Waste 1986, 13–25).

The debate between these opposing views became relatively fruitless by the 1970s. Both sides had recognized that a small percentage of community residents were involved in community governance, and they disagreed primarily on the question of a cohesive and closed versus fragmented and accessible leadership. In the 1970s and 1980s new ideas emerged to expand and renew the study of community politics. There are any number of ways to categorize these concepts. For example, Harrigan grouped them into five areas: neo-Marxist and structural approaches; growth machine theory; the unitary interest theory of Paul Peterson; the systemic power and regime paradigm theories of Clarence Stone; and the pluralist counterattack (Harrigan 1989, 191).

Bachrach and Baratz (1962) noted ways in which local elites could exclude the public from the governance process by keeping important issues from reaching the point of open discussion. They could do this by controlling the agenda of public discussion while making the really important decisions in quiet consensus among themselves; Bachrach and Baratz called this "nondecisionmaking." Verba and Nie (1972) explored the concept of *concurrence,* the degree of agreement about public issues between citizens and community leaders. They found that people of higher socioeconomic status participated more in community politics and community leaders were more responsive to them. It seems intuitively logical that political and economic elites would be dominant in shaping the community policy agenda, and research supports this assumption.

## CITY LIMITS

Paul Peterson's book *City Limits* (1981) was noted previously for its emphasis on the community politics of land speculation and development. Peterson argued that, as local decision makers make choices about policy, they do so in three broad areas: developmental policy, redistributive policy, and allocational policy (Peterson 1981, 41–46). Developmental policy is concerned with generating economic growth in the community, whether by luring businesses to the area, helping local firms to grow, or improving local public services like schools or infrastructure so the community is more attractive to businesses that need infrastructure to operate and schools to provide an educated workforce. Allocational policy is concerned with "housekeeping" functions such as police and fire protection and garbage collection. Such functions have relatively little impact on intercommunity competition, because most communities provide such services.

Redistributive policies "help the needy and unfortunate" (43) and detract from the overall objective of improving the economic health of the community. This is because public resources are used to assist less productive citizens instead of to promote the economic advancement of the community in relation to other communities. For this reason, Peterson argued that "the competition among local communities all but precludes a concern for redistribution" (38). Redistribution is seen as consisting of policies that are "not only unproductive but actually damage the city's economic position" (43).

According to Peterson, when the community successfully creates new wealth in the process of competing with its neighbors for economic growth, everyone in the community is made better off; entrepreneurs, owners, and managers benefit through greater return on their investments and they are able to employ more people and pay them well. In this way, a "unitary interest" is created in which the success of the governing group makes everyone better off, and all community residents share the same goals (interests).

The reader may notice a parallel between Peterson's theory and the debate in American national politics about "trickle-down economics." Peterson's unitary interest concept is much like trickle-down (supply-side) economic policy. Both hypothesize that the way to make everyone better off is to put more money into the hands of entrepreneurs, capitalists, who will then create jobs by investing in new businesses, production facilities, job training, and so on. Thus a decrease in taxes on the wealthy, funded in part by cutting social welfare expenditures, will theoretically result in economic betterment for all. The reader is left to make an individual judgment about whether this has worked on the national level. The point here is that Peterson's unitary theory is subject to the criticisms normally associated with trickle-down economics, especially that it may make some wealthy people even wealthier at the expense of some who have relatively little.

In addition, Peterson's theory has been criticized for being too absolute and mechanistic to accurately reflect the on-the-ground reality of local policy making. Robert Waste (1993) noted that communities sometimes do enact redistributive policies and that policies that appear to the residents of one city to redistribute wealth inappropriately from rich to poor, such as busing of school children to achieve better education of students of low socioeconomic status, might be viewed as developmental in another community, where a skilled and prosperous labor force is thought to be a good thing. Local politics are varied and complex, so that it is difficult to predict the responses of community residents to policy challenges. As Waste put it, "local conditions, local actors, local policyframers or policy entrepreneurs matter" (452).

## THE GROWTH MACHINE

Of the theories of community politics that emerged in the 1970s and 1980s, I find the *growth machine* model advanced by Harvey Molotch in 1976 and elaborated by John Logan and Molotch

(1987) to be a good fit with the daily reality of local government. This model incorporates market forces, elite theory, the importance of land in the local economy, and the impact of non-decision making. It goes beyond describing the membership of leadership groups in the manner of elite and pluralist theorists in an attempt to identify the underlying mechanisms of community politics.

Molotch built on the idea that land is the dominant factor in local politics and economics. The basic premise is that the important players in community governance are those who have the most to gain or lose from changes in the rate of return from the use, development, and speculation in land. These people include those most immediately connected with land, such as landowners, local businesspeople, investors in locally owned financial institutions, lawyers, realtors, and so on, in addition to others who depend on growth for increases in their economic well-being (Molotch 1976, 314). *Growth* may mean development of vacant land or it may mean the improvement or redevelopment of land that is already being used. Not every community has significant amounts of vacant land that can be developed, but the effort to make money from the ownership and use of land can take several different forms.

Through their voluntary associations like the Chamber of Commerce and their efforts to influence the activities of local government, these people work to create a "we-feeling" of community, using athletic teams and community events to instill "a spirit of civic jingoism regarding the 'progress' of the locality" (Molotch 1976, 315). It is these leaders of community opinion and decision making who set the course of the city, and "the city is, for those who count, a growth machine" (Molotch 1976, 310).

In developing his model, Molotch took a relatively extreme, unicausal position about the linkage between the market-driven individual imperative to profit from the use of land and the nature of community leadership. The model assumes that, "People dreaming, planning, and organizing themselves to make money from property are the agents through which accumulation does its work at the level of the urban place" (Logan and Molotch 1987, 12). Because economic activity shapes community politics, Molotch finds that "this organized effort to affect the outcome of growth distribution is the essence of local government as a dynamic political force" (Molotch 1976, 313). The model is less concerned with the structural question of whether community leaders form a cohesive elite or shifting issue-based coalitions than with the underlying economic dynamic of urban politics. The desire to make money from land is pervasive at the local level. So is the impulse to use the financial and regulatory powers of local government to gain advantage for particular individuals and groups. The financial powers are those of taxation and debt issuance to build infrastructure; planning and zoning regulations are used to provide advantages to some people and deny them to others (Burns 1994, 54–7).

In a community in which the growth machine is operative, there will be people in favor of the machine's objectives and those who are opposed. Molotch's model is an elite model, because it depicts a strong difference of interests between those who benefit from growth and those whose interests might be damaged by it. Much like the distinction made by William's and Adrian's in the section that follows, between the work environment and living environment, Logan and Molotch contrast *exchange* values and *use* values (1987, 17–49). *Exchange values* are those of the marketplace, of people whose interest in land is primarily to make money from it. *Use values* are those of people whose primary interest in land is its use for creating a peaceful and pleasing living environment for themselves and their families.

There is a natural clash between these two sets of interests, because living space values may constrain those who wish to use the land for commercial enterprise, and marketplace values may threaten the living environment. It is for this reason that those involved in the growth machine attempt to control the agenda of community decision making, whether through influencing decision makers or keeping key growth-related issues off the public agenda (non-decision making). While the public is occupied with sports teams and other community events, the people who make decisions about and benefit from growth are holding a "dull round of meetings of water and sewer districts, bridge authorities, and industrial development bonding agencies" (Logan and Molotch 1987, 64). Because these meetings are indeed dull and the decision process extends over months and years, few people other than those directly involved attend and the media pay little attention.

Community residents will come to meetings when a major decision point is reached, at least if it directly affects them. Sometimes people will organize to resist the growth machine, though it is difficult to do so on a sustained basis. There are places where resistance to the growth machine is prolonged and well organized. People may band together to protect their neighborhood from development they regard as damaging, local business people may resist commercial or industrial development that could detract from their businesses, affluent suburbs may resist development that might lower property values, and certain kinds of cities, like university towns, may contain groups of people who are highly active and vocal about the living environment.

Despite all this, the underlying concept of the growth machine model, that people naturally want to make money from the use of land, means that it is difficult for citizens to resist the growth machine phenomenon and that it takes sustained effort to do so. For public professionals, the pervasiveness of the growth machine sets clear limits to action, as the "growth machine elite" controls the political power in a community and thus can influence hiring and retention decisions affecting professional careers.

In recent years, many researchers have explored aspects of the growth machine hypothesis. They have found, for example, that growth machine elites use arts and cultural organizations to improve the market attractiveness of their properties (Whitt and Lammers 1991), that the growth machine phenomenon can be observed in established industrial cities such Flint, Michigan (Lord and Price 1992) as well as in newer, rapidly growing places, and that progrowth and antigrowth "entrepreneurs" may emerge to promote or discourage growth in a community (Schneider and Teske 1993a; 1993b). As with Peterson's theory, we need to take into account other political and economic factors, but it appears that the growth machine is indeed a powerful force shaping the environment of community governance.

## FOUR CITIES

Another way to describe communities is to examine the choices residents have made about the role of community government in local affairs. Oliver Williams and Charles Adrian, in their 1963 book, *Four Cities: A Study in Comparative Policy Making,* identified four basic orientations in communities; the first is a community in which the primary concern of the government is the promotion of economic growth. This concern in *promotion* cities is driven by "speculative hopes" (23) and a desire for increases in population and wealth. In such a community, "the merchant, the supplier, the banker, the editor, and the city bureaucrats see each new citizen as a potential customer, taxpayer, or contributor to the enlargement of his enterprise, and they form the first rank of the civic boosters" (24).

The second type of community is one in which the primary goal of government is to provide and secure life's amenities, the "home environment rather than the working environment" (25). In such places, growth is often seen as a threat to the living environment. Emphasizing values of the living environment as opposed to those of the work environment can be expensive, either in missed economic opportunities for residents or in provision of attractive physical facilities. *Amenities* communities tend to have relatively homogeneous populations—that is, their residents are sufficiently similar in socioeconomic status and desire for a certain type of community that agreement on community goals can be reached and kept. The well-to-do suburban enclave where people move when they wish to escape the problems of the central city is typical of this community type, as is the planned residential community developed privately within a city.

The third community type described by Williams and Adrian is one in which maintenance of traditional services is the primary goal of government. Residents of this *caretaker* city wish to keep taxes low, minimize land-use planning and other restrictions on the use of private property, depend on the "freedom and self-reliance of the individual" (27), and provide only basic and essential services through the local government. This laissez-faire approach to local government may leave the state or national government to solve problems the community wishes to avoid dealing with.

The fourth type of community is very diverse, with many interest groups competing for political advantage. The function of local government in this community type is to serve as *arbiter* between the competing groups. In this hyperpluralist environment the highest value is placed on political responsiveness.

Williams and Adrian found that promotion and amenities communities have "unitary," broad agreement on goals. In these places, centralized and professional structures such as the strong mayor or council-manager system work well. But in arbiter and caretaker communities, there is a pluralistic diversity of interests that is best served by decentralized structures that distribute power, such as the weak mayor form and ward elections for council members (29–31).

Williams and Adrian's typology of community orientations is similar to more recent work by Clarence Stone. Stone (1993) created a typology that included four types of communities as well. The first is the *maintenance regime* (similar to Williams and Adrian's caretaker community) that preserves the status quo and introduces few changes, and the second is the *development regime* that is "concerned primarily with changing land use in order to promote growth or counter decline" (18), much like the promotion community. The third regime type is the *middle-class progressive regime* that focuses on "such measures as environmental protection, historic preservation, affordable housing, the quality of design, affirmative action, and linkage funds for various social purposes" (19); this is analogous to the amenities community. Stone's fourth regime type is the *Regime Devoted to Lower-Class Opportunity Expansion,* through programs such as "enriched education and job training, improved transportation access, and enlarged opportunities for business and home ownership" (20). Stone sees this fourth type as being "largely hypothetical," though there are hints of it in certain places at certain times. It is different than Williams and Adrian's arbiter community, in which there are many competing interests, because the arbiter community may or may not emphasize a particular policy orientation, such as expansion of lower-class opportunity.

We might extend Williams and Adrian's work to speculate further about how professional administration is regarded in different communities. In the promotion and amenities communities where many people share the same goals and want government to play an active role in community life, professionalism may be especially valued. But where people wish to limit the role of government as they do in caretaker communities, or where political skills are needed as in the arbiter community, professional skills may be valued in the technical process of implementation but not in the process of policy formulation. Thus we may expect greater reliance on creative and innovative public administrators in the promotion and amenities communities and greater intrusion into administrative matters, with significant limits on administrative discretion, in the caretaker and arbiter communities.

The promotion community may encourage an administrative orientation toward adapting to rapid expansion, accommodating the needs of developers and business people, and showing a desire to "market" the community. Citizens in such a community may expect their public service practitioners to encourage innovation, activity, and change, although of a particular character. Where this is true, the administrative philosophy needs to include a willingness to find solutions that enable people to make money from land speculation and development, because the political environment may not be receptive to administrative action that puts aesthetics or environmental protection ahead of promotion and growth. Where political leaders come to see administrators as standing in the way of development, they may take action to decrease administrative discretion or move decision-making authority to higher levels, for example from a department head to a mayor or city manager.

In the amenities community, professionals in development-related services may be expected to be experienced in the areas of architectural review, historic preservation, requirements for developers to bear the lion's share of infrastructure costs, and regulations that protect the residential environment from the noise, traffic, and appearance of commercial and industrial land uses. With the greater emphasis on residential amenities, park and recreation professionals in the amenities community may have a clientele eager for attractive open spaces. The police department in such a community may focus on keeping residential areas free from disturbances and petty crime, perceived as largely perpetrated by outsiders.

Practitioners in the caretaker community may concentrate on cutting costs and finding new ways to do the same amount of work with fewer resources. They might be expected to support the concept of limited government, a limited sphere of discretion for administrators, and a minimum of change and innovation aside from that which saves money. Program or policy development skills may not be highly valued and administrators may be rewarded for keeping things on a quiet and even course.

The political environment in the hyperpluralist arbiter community might force administrators to be aware that most decisions or actions of significance will offend *someone,* given the range of competing interests. The practitioner might need to carefully evaluate her or his motivations and objectives, because public expectations and potential reactions to administrative action would often be unpredictable and may change from one day to the next. Technical, professional skills might be valued, but interpersonal and conflict resolution skills would also be very important.

Of course, care must be taken in applying such generalizations to a specific community. The four broad descriptions of communities used by Williams and Adrian may not explain some important differences between places. For example, though we might expect the caretaker city to be run by elected representatives who have a low opinion of professionals and government, it might be the case in a particular caretaker community that leaders value professional competence and administrative efficiency and at the same time hold to values of limiting the size and intrusiveness of government. Leaders in promotion communities might often care more for maximizing the profit of speculators and developers than for the aesthetics and environmental impact of development, but it is also possible to combine enthusiasm for growth with concern for amenities.

Two concerns common to communities in this model are the future of the community and the role of government in shaping that future. In local government there are a number of administrators who are involved in helping elected representatives make decisions about the future and the role of government. They include city or county managers, administrators and their assistants (where there is a city or county manager or administrator), and department heads and their subordinate professional staff in the areas of financial management, planning, public works ("public works" covers many functional areas, such as water, sewer, streets, airports, sometimes gas or electric services, and public buildings), parks, human services, police, fire, and others.

School districts face these questions as they grapple with issues such as whether to build new buildings or remodel old ones, where to expand and where to economize, and how to deal with calls for greater parental control or privatization of schools. Schools are not explicitly included in this narrative because they are usually administratively separate from general-purpose local governments (cities and counties) and are organized as single-purpose special districts. However, the concepts discussed in this book also apply to schools, as well as all manner of special-purpose districts (for example, districts that deal with air or water quality, libraries, parks and recreation, mass transit, animal control, etc.).

Though generalizations must be used cautiously and general concepts may prove inaccurate in specific situations, it is clear that there are important and identifiable differences in community orientation. For citizens and administrators directly involved in the debates over the future of communities or the role of government in shaping that future, differences in community types as described by Williams and Adrian may make a significant difference in how they choose to approach their work.

## VARYING RESPONSES TO COMMUNITY POLITICS: A CASE STUDY

Molotch and Peterson's models are subject to the criticism that they are based on only one thing, the response of community residents to pressures for economic growth, the use of the community's land such that individual (Molotch's model) or general (Peterson's model) wealth is maximized. Despite this potentially damaging criticism, the growth machine and unitary interest models have much to tell us about how community politics and power work.

Williams and Adrian's (and Stone's) typology of community values has the advantage of being based on contrasts between fundamental philosophical approaches to local government. In their typology there is no single issue that motivates local decision makers, but instead there is a general orientation toward achieving a certain kind of community, an orientation that develops over time as a result of the characteristics of community residents and their interactions. Some places are more interested in growth, some in balancing competing political interests, and so on.

These models are not necessarily contradictory. They provide different perspectives on community politics, politics that vary between cities and within the same city over time. I found this to be true during a research project in two Oregon cities in 1990. As described by the Molotch and Peterson models, land use was the most important issue in both communities, whereas each displayed characteristics that fit Williams and Adrian's typology as well. These two cities are about 10 miles apart geographically but worlds apart in orientation to community governance. City A was the largest in the region, a community of 50,000 that was the commercial and governmental center for the county. City B was a community of 16,000 that thrived on tourism, the arts, and higher education. Both communities were faced with substantial growth pressures during the study period, as well-to-do Californians and others seeking to escape urban centers migrated to Oregon. By comparison with California, prices for land and homes in Oregon were very low and the real-estate boom brought on by the new residents drove land prices up dramatically, making it hard for many long-time residents to pay their property taxes or buy a home. In City B, assessed property values had increased 25 percent in the year prior to my fieldwork.

The response to growth in City A had remained consistent over almost fifteen years, even though there were times when very little growth was occurring; in public meetings and in mayoral campaigns, there was a pattern of emphasizing growth as a good thing, a way to improve living standards for everyone and to provide jobs. City officials were aware of the need to manage growth to create a relatively pleasant living environment, one that could be served efficiently by city infrastructure. Even so, it had always been clear that citizens or professionals who asked uncomfortable questions about who benefited from growth and who paid for it were regarded as obstacles to be overcome and that growth was to be encouraged whenever possible.

Things were much different in City B over the same time period. In the 1970s and into the early 1980s, relatively conservative business people controlled public discussion of policies that affected land use. They were in favor of attracting development in much the same way as in City A. This was the case even though most members of the city council were, by the late 1970s, in favor of a more careful and restrictive policy toward growth. But by the mid- to late 1980s, the policy process had come to be dominated by a younger group. Though many of these people were also involved with the business community, most of them were concerned about preservation of what they saw as the character of the community in the face of pressures for change through development. By the time of my work in 1990, resistance to growth was not only a dominant value, the city council and planning commission had publicly and consciously decided that the normal market processes that determine the type and extent of growth in most places were not to be determinative in City B. Instead, decisions about growth would be made on the basis of impact on the aesthetics of the living environment.

Looking at what had taken place in these communities over the period studied, I came to realize that there is a continuum of responses to pressures for growth. In Williams and Adrian's typology, City A was and had been a promotion city throughout the period studied. City B had been a mixture of the caretaker and promotion types, welcoming growth but working to keep taxes and services to a minimum; then it changed to an amenities community. An interesting study by King and Harris (1989) examined attitudes toward growth of people serving on planning boards in rural towns in New York and Vermont. They found that boards in towns faced with growth pressures wanted to control or stop it through "rigid adherence to zoning bylaws and by reference to land suitability maps" (186). In places where little growth was occurring, the boards encouraged any potential development and approved almost any proposal. This characterization fit City B over the study period; when there was little growth the progrowth faction was in control, and with later rapid growth there was a transition to a strong growth resistance attitude.

However, this change did not occur immediately and the causes of the change were more complex than just the changes in external pressures for growth. For several years in the late 1970s and into the early 1980s the progrowth faction remained dominant even though they were no longer in the majority on the city council and growth pressures were increasing. Progrowth people were heavily represented on the planning commission and the people who spoke to decision makers on a regular basis about issues of development were to a large extent the business people who had a direct financial stake in the outcomes of the policy process. Because the economy had been poor in the mid-1970s, these people had not yet realized they could do even better financially by making the city's growth attitude appear environmentally sensitive, thus increasing demand for property and maintaining a strong rate of growth while property values went up.

This "conservative" group was largely replaced by new people in the 1980s. Some of the new decision makers were long-time residents (though relatively young, in their 30s and 40s), and some were people who had moved to the area in the last few years. These recent migrants tended to be from large urban areas to the south and they were often *drawbridge* advocates—that is, they had arrived, they liked the living environment, and they wanted to keep it the same by pulling up the metaphorical drawbridge to keep other people from coming to the community.

During debates on several proposed developments and during a key mayoral campaign in 1988, the faction that wished to discard the market as a guide to growth decisions became dominant in the policy process. So, in City B the change from active growth promotion to resistance took several years. External growth pressures set the stage for the change, but it took time for slow-growth advocates to build support, take over the mayor's position, and begin putting their people on the planning commission. The mayoral election was the point at which it became clear to everyone that the antigrowth sentiment that had been building in the community had come to political maturity and domination of the policy process.

As I thought about these two cities it seemed that the response to growth could be portrayed along a continuum that stretched from growth machine dominance to a weak growth machine. City A had a very strong, almost dominant growth machine; there were people who questioned the desirability of unlimited growth but they were few and far between. Specific projects would sometimes be challenged by people living in the immediate vicinity but development proposals would usually prevail in the political arena. City B had moved through a period of turmoil and conflict from support of growth to resistance to it. The growth machine in City B seemed virtually nonexistent at the time of my fieldwork given the proenvironment orientation of the mayor, council, and planning commission.

In between the dominant and weak polar opposites on the growth machine continuum are intermediate responses to growth. The idea of a dominant growth machine in full control of the governance process is similar to Williams and Adrian's promotion type, and it fits Peterson's description of the city with unitary interests. But it does not fit Molotch's description of the growth machine, which tries hard to convince people of the desirability of growth, distracts them with civic boosterism, and makes the crucial decisions quietly and unobtrusively. In Molotch's community the growth machine is vigorous but not dominant, so an intermediate step in the continuum from dominant to weak growth machine types could be called the *strong* growth machine community.

These descriptions of communities are really descriptions of local public opinion. A "community" cannot become a dominant, strong, or weak growth machine place; after we study and understand how the residents of a place feel about growth, we take an aggregate estimate of opinion and label the community on one of our continuum steps. What has been labeled is not the community as if a community could have a certain character separate from the attitudes of its residents. Instead, what has been labeled is our best guess about what the majority of residents would like to see happen with their community in the future. Residents form such opinions based on their past personal experience and their observations about what has been happening in the community.

City B passed through a stage of political conflict in the 1980s, from what appeared to be a strong growth machine community type to a weak growth machine type. During this conflict stage, community opinion was divided on growth issues with the opposing groups in some rough sort of balance; logically we could label a community like this a conflict type, adding it to the continuum between the weak and strong communities. There are any number of possible types of public opinion about growth that fall in between the weak, conflict, strong, and dominant types. Rather than clutter the model with lots of interesting but nonessential intermediate points, it may be better to highlight those public opinion characteristics that help us understand that differences exist between communities and in communities over time. The dominant and weak growth machine types are the extremes of community opinion about land use, the strong type is the community described by Molotch, and the conflict type, between the strong and weak types, reflects a community in which a lively debate is taking place about the future. The result is a four-part "expansion" of Molotch's growth machine model.

The growth machine phenomenon can take place in communities with physical room to grow and it can take place in communities with no vacant land through speculation, redevelopment, or intensification of the density of development. In addition, positive attitudes about growth are not only found in communities that are growing. In fact, some studies have shown that communities with little growth pressure are more likely than those experiencing significant growth to view growth favorably (Anglin 1990; King and Harris 1989). If this is true, it would be reasonable to expect to find places with stagnant or declining economies or population sizes in which the political climate is strongly progrowth. This makes sense, as people become concerned when the economy is bad and they worry about the living environment when the rate or amount of change creates obvious environmental impacts. For this reason, the impact of the growth machine on community politics and administration can be significant even where it seems unrelated to what is happening in the economic environment. A community in economic decline may be governed by people whose values are those of the growth machine.

There are a number of issues facing any community that do not seem to be directly connected with growth, such as how to treat questions of racial equity, whether to give public employees a large or small raise in the annual budget, or whether to build new parks. But issues like these are indeed affected by the community's economic situation. A community in good economic circumstances may be able to require developers to include affordable housing in their developments, operate a small business loan program in minority-dominated areas of town, attract top-quality employees with aggressive recruitment and high salaries, and improve on parks and open space amenities.

If Molotch is correct in believing that there is a linkage between the market for land and the content of local politics, the character of public opinion about developmental issues is very important to citizens and administrators. In the long term, the nature of land use and economic development has a significant impact on the needs and challenges the community faces. Choices made about who bears the costs of development and how private sector activity is regulated say much about the values of community residents and their government. Such choices constrain citizens and professional administrators in the daily work of making policy recommendations and carrying out policy directives.

## COMMUNITY POLICY ORIENTATIONS

In this [reading], we have examined the development of American communities, from early forms of governance to the current search for greater citizen self-determination. We have found that community elites often use the powers of local government, the ability to provide infrastructure and control the use of land, to benefit themselves. And, there is concern over whether this provides benefits for everyone in the community or transfers wealth from the poor to the rich and maximizes marketplace values at the expense of aesthetic values of the community as a living space.

The range of issues of importance to people in American communities is very large. This is not the place to discuss the content of the more common substantive issues or to take positions on how to deal with them. We must accord equal respect to all local interests and issues, focusing on the problem of enabling citizens to govern communities rather than the specific substantive concerns on the policy agenda of each community. In short, we are concerned with access and process, leaving issues such as economic development, race, income redistribution, how to fund streets, parks, and schools, and a host of other matters, to each local policy dialogue.

Even in the face of the complexity of substantive community issues, it is possible to distill into a few key points the features of community life that most affect the ability of citizens and practitioners to influence public policy. We have seen that visions of the future of the community and the role of local government in the community's future create a variety of political environments. Though a particular community is likely to be a mixture rather than a pure example of one of Williams and Adrian's four types of community orientation or one of the four types of community response to the growth machine in the "expanded" growth machine typology, there are very real differences in local political environments. The expectations of community residents and leaders about how citizens can become involved in the policy-making process and what role a professional administrator should play are important features of the local political setting.

In the dominant and strong growth machine communities, the conditions for citizen or professional action on policies affecting land and economic conditions are similar to those in Williams and Adrian's promotion city. There is significant flexibility and room for innovation as long as actions taken enhance the economic gains of local elites. In the conflict community there is a broad range of opinion about the future of the community, allowing citizens greater area for potential action. The public professional in the conflict community may find it difficult to take a position on a given issue without alienating individuals or groups because of deep divisions in community opinion. Paradoxically, failure to express a professional opinion or assist in policy formulation may also create an adverse public reaction. In this political setting it is understandable that some professionals may choose to keep their heads down, trying to avoid controversy.

Citizens and practitioners in a weak growth machine community are in a mirror-image of the situation of those in the dominant or strong machine community. They may have plenty of discretion to innovate but only where they conform to community values of protecting and enhancing the living, rather than the commercial, environment.

To the extent these ideas accurately portray community politics, the possibilities for successful citizen or professional action are shaped by attitudes toward economic expansion and the use of land to provide environmental amenities. This is not much different from Williams and Adrian's focus on the future of the community and the role of government in shaping that future. Or more precisely, the growth machine view of what is important in communities is a subset of the four cities typology, a part of the question of the future of the community. Williams and Adrian's typology adds the dimension of the appropriate role of government to the concern about land use. We can bring together many of the ideas from our discussion to this point in a four-part typology of community orientations to the creation and implementation of public policy. Each of the community policy orientations described next contains a continuum of thought (for example, from preference for a large role for government to a restricted role), so that the four orientations are not mutually exclusive.

1. *Accessible and Open or Excluding and Closed Governance System.* These opposing views of the degree to which citizens should be able to take part in the governance process correspond to the pluralist and elite views of community power. If a citizen who is not part of the "in-group" of powerful or wealthy people is able to enter the decision-making process and make a meaningful contribution, the governance process is open and membership in the governing "elite" is changeable and accessible to new members. If an interested citizen is kept on the fringes of the process, allowed to attend meetings but not to make a significant difference in local policy, the governance process is closed and the governing elite group is stable and difficult to penetrate. Becoming a member of this group or being trusted by them as an outsider may take years of volunteer involvement in community activities to demonstrate that the citizen is not a threat to the values and financial interests of the elite.

2. *Community as Marketplace or Community as Living Space.* This is a measure of the degree to which the local policy process is controlled by people who favor growth and economic development because they view the community as a marketplace, or those who favor environmental amenities because they view the community as a living space. A community that emphasizes the marketplace view may be one in which the growth machine is dominant or strong. One that emphasizes the living space view has a weak growth machine, and the conflict community lies in between, with polarized views of the desired future.

3. *A Desire for a Large or Restricted Role for Government.* This is the macro-level scope-of-government question. Those who favor a large role for government may have a variety of policy goals in mind, such as social welfare or racial equity, environmental enhancement, or providing incentives for economic development. Those who favor a limited role press for low levels of taxation and provision of only basic and essential public services.

4. *Acceptance of, or Resistance to, Public Professionalism.* This is a measure of community opinion about the appropriate extent of influence to be exerted by the professional knowledge and values of the public service practitioner. Highly politicized arbiter, or conservative caretaker, communities are wary of the influence of professionals, whereas communities with relatively consensual politics or those facing serious problems of finance or infrastructure may depend on professional knowledge. The degree of acceptance or resistance to professionalism may vary over time in a community, depending on circumstances and personalities.

Each of these policy orientations can have an impact on the principles of Citizen Governance. A closed governance system damages the democracy principle, resistance to professionalism could make it harder to fulfill the rationality principle, and so on. The ways in which community policy orientations affect the principles depends on the specific conditions in a community at a particular point in time. Taken together, the concepts discussed [here] and the community policy orientations supply the citizen, representative, and practitioner with tools for understanding the nature of community governance. . . .

## REFERENCES

Adrian, Charles R., and Ernest S. Griffith. 1976. *A History of American City Government. Vol. 2, The Formation of Traditions, 1775–1870.* New York: Praeger.

Anglin, Roland. 1990. "Diminishing Utility: The Effect on Citizen Preferences for Local Growth." *Urban Affairs Quarterly* 25 (June): 684–96.

Bachrach, Peter, and Morton S. Baratz. 1962. "The Two Faces of Power." *American Political Science Review* 56 (December): 947–52.

Blodgett, Terrell. 1994. "Beware the Lure of the 'Strong' Mayor." *Public Management* 76 (January): 6–11.

Box, Richard C. 1993. "Resistance to Professional Managers in American Local Government." *American Review of Public Administration* 23 (December): 403–18.

———. 1995. "Searching for the Best Structure for American Local Government." *International Journal of Public Administration* 18(4): 711–41.

Burns, Nancy. 1994. *The Formation of American Local Governments: Private Values in Public Institutions.* Oxford: Oxford University Press.

Cook, Edward M. 1976. *The Fathers of the Towns: Leadership and Community Structure in Eighteenth-Century New England.* Baltimore: Johns Hopkins University Press.

Dahl, Robert A. 1961. *Who Governs? Democracy and Power in an American City.* New Haven, CT: Yale University Press.

Elder, Shirley. 1992. "Running a Town the 17th-Century Way." *Governing* 5 (March): 29–30.

Fox, Charles J., and Hugh T. Miller. 1995. *Postmodern Public Administration: Toward Discourse.* Thousand Oaks, CA: Sage.

Giddens, Anthony. 1984. *The Constitution of Society: Outline of the Theory of Structuration.* Berkeley: University of California Press.

Goodnow, Frank J. [1904] 1991. *City Government in the United States.* Holmes Beach, FL: Wm. W. Gaunt and Sons.

Griffith, Ernest S. 1938. *History of American City Government. Vol. 1, The Colonial Period.* New York: Oxford University Press.

———. 1974. *A History of American City Government. Vol. 3, The Conspicuous Failure, 1870–1900.* New York: Praeger.

Gurwitt, Rob. 1993. "The Lure of the Strong Mayor." *Governing* 6 (July): 36–41.

Harrigan, John J. 1989. *Political Change in the Metropolis.* 4th ed. Glenview, IL: Scott, Foresman.

Hunter, Floyd. 1953. *Community Power Structure.* Chapel Hill: University of North Carolina Press.

King, Leslie, and Glenn Harris. 1989. "Local Responses to Rapid Rural Growth." *Journal of the American Planning Association* 55 (Spring): 181–91.

Lappe, Frances Moore, and Paul Martin Du Bois. 1994. *The Quickening of America: Rebuilding Our Nation, Remaking Our Lives.* San Francisco: Jossey-Bass.

Lockridge, Kenneth A. [1970] 1985. *A New England Town the First Hundred Years: Dedham, Massachusetts, 1636–1736.* New York: W.W. Norton.

Logan, John R., and Harvey L. Molotch. 1987. *Urban Fortunes: The Political Economy of Place.* Berkeley: University of California Press.

Lord, George F., and Albert C. Price. 1992. "Growth Ideology in a Period of Decline: Deindustrialization and Restructuring, Flint Style." *Social Problems* 39 (May): 155–69.

Martin, Lawrence L. 1993. "American County Government: An Historical Perspective." In *County Governments in an Era of Change,* ed. David R. Berman, 1–13. Westport, CT: Greenwood Press.

Massialis, Byron G. 1990. "Education Students for Conflict Resolution and Democratic Decision Making." *The Social Studies* 81 (September/October): 202–5.

Molotch, Harvey L. 1976. "The City as a Growth Machine: Toward a Political Economy of Place." *American Journal of Sociology* 82 (September): 309–32.

Morison, Samuel Eliot. 1965. *The Oxford History of the American People.* New York: Oxford University Press.

Peterson, Paul E. 1981. *City Limits.* Chicago: University of Chicago Press.

Schattsehneider, E.E. 1975. *The Semisovereign People: A Realist's View of Democracy in America.* Hinsdale, IL: Dryden Press.

Schneider, Mark, and Paul Teske. 1993a. "The Antigrowth Entrepreneur: Challenging the 'Equilibrium' of the Growth Machine." *The Journal of Politics* 55 (August): 720–36.

Schneider, Mark, and Paul Teske. 1993b. "The Progrowth Entrepreneur in Local Government." *Urban Affairs Quarterly* 29 (December): 316–27.

Selznick, Philip. 1992. *The Moral Commonwealth: Social Theory and the Promise of Community.* Berkeley: University of California Press.

Smith, Page. 1966. *As a City Upon a Hill: The Town in American History.* New York: Alfred A. Knopf.

Stone, Clarence N. 1993. "Urban Regimes and the Capacity to Govern: A Political Economy Approach." *Journal of Urban Affairs* 15(1): 1–28.

Verba, Sidney, and Norman H. Nie. 1972. *Participation in America: Political Democracy and Social Equality.* New York: Harper and Row.

Waste, Robert J. 1989. *The Ecology of Policy Making.* Oxford: Oxford University Press.

———. 1993. "City Limits, Puralism, and Urban Political Economy." *Journal of Urban Affairs* 15(5): 445–55

Waste, Robert J., ed. 1986. *Community Power: Directions for Future Research.* Beverly Hills: Sage.

Whitt, J. Allen, and John C. Lammers. 1991. "The Art of Growth: Ties Between Development Organizations and the Performing Arts." *Urban Affairs Quarterly* 26 (March): 376–93.

Williams, Oliver P., and Charles R. Adrian. 1963. *Four Cities: A Study in Comparative Policy Making.* Philadelphia: University of Pennsylvania Press.

Zuckerman, Michael. 1970. *Peaceable Kingdoms: New England Towns in the Eighteenth Century.* New York: Alfred A. Knopf.

READING 3.2

# "THE QUEST FOR A CIVIL SOCIETY"

## DON E. EBERLY

We are all recognizing that the kinds of rationalistic, scientific, technical, organizational responses to human needs in the past several centuries are not sufficient to respond to people's deeper yearnings, their spiritual desires, and the way they treat one another.
—*First Lady Hillary Rodham Clinton*[1]

There has to be some core of shared values. Of all the ingredients of community this is the most important. The values may be reflected in written laws and rules, in a shared framework of meaning, in unwritten customs, in a shared vision of what constitutes the common good and the future.
—*Common Cause Founder, John W. Gardner*[2]

The alternative to the naked public square is the reconstitution of civil society in America. And what is "civil society?" Civil society is the achievement of a genuine pluralism in which creeds are intelligibly in conflict.
—*Conservative think tank President George Weigel*[3]

The communitarian perspective recognizes that the preservation of individual liberty depends on the active maintenance of the institutions of civil society where citizens learn respect for others as well as self-respect; where we acquire a lively sense of our personal and civic responsibilities, along with an appreciation of our own rights and the rights of others; where we develop the skills of self-government as well as the habit of governing ourselves, and learn to serve others—not just self.
—*From the Communitarian Platform*[4]

We are increasingly living in a political society at the expense of a civil society. The challenge . . . is to stand up for the principles of a civil society—one based on voluntarism—while standing in the midst of a statist conflagration . . . the level of taxation is the measure of our failure to civilize our society.
—*Libertarian advocate, Edward H. Crane, President, The Cato Institute*[5]

The American political party that best gives life and breath and amplitude to civil society will not only thrive in the twenty-first century. It will win popular gratitude and it will govern.
—*Catholic theologian Michael Novak*[6]

There seems to be a great deal of agreement, from conservatives to liberals, about the need to improve civil society. Despite this seeming consensus across the political spectrum, however, the search for a humane, civil, and well-ordered society remains elusive as the 21st century approaches. As the title suggests, this volume explores the condition of America's civic and democratic life and seeks a foundation for progress in the 21st century.

America is in the midst of social convulsions at home and global change of a scale, velocity, and uncertainty that futurists say may be unprecedented in human experience. In the midst of a surging economy, large majorities from virtually every sector of society worry that something very basic has gone wrong at the core of society, something that cannot be measured in the traditional terms of Gross National Product (GNP) growth rates and unemployment figures. Public worries increasingly concern cultural indicators, such as rising teen pregnancy, violence, and declining Scholastic Achievement Test (SAT) scores, factors as important to national advancement as a growing economy.

But even this social data reveals only the measurable indicators of societal regression. The quality of life is more and more shaped by immeasurable things that have weakened, like basic civility and manners. Many report finding life coarser, their culture cruder, the public debate angrier, and the treatment of individuals less respectful than when they grew up. There is a sense that Americans have become shortsighted and selfish, obsessed with rights and entitlements, and that quality, excellence, and commitment to work have all waned.

The mood is peculiar in light of the nation's successes. For one, America presently stands taller than perhaps any nation in history. By defeating the forces of totalitarianism, the United States occupies the position of undisputed heavyweight: the only true military and economic superpower. America is still the most coveted destination for immigrants; it is exporting more goods and services than any nation in history; it is employing more of its population than ever; and its technological genius is still unmatched around the world. In sum, it is the richest, mightiest, and most magnetic of any nation on the globe.

For decades during the cold war, the need for a well-defined identity and moral purpose was reinforced as America led a unified democratic front against Communism. Her core principles needed little further articulation. That the East-West conflict and the national resolve it produced held things in equilibrium cannot be denied. With the East-West conflict gone, however, America no longer leads out of necessity. If it is to continue to lead other nations, it will do so less out of moral necessity and more by moral example.

The entire world is free to decide which sociopolitical system to adopt. America is increasingly being judged globally, not by its military prowess, but by social conditions at home. It is humbling for the world's richest industrial nation to have a poverty rate twice that of any other industrial nation and to be singled out by international agencies as a world leader in child poverty and youth homicides.

While no ideology appears to compete today with the dominance and increasing popularity of democratic capitalism as the best system to satisfy human aspirations, the picture is by no means that simple. The factors leading to human conflict that have reemerged around the globe, and even at home in recent years, are the ancient forces of race, ethnicity, religion, and nationalism. They represent some of the most vexing fissures that have marked the human landscape from the beginning of time and are the kind that democracies have proven weak to manage.

Democratic capitalism has vindicated itself in surviving repeated challenges in the 20th century, whether depression at home or ideological challenge abroad. But how sufficient is economic advancement, many are asking, if our schools do not function, if crime defies control, and if children have lost their innocence in an adversarial culture of violence and banality? When a society becomes completely indifferent to the need to safeguard its own children from harm and lets one in four fall into poverty, it risks losing its status as a world leader.

Voicing worry about the status of American society involves risk. One can easily be tagged a pessimist, a declinist, or a nostalgist. Such talk seems out of place in America for the simple reason that we have always been an optimistic nation; pessimism is almost unAmerican. Social historians have described America as a country that fundamentally lacks a tragic sense.

The United States has always passed through difficult transitions with unbroken resolve, confident that our civilization stands above the immutable laws of history which seem to assure that nations wane as surely as they wax. Americans have been reminded again and again by their leaders that, come what may, we are a people of destiny; economic prosperity at home and privileged status in the world are assured almost as a matter of birthright. But growing numbers of American leaders are apparently taking exception to this American "exceptionalism." In fact, some fear that if our internal vulnerabilities are left unattended, we may succumb to those laws of history America has defied for over two centuries. Unlike some of her European counterparts, countries held together by racial or ethnic homogeneity, long histories, or strong ancestry, the United States is held together by a set of ideas and values. The glue that holds the United States together is of the kind that, with neglect, could grow fragile.

As we near the end of a century and a millennium, America is entering a collective search for its national soul. The country is in the midst of profound changes in demographics, technology, and the structure of society. The great projects that stirred nationalist spirits in the past—whether settling frontiers, defeating international communism, or launching "great societies"—have either been accomplished or were tried and failed.

The debate for the balance of this decade, and perhaps well into the next century, will focus on a modern paradox: how can a society that has produced more freedom and prosperity than any other in history, and has been so generous in its distribution, also increasingly lead the world in so many categories of social pathology? Many realize that politics alone has not effected what Americans prize the most—a humane and civilized society. Neither the welfare state nor a surging capitalism has solved many of society's persistent problems; in fact, each has contributed in its own way to the corrosion of civil society and its institutions. When the mediating structures of society—families, churches, communities, and voluntary associations—are weakened to the breaking point, individuals are increasingly left isolated and vulnerable within an ever expanding state. Neither the conservative mantra of "more markets" nor the liberal song of stronger "safety nets" in an expanded state has proven adequate.

Forces of modernity, more than politics, have produced much of society's fragmentation and rootlessness. Social thinkers have long recognized the precariousness of society. The magnetic forces of cooperation and solidarity can quickly be replaced by the centrifugal tendency to abandon attachment and obligation. These pressures have been particularly acute during our era, a time sociologist Daniel Bell has characterized as "a rage against order" because of its steady undermining of voluntary institutions and restraints.

## THE SEARCH FOR NORMATIVE VALUES

America, the undisputed military and economic leader of the globe, shows every sign of being under immense social strain, as evidenced by her cities, her schools, her families, and her youth, problems which seem largely unaffected by changes in politics or the national economy.

Many observers argue that the crisis that has come to former totalitarian countries may come to the liberal West as well. Futurist Richard Eckersley attributes social disintegration throughout the West to a failure to provide a sense of "meaning, belonging and purpose in our lives as well as a framework of values. Robbed of a broader meaning to our lives, we have entered an era of often pathological self-preoccupation."[7]

What happens to the society that has been severed from its underpinnings, in which faith, culture, and politics have become fragmented and devoid of meaning and citizens have lost a shared basis for a common life together? The result is the loss of community, a declining social order, the erosion of trust in authority, and the increased assertion of human passion through power rather than reasoned judgment.

As sociologist Peter Berger has said, excessive relativism produces a painful sense of impermanence and uncertainty. When relativism reaches a certain intensity, "absolutism becomes very attractive again" as people "seek liberation from relativism."[8] For growing numbers throughout the Western world, the desperate search for meaning and belonging ends in the total subjugation of the self in fanatical nationalism and fundamentalism.

Liberal scholar John Gardner states starkly: "Without the continuity of shared values that community provides, freedom cannot survive. Undifferentiated masses never have and never will preserve freedom against usurping powers." The answer, according to Gardner, is to strengthen the intermediary structures of society in order to close the gap between the individual and state and fortify families and communities which are the ground-level generators and preservers of values and ethical systems.[9]

The challenge is to restore an ordered freedom and to resist abuses of either freedom or order. Some have used freedom to advance a radical expressive individualism that completely dismisses concern for the moral ecology of society. Conversely, the abusive reach for order has made its debut in growing cases of warrantless searches, vigilantism, and the National Guard being summoned to police urban housing projects.

This project of restoring an ordered freedom will require shoring up personal and civic values and treating the stress fractures on the institutions of civil society. If America suffers from a poverty of spirit, it suffers even more from a loss of language to address declining public life and common values. The language embodying the concepts of personal and public virtue has been eradicated. Traditionalist scholar John Howard has documented the broad segment of vocabulary along with the concepts they embodied that have simply been rendered obsolete on American soil. Modesty, decency, probity, rectitude, honor, politeness, virtue, magnanimity, and propriety are words that—along with their opposites such as shame and disgrace—have disappeared from current use. "They don't even enter into the calculus of public discussion and decisionmaking."[10]

Perhaps words like probity and rectitude bespoke the mentality of an era that is now gone forever and one to which few would prefer to return if given the choice. But they were words that conferred approval or disapproval on behavior that was thought good or bad for a healthy society. And they embodied respect for others.

The language of the latter 20th century, by contrast, treats personal and civic virtue as though they are purely private concerns. The language of public life is the dialect of the quantifiable, the rational, the scientific, and the technical—language of calculation and control but not language of values and meaning. It is a dialect that suits well the designers and managers of the paradigm of the present age.

Americans of good will who seek escape from the sterility of modern secular society are handicapped by the absence of a shared framework for discussing core American values. While secularism has not supplied public life in America with a sense of meaning and purpose, sectarianism has framed spiritual restoration largely in terms of politics and power. Secularists and sectarians alike have come to understand and explain social reality in the context of power and, thus, resist attempts to transcend politics and to find common ground in building a better, less political society.

The emergence of religious and ideological wars in America only points to the collapse of the American public philosophy. In reality, Americanism is a matter of the mind and heart. G.K. Chesterton said that America was built upon a unique "creed," a creed so dominant that it was set forth "with dogmatic and even theological lucidity" in our founding documents.

Nations are said to live by their myths. What binds America together are the ideas that live on—its myths, its history, and its still unfolding story. Is there an American creed around which Americans can unify in the late 20th century, or will centrifugal forces continue to gain strength? What are the myths and stories of America as a land of imagination and opportunity? What informs our attitudes and shapes our political sentiments? Is there an American memory that stretches, as Abraham Lincoln said, like a "mystical chord" from "every patriot grave and battlefield to every heart and hearthstone in America?" Or have Americans, as is often alleged, become self-absorbed, animated mostly by the promise of more rights and entitlements than calls to civic duty?

The anxiety about the current rush to debunk the myths that nurture a nation come from across the political spectrum. Conservative Reagan speechwriter Peggy Noonan decries the "compulsive skepticism" of the modern mind that only feeds cynicism. Liberal scholar Arthur Schlesinger Jr. worries that "the historic idea of a unifying American identity is now in peril in many arenas—in our politics, our voluntary organizations, our churches, our language." He believes the end result will be "the fragmentation, resegregation and tribalization of American life."[11] Free societies must be replenished with things that classical philosophers would describe as "pre-political," those things that are more important than and prior to politics and economics. Given the nature of our dilemma, America appears to be entering one of those phases again.

Even though the visible signs continue to point to social balkanization, a new paradigm which will draw people in from across the political spectrum is slowly being born. The ancient Greek word paradigm means a new model, or framework for how we "see" problems in society. And so it is with the paradigm that is emerging: it is being constructed almost entirely on the basis of how Americans and their leaders choose to see and understand root causes of America's social and democratic problems. People are not moving so much to the Left or to the Right as they are moving out of old categories and old ways of looking at problems. Civic renewal is a different enterprise than winning elections for one's ideological or partisan point of view.

We may assume that in the century immediately ahead, the forces for synthesis will gain strength and steadily compete with the forces of polarization. Growing numbers will turn away from the false choices that are so frequently presented in fragmented, ideologically driven politics: extreme moralism versus extreme secularism, feminism versus traditionalism, individualism versus communitarianism, and so on.

When the realization emerges that what America needs, above all else, is civic revitalization, then growing numbers will also come to accept that what America does not need [is] a lot more of what is currently being offered in public debate—more capitalism or anti-capitalism, more political Left or political Right, more government or less government. Americans can afford to be partial on all of these points, . . . and still conclude that what America really needs is an all-fronts mobilization of individuals to improve the social and moral infrastructure of America.

Chief among the objectives of this movement will be forging a new consensus on the basic values upon which a free society rests. Citizens and leaders from all sectors of society will be needed to rebuild American greatness around the tripod of character, community, and culture.

The task will be difficult. It will, of necessity, involve a debate about values: a debate which can either divide or unify, attract or repel, depending upon whether the antagonists in that debate have in mind forging a new American public philosophy or simply winning partisan squabbles.

The debate about values is not about single issues, or even predominantly about legislative conflicts. Declining values either cause or compound some of society's most pressing problems—its litigiousness, its runaway entitlements, its special interest politics, and its insatiable demand for new rights. But legislative correction alone will do little to change the deeper cultural roots.

By definition, asserting values cannot be confused with a call to simply assert power. Attempts to restore declining cultural values and order through power often only provoke and exacerbate the conflict. The American renewal movement that is coming will not focus predominantly on political ideology or partisanship for the simple reason that neither is capable of social regeneration, and both often undermine it.

The objective of ethical renewal is the renewal of people and their social institutions, not just government. It will require reasserting certain core values, which will necessitate doing away with a radical, ethical pluralism which holds that no ideal is superior to another and that building the good society can be done without any basic agreement on the rules. This project will not be accomplished through the heavy-handed power of the state, but through voluntary value-shaping institutions of America.

America needs a new venue for discussing values, one less dominated by partisan politicians, and a new language, more civic and less sectarian, more civil and less belligerent. Anxiety about public and private morality runs across the political spectrum, and there is much at stake for everyone, regardless of religion or ideology. The debate concerns the kind of society all Americans, conservative or liberal, live in.

Since this [reading] is about the renewal of community, of citizenship, and of civil society, some definition of concepts is imperative.

## WHAT IS COMMUNITY?

Man is a social creature by nature and is dependent upon society for satisfying his basic needs. The citizen in a liberal society is one who can rise above purely private calculation that comes with competitive individualism and can dwell cooperatively in community.

Politics has done little to call Americans to higher ends than material striving or the pursuit of rights. Neither of today's competing political philosophies of liberal egalitarianism or conservative libertarianism present a framework for human progress outside of man as an autonomous, rights-bearing consumer. One envisions man as the bearer of ever multiplying legal rights; the other pictures man mostly as a consumer in an unfettered marketplace. In neither framework is the individual encumbered with a duty to serve something one would call society. American society has invented more rights and provided more consumer choices than any in human history. But neither legal rights nor consumer opportunities are sufficient by themselves to hold a society together.

The alternative to rights-based individualism is what Michael Sandel describes as the "classical republican tradition" in which private interests are "subordinated to the public good and in which community life takes precedence over individual pursuits."[12] The classical republican tradition offers a vision for the individual operating, not just in competition with others or in conflict with the state, but in relationship to community.

Periodically, the American tide turns against individualism's excesses. Guardians of the American experiment, from the Founding on, were conscious of the need to balance untrammeled self-interest with the solicitation of virtue. It is not from the benevolence of the butcher, the brewer, or the baker which we can thank for our dinner, as Adam Smith said, but from their regard for their own interest. But the pursuit of interest alone does little to ensure that one's dinner will be enjoyed in a society of safety and civility.

Alexis de Tocqueville, the Frenchman who traveled the United States in 1833, is broadly regarded as one of the most astute observers of America's unique social and political culture. . . .

Tocqueville noted the creative tension in American society that exists between the power of self-interest and the pull of the public interest. He said Americans tended to explain almost all the actions of their lives by "the principle of self-interest rightly understood." An enlightened regard for themselves constantly prompted them, according to Tocqueville, "to assist one another" and to be willing "to sacrifice a portion of their time and property to the welfare of the state."

This was American self-interest, enlightened and pragmatic, in full and unique display. Self-interest develops continents and fuels enormous energies toward industrial and technological innovation. The impulse to rise above the narrow pursuit of private interests is equally important to preserve well-ordered human society in which the fruits of one's labor can be enjoyed. Today, as has long been the pattern, new attempts are being made to invent a social philosophy that issues a call to community.

Communitarianism, for example, guided by sociologists Robert Bellah and Amitai Etzioni, is moving to erect a new sociopolitical framework that would be, in Bellah's words, "less trapped in the cliches of rugged individualism" and "more open to an invigorating, fulfilling sense of social responsibility."[13] This is a movement, like those that came before, that is searching for a new social equilibrium. It seeks to provide a greater balance between rights and responsibilities, to establish checks on the deepening impulse toward moral license, and to apply brakes to run-away individualism.

The key to community is understanding its voluntary nature. Community cannot be ordered through legislation or summoned forth by verbal admonition—it springs from the habits of a people's heart. Calls to community can quickly degenerate into a reach for something that is far removed from, rather than firmly rooted within, the American heritage of civic republicanism. The answer to the excesses of liberal individualism is not coercive communitarianism.

If the concern is repairing social order and moral connectedness, then community is not and cannot be the work of the national government or even massive social movements. Indeed, the idea of creating a national community, as some are prone to suggest, may be a contradiction in terms, particularly on the ultra-pluralistic soil of America.

Rebuilding community cannot be achieved by absorbing the idea of community into the massive organizational structures of modern society. Small indigenous communities face constant threats from the homogenizing power of the large modern organizations. Indeed, it is large omnipresent structures—such as the central state, ideological movements, or the national media culture—along with their totalizing effects on society, that may be the most destructive of community.

Each structure, in its own way, presents the paradox of modern individualism: the individual exalted, the individual overshadowed and abandoned.

The tendency in a nation driven to ever greater fragmentation by political interest groups and racial and ethnic subgroups is to resist groupings in the name of asserting a common civil life. Yet, properly understood, these groupings may represent an effort to reconnect atomized individuals. These subcommunities may subordinate the individual to a local voluntary society without wholly submerging him or her in a national community thus, causing further isolation.

If Americans are to build "a community of citizens," . . . they will have to do so within, not outside of, American philosophical traditions. Rediscovering the American public philosophy will aid in asserting new definitions of the common good without entirely submerging individual or group identity, or absorbing the private sphere into the already overbearing public realm.

The call to community and greater social cohesion in response to rapid and often reckless social change can also degenerate into a nostalgic and reactionary yearning for a lost golden age, one that was perhaps more carefree, harmonious, and virtuous, but also more stratified by race, gender, and class.

Few would seriously want to trade off all of the progress—whether economic, technological, or social—that has been won in recent decades for a return to the 1950s. The need for shared identity and belonging that most Americans now sense will have to be achieved by accommodating, not reversing, the social advances of recent decades. The frontiers of the 1990s will involve a search for improvements in our social life by moderating the excesses of social movements.

## WHAT IS CITIZENSHIP?

The defining characteristic of America's unique experiment in democracy is reducible to one core principle: self-government. In early America, the people were considered sovereigns, not subjects, because they were individually self-regulating. When Chesterton spoke of the creed that animated American life he described its core component as the pure classic notion that "no man must aspire to be anything more than a citizen, and that no man shall endure to be anything less."

There was a clear sense among the Founders that the size, scope, and cost of external government was directly and inversely related to the presence of internal self-regulation among the people. Maintaining a capacity for self-government based in an ample supply of individual character, knowledge, and commitment to democratic participation was seen as the most urgent project of democracy's guardians. Anything less than an informed and active citizenry could lead to the abuse of government power. The call to civic action, issued then and periodically since, has also been presented as a direct challenge to the American cult of egoism, individualism, and privatism.

Today's understanding of citizenship is defined almost entirely in terms of political activity, especially voting. . . . When politicians appeal to individuals to exercise citizenship, they usually mean "come out and vote for us so we can do something for you."

Empowerment, to use the term of current vogue, is mostly a political term. When used in politics, it is frequently used to build support for legislative action by politicians, policy professionals, and social service providers. It means turning control of the work of public life over to qualified experts who, then, design and launch public programs on an affected group's behalf. Democracy ceases to be about participation and becomes about programs.

To be a citizen is to be socially engaged. David Green, author of *Reinventing Civil Society*, suggests that since the contest between socialism and capitalism is over, advocates of freedom must work to build a new framework he describes as "communal liberalism," which aims to rebuild "a sense of community or solidarity compatible with freedom."[14] A society strong in communal solidarity is a society rich in citizenship, where common people take it upon themselves to nourish the institutions, habits, and morals upon which a humane society rests. The civic ethic includes meeting human need directly and voluntarily. It means turning acts of benevolence into opportunities to treat the whole man by strengthening the ideals of self-improvement and character development.

This vision for wellness and vitality is one on which free societies depend for progress, yet it is conspicuously absent from much publicly funded and administered charity. Vast resources in both the public and private sectors are now controlled by professional caregivers who exhibit a morbid preoccupation with illness, abuse, and victimization.

The concept of the individual, not as citizen, but as client is now pervasive. The provision of public assistance to tens of millions of Americans is often accompanied by the subtle message that recipients are hopelessly trapped in conditions that require the permanent help of advocates, interest groups, and government workers. It produces a mindset that dwarfs citizenship for poor and nonpoor alike. It leaves the poor feeling justified in doing little to reclaim control of their lives and the nonpoor feeling no obligation to intervene with neighborly aid.

The language of social action today is not the common language of a caring citizenry but rather the technical jargon of a professionalized therapeutic state. It is the language of programs, policies, and healing through clinically certified experts. Our current public language creates a picture of society in which most institutions except the state are dysfunctional and where human relationships are frequently abusive and, thus, in constant need of trained mediators. When society is motivated less by a vision for wellness and progress than for identifying new forms of disorder, politics is turned into what neighborhood empowerment activist Robert Woodson calls "grievance-based" politics. Citizens are reduced to clients; politicians are reduced to voicing grievance-based claims. The result, Woodson says, is "an inverse relationship between the leader and the people; his success depends on their continued failure."

Too many leaders, activists, and professional caregivers have a vested interest in preserving rather than solving human problems. Many social and political organizations across the political spectrum would face precipitously declining budgets if the individuals, families, and neighborhoods they presume to represent were suddenly restored to strength.

The politics of grievance and victimization have not made Americans any more tolerant, any more compassionate, or any more sensitive to the volatile issues that so quickly erupt into nasty recriminations when social problems become entirely the concern of the political class.

The key ingredient of citizenship is fostering a genuine social sympathy among the people. The real objective of a renewed citizenship would be to reclaim the individual by refusing to delegate the business of public life and public aid entirely to government professionals. Programs would seek to strengthen mediating structures and would view citizenship as the very lifeblood of society. Programs would seek to join the poor and non-poor together as neighbors, volunteers, and partners in community renewal.

Individuals must be restored to full participation in the whole of public life. Whatever government does, or does not do, the job of restoration and renewal will fall heavily to citizens. Citizenship will be restored when people recognize that the reach for political power to solve our deepest social problems has been a Faustian bargain. If empowerment means anything it must mean that individuals, including the family and neighborhood structures that form one's social habitat, are granted greater independence from enlightened bureaucratic caregivers and from the permanent political class that presumes to speak for them.

## WHAT IS CIVIL SOCIETY?

Citizenship entails a range of social duties that are carried out through the associational life of a community. Civil society is the realm of volunteer networks and informal associations in which individuals conduct much of their lives.

A well-functioning society is made up of what Edmund Burke called "little platoons"[15]; Emile Durkheim called "the little aggregations"[16]; and Christopher Dawson called the "interpenetrating orders—political, economic, cultural and religious."[17] Each order should possess a considerable amount of independence.

Many of these vital institutions of society have been weakened. Robert Bellah, author of the critically acclaimed book *The Good Society*, points to the collapse of institutions, particularly the family, that nurture and socialize children. In recent years, "the local communities, extended kin groups and religious organizations, as well as the many economic, legal and political functions" which support the family have been sapped of strength and vitality.

The essence of civil society was established most effectively in our own generation by Peter Berger and Richard John Neuhaus in the classic *To Empower People,* which entered the term "mediating structures" into the vernacular. By mediating structures they meant, above all, the institutions such as families, churches and synagogues, voluntary associations, and neighborhoods, that come between the individual and the state. "Mediating structures," they argued "are essential for a vital democratic society."

Public policy, Berger and Neuhaus said, should protect and foster mediating structures and, whenever possible, should "utilize mediating structures for the realization of social purposes."[18] This would result in true empowerment, not just for individuals, but for the institutions and associational networks that are necessary for sustaining individuals in community.

Recently politicians, think tanks, and political interest groups have rediscovered the notion of civil society, and their interest is welcomed. But, it is very doubtful that reconstruction of the civic order will be led by politicians. Says scholar Robert Royal, "Barring a truly great spirit in the White House—someone like Lincoln, who would understand the shape and limits of what the state can do to encourage a healthy civil society—we must look elsewhere for moral and civic reconstruction."[19]

Civil society is not about politics and it is not, according to Royal, about "democratic machinery or the well-managed bureaucratic state. Civil society is precisely a human order that is larger and richer than the state."[20] The institutions that make up civil society thrive in the absence of political encroachments and frequently falter when supplanted by government. . . .

## WHAT IS CULTURE?

At its deepest level, culture is the values, beliefs, and habits by which individuals order their lives and which are then manifested and mirrored in public life in hundreds of forms, from the condition of our universities to the quality of our prime time television, to the health of our neighborhood institutions, to our civility in public debate. For culture to be culture it must have the effect of civilizing humanity.

Webster presents culture as a concept indicating progress, not regress. It is not a term that would be appropriate to use to describe a group of people who are brutal, barbaric, and anarchic.

Culture is the soil in which the seeds of a healthy human society are planted and replanted. The civilizing work of culture is accomplished primarily through the cultivation of character and intellect, and the refinement of aesthetic tastes. It is about our manners and our morals. . . .

Culture may be more determinative of the quality of our common life, and over the long term of our national success, than our public policies, advanced legal system, or vibrant economy. The life of culture is deeper and more important than government because it reflects what individuals value. As Plato said, "Give me the songs of a nation and it matters not who writes its laws." Burke said: "Manners are more important than laws. Upon them, in a great measure, the laws depend."

In addressing the issue of culture on the American continent, one must be especially careful. Unlike cultures having ancient roots such as on the European continent, many find the very thought of an American culture oppressive, old-fashioned, or simply irrelevant. Any attempt to reassert shared values runs against a powerful grain of individualism, which is perhaps the most distinguishing mark of contemporary American culture.

America's acceptance of rampant pluralism militates against attempts to create an undergirding culture or reinforce common values. There appears an almost inbred native resistance to anything monolithic or uniform. American individualism and pluralism run deep and, in recent decades, have become excessive. Americans display a remarkable tolerance for the conflict and disorder that frequently result from a lack of commitment to the common good. Ralph Waldo Emerson described the idea that society would issue a claim on the individual as a conspiracy against "the manhood of every one of its members."

We have a culture, then, with a penchant for negating culture. In America, perhaps like nowhere else, the very definition of freedom means the freedom to be different, the freedom to live by one's own lights, and to find happiness in one's own ethnic identity and cultural experience—in other words, to escape the constraints that often operate on individuals in less dynamic cultures. Rugged individualism and cultural pluralism have not been without their benefits. For centuries, immigrants have come to America seeking escape from cultures that suffocate individual expression and achievement through rigid social stratification. The spirit of adventure and innovation that has animated Americans for centuries drew its strength, not from conformity to a reigning culture, but from throwing off the boundaries of social class and old sociopolitical arrangements, whether mercantilism or state-sponsored religion.

Is a common culture necessary to maintain a liberal civil society? Calls to restore traditional American values are not likely to succeed if what is intended is the denial of freedom to individuals as well as ethnic groups to assert their own uniqueness and escape the boundaries that exist in the countries from which they emigrated. Such an agenda would be neither realistic nor compatible with over 200 years of American experience.

Still, this wholly privatized concept of society can be very destructive of public life. Sociologists dating back to Emile Durkheim have warned about societal disintegration resulting from the erosion of shared beliefs and common purpose. Tocqueville worried about excessive individualism and materialism producing an atomistic society. What Tocqueville was describing was not a society at all, but a collection of individuals living in resistance to the notion of enduring common interests, pursuits, traditions or beliefs that make up a society. Tocqueville predicted "democratic despotism" would result for a society with weakened civic institutions in which the individual stands alone and alongside a powerful state.

## EDUCATING FOR A CIVIL SOCIETY

The cultural transmission belt of a free society is education, which probably explains why education policy has become such a battle zone. Education transmits a free society's highest ideals and imparts the knowledge necessary for democratic participation. The ultimate end of education is not simply preparation for competition in a technological world; it is nourishing the habits of virtue and character needed for cooperation in human society.

American public education was forged out of a desire to draw together many separate identities into one national quilt. Its founders constructed it, according to Arthur Schlesinger Jr., as "the great instrument of assimilation" whose purpose was the creation of "a unifying American identity." Today's educational system is paralyzed by the thought of assimilation into Americanism.

Schlesinger maintains that the "militants of ethnicity" now view the main objective of public education, not as assimilation, but "the protection, strengthening, celebration, and perpetuation of ethnic origins and identities." But this separatism, he says, "nourishes prejudices, magnifies differences, and stirs antagonisms."[21]

In the struggle to build student self-esteem and achievement around ethnic identity, the search for shared civic values and common democratic interests can easily be defeated. The late Christopher Lasch argued that the loss of historical continuity in America has created a culture of competitive individualism which is degenerating into a narcissistic preoccupation with self.

Learning respect for the accomplishments of a nation's past is necessary to preserve a society's identity and heritage. The literature of earlier periods emphasizing fortitude, perseverance, and self-control gave way to the pursuit of success and self-fulfillment. The narcissist, says Lasch, lives "only in, and for, the present."

The cumulative effect, according to educator Peter Gibbon, is that our children "are not being raised by exemplary lives and confident schools; nor by high culture, vigilant communities, families, churches and temples, but rather by an all-enveloping enemy culture interested in amusement, titillation, and consumerism."[22]

Schools cannot be blamed for the collapse of other value-shaping institutions, such as the family, nor for an epidemic of ethical relativism that is society-wide. Nevertheless, no movement to strengthen American democracy, civil society, or ethical standards will be effective if it fails to counter the American cult of the unrestricted self.

The pursuit of the moral life requires the pursuit of ends and ideals larger and higher than the subjective self. Teaching the virtuous life is difficult when man is viewed as a completely autonomous decision maker. When morality is contingent, heroic acts and virtuous deeds are not regarded as universally valued services to man but merely as expressions of subjective desires. . . . Few would suggest that schools return to moral exhortation or to the teaching of religious doctrine. The key is in understanding what is wrong with the false extremes that currently present themselves and to build consensus around values that are still nearly universally held.

Philosopher Alasdair MacIntyre argues that the chief means of moral education in classical societies was storytelling. In classical antiquity, students were tutored by the heroes of ancient Greece and Rome.

Storytelling is a lost tradition in the modern era. Lessons meant to encourage moral reasoning often involve perplexing and unresolvable dilemmas. When moral lessons are filled with ambiguities, the suggestion is that all morality is in a state of constant flux. Rather than aiming to strengthen ethical awareness, students are trained in procedural skills in order to arbitrate differences in a moral environment characterized by constant disagreement.

Societies committed to preserving a sense of heroism and the pursuit of virtue do not treat morality as though it consists of endless argument over rules. They do not give children the impression that the pursuit of the moral life is completely subjective, private, and futile. The result of encouraging endless criticism in the name of promoting cultural pluralism is weakened allegiance to any common heritage or shared values.

## POLITICS, LAW, AND AMERICA'S PUBLIC DISCOURSE

### The Decline of Politics

The state of American democracy, by broad agreement, is not good. It has been a long while since politics was given credit for having made life fundamentally better for Americans. The combined effect of a spiraling national debt, unrestrained entitlements, a failing educational system, and an ineffective criminal justice system has most Americans charging government with extortionate incompetence.

The linkage between citizen participation and decision making in a representative government is indispensable to preserving democratic legitimacy and vitality. In America, that sacred connection between elections and what government does is an ever more tenuous one, making it harder to convince an already cynical public that "every vote counts."

First, there is a deepening realization that political debates have become disconnected from the real life experiences and concerns of ordinary Americans. Debates have turned sterile, repetitive, and, to many citizens, simply irrelevant.

Second, citizens are mostly separated from any meaningful role in shaping the government they are asked to finance. Their role is mostly one of following, not affecting, the formation of major decisions. The manipulation of the electoral process by interest groups, lobbyists, and political professionals has made the process seem remote to ordinary individuals. Liberal Tom Hayden observes:

> The real crisis is the emergence of the special interest state, a permanent, insulated state within the democratic state. It's defining obsession is with immediate interests at the expense of future welfare.

What happens to a democracy when its basic institutions are perceived to lack credibility, even legitimacy? Part of the problem, according to growing numbers of observers, is that our institutions of democracy were not set up to handle the scope of functions and conflicts currently weighing upon it. As currently organized, the American political process can only be seen as inefficient and ineffective because managing public affairs in a society as complex as America's, whose problems are often beyond managing, is nearly impossible.

. . . Politics cannot possibly meet the expectations it has built up in the people because few politically viable answers exist for the public problems that trouble Americans the most. America's most vexing social problems—illegitimacy, father-absent families, crime and violence—lie largely beyond the capacity of political agendas of either the Left or Right to solve. For the foreseeable future, academic and policy professionals will be asked more frequently to document the failures of public programs than to map believable plans for new ones. The best thing for politics to do is tackle the tough problems that only governments can solve and candidly admit what politics cannot do.

## Politics As Governmentalism

Most discussions of politics immediately assume that the object of "political" activity is always to order the affairs of government. The central feature of modern politics is its peculiar preoccupation with simply managing government—tinkering with bureaucracy, adjusting tax codes, and tightening criminal statutes. The ancients viewed politics as a means of improving man's capacity to better himself and to practice self-government. Politics entailed the elevation and ennoblement of man through broader engagement in the life of the polity. Politics was rarely separated from the deeper normative questions regarding the purposes for which citizens lived, individually and corporately. Statecraft meant more than simply advancing individual interests within the realm of government; in large measure it was soulcraft, to borrow columnist George Will's term.

Politics, practiced as soulcraft, concerned itself with the condition of a person's mind and soul, his or her habits of industry, moderation, courage, public spiritedness and capacity for sound democratic judgment—characteristics which republican forms of government were thought to depend upon for survival. It sought not to overwhelm public managers with insoluble problems, but to summon forth from the people a willingness to engage fully in the resolution of public concerns.

. . . democratic institutions, including parties, often undermine civic involvement. . . . A political system that has given up on the possibility of individuals governing themselves has lapsed, whether it wants to admit it or not, into some form of paternalistic authoritarianism, be it oligarchy, aristocracy, or simply a benign system of central control in which enlightened public managers are charged with governing a compliant and docile citizenry. Little about politics today seeks to fortify public-spirited citizenship. Politics reinforces a vision of society inhabited by unencumbered private individuals, pampered with promises, fortified with multiplying legal rights, and awash in consumer choices, yet paradoxically, more subject than citizen. Political debates which once entertained public reflection on the higher ends of life in a just and good society have been replaced by politicians reciting their accomplishments in ladling out pork-barrel projects.

The corroding effects of modernity and materialism have eroded the deeper linkages of the citizen and his or her responsibilities to transcendent principles. Politics has lost ground because it is about little more than governmentalism, particularly the crass angling for leverage over the appropriations process. As author Jonathan Rauch laments, today's politicians "have turned every social argument into an argument about government, politics or law. Libertarians, anti-government conservatives and left-liberals disagree on many things, but they are all governmentalists."[23]

Can we find an approach to thinking about ourselves, developing rules for conduct, and reclaiming normative values without ritual deference to the government and its alphabet soup of program offerings? Can individuals and communities be called upon to debate how to govern themselves in any meaningful sense?

Rauch recommends a rather modest experiment that could put the question to the test. He suggests planting the brainiest policy wonks from the political Left, Right, and Center in a three day conference. They would be asked to address America's social problems with only one basic condition: no one would be permitted to promise a single change in federal or state statutes. At first, they would be unhappy, confused, at a loss. Then, after a while, he predicts, "they will start to think about the very large portion of America that is not the government. They will hatch ideas, some of them possibly good. They will begin to think a bit differently." That, says Rauch, is where the next quarter century begins—"with the realization that governmentalism has long since reached its limits.[24]

The business of democracy is enabling the people to govern their own lives as well as to draw citizens into the process of shaping public life across a range of functions. The good society can *not* be doled out like just another entitlement; it can *not* be pieced together through programs, or stimulated into existence by more tax cuts. It must be achieved through the cooperative efforts of individual Americans from all walks of life.

### Personal Manners and the Public Discourse

If democracy were somehow hoisted upon the examination table and its contents placed under clinical examination, its words would reveal a body riddled with disabling viruses.

In no area are society's extremes of individualism more evident than in its public debate. Rarely is there the faintest reference to any concept of the common good. *Washington Post* columnist William Raspberry describes a process that has moved beyond the normal give and take among various competing sectors and ideologies into "a near-total breakdown of the American society into warring component parts." Raspberry questions whether anyone speaks the language of national interest any more.

Civil society involves the exchange of ideas—even vigorous arguments over competing ideas and values. But to disagree in a civil society requires sharing enough in common to have something to disagree about. According to the late John Courtney Murray:

> The whole premise of the public argument, if it is to be civilized and civilizing, is that consensus is real, that among the people, everything is not in doubt, but that there is a core of agreement, accord, concurrence, acquiescence. We hold certain truths; therefore we can argue about them.

Without this sense of acknowledging the other's humanity, says Murray, debates degenerate into shrill mutterings. Whoever wins, society loses, and "the barbarian is at the gates of the city."[25]

The classical philosophers were concerned about rhetoric and the public discourse. The story of the Roman Empire is a record of one monumental success heaped upon another only to be followed by internal collapse of civilization into disorder. Disorder was fueled in part by the endless polemics that broke out between warring factions during the latter stages of the republic, producing bitter and eventually violent recriminations.

Strong republics are not shaken by bitter polemics and recriminations, but then neither are such tactics necessary if the political and social order is healthy. Public rhetoric is a barometer of a political regime's health and reflects whether society has given way to a tacit state of nature in which individuals and groups are determined to go their own way, come what may.

Thomas Hobbes, a philosopher familiar with the notion of the state of nature, believed a well-ordered society required broadly shared rules of debate. Writing in *Leviathan,* Hobbes established 19 rules, or "natural laws" for preserving peaceful and dignified conduct in debate. These "articles of peace," as they were called, essentially established the "golden rule" of public life—"do not that to another, which thou wouldst not have done to thyself." To Hobbes, the one irreducible requirement for civil public debate was possessing genuine respect for the rights and dignity of one's fellows. Article 10 insisted that "everyone acknowledge another as one's equal by nature." Article 8 admonishes that no practitioner of democracy "by deed, word, countenance, or gesture declare hatred or contempt for others."[26]

## Rights, Wrongs, and Responsibilities

The well-ordered civil society and the political society are in dialectical tension. When government becomes the ally of an extreme rights-based individualism cut off from any connection to a higher morality, it is forced to accommodate greater demands for individual legal gains. This restless search for human progress through legal reforms is the very root of the politicized society. When only the law and politics arbitrate human affairs, everything becomes political—even the most basic human relations. What follows is a state that expands radically even as its competence and legitimacy ebb.

The American legal system has developed the tendency to translate every dispute into the language of untrammeled individual entitlement, according to Harvard law professor Mary Ann Glendon. The result is the replacement of substantive constitutional principles for what political scientist Michael Sandel calls a "procedural republic" in which individuals are endowed with rights and entitlements but have little consciousness beyond that.

When any concept of a higher law or any notion of self-evident truths has been destroyed, the only real basis for asserting the inestimable moral worth of all of human life is also extinguished. Without the aid of transcendent principles, morality becomes contingent on human choice, and the human person is subject to the arbitrary whims of democratic majorities.

The result is what Philip Johnson calls "the modernist impasse." When people have concluded that man gets to decide everything on the basis of democratic majorities, people are subjected to "the whims of whoever controls the law-making apparatus."[27] The law degenerates into an arbitrary tool of the politically organized. A right conferred on one group becomes an obligation imposed on another. One person's gain is another's loss. The legal system is forced to find ever more perfect balances and boundaries between conflicting parties and claims. People expect the law to simultaneously confer the right to sexual freedom as well as freedom from sexual assault; to guarantee gender and racial advantage for some and the protection against reverse discrimination for others; to protect the rights of criminal offenders and the rights of those offended; to guard the rights of free speech but initiate new rights against the insult of hateful speech; to defend both the rights of individuals and communities, and so on.

The law has always been expected to strike careful balances in these areas, but never before has it been called on to split conflicting demands with such exasperating precision. This degree of legal harmony and balance is, of course, beyond the capacity of the law and state agencies to achieve. The law begins to resemble a harried referee who has the impossible task of policing a sport that is both choked by rules and overwhelmed by rule infractions. The players of society are reduced to fighting over the rules.

## Beyond Left and Right

On the political front, politics is about to yield to new coalitions and new themes emphasizing quality of life in the context of neighborhood. These themes will include a new citizenship that views individuals, not just those speaking for them, as the agents of change and a new localism that aims to expand the capacity of private and public institutions closest to the people to solve problems.

Most solutions are local, which means they are beyond affecting by national ideological movements operating at the top of the political system. Labels are relevant then only to the extent that political ideology is relevant.

The old paradigms of Left and Right are increasingly less useful and more internally conflicted. . . . Lawrence Chickering argues that words like "conservative" and "liberal" have come to mean so many conflicting things "that coherent talk about politics has become all but impossible." Liberalism and conservatism each has its freedom and order wing, one emphasizing modernist values of individualism, freedom, and reason, the other emphasizing order, community, and values.

Each distinction is further reduced by its common assumptions and attitudes about government. According to Chickering, "The left and right argue opposite positions from the same basic assumption: that institutional reform—change in the powers and actions of the central government—is necessary to correct all fundamental problems. Liberalism and conservatism alike are preoccupied with the central state. Almost no one, says Chickering, "is concerned with building positive forms of individual and community self-governance."[28] He wonders why there is no greater debate about how to empower individuals in small communities to solve economic and social problems.

If individuals are to be made not subjects but citizens again in small self-governing communities, new institutional arrangements will have to be devised to provide for greater participation in governing.

American politics pays lip service to the ideal that "all politics is local." But the greatest intellectual thought is given, not to the 82,000 subnational units of government, but to the presidency and Congress. The symbol of American democracy has become less the town hall meeting and more the presidential pen. The prize for a successful political party or ideological movement is the presidency or control of Congress.

## Federalism

The tendency in modern times is to assume that the more difficult the problem, the more desirable it is to turn to specialists and governmental authorities at a higher and more remote level. The federal government's design still reflects that of the 20th century's architect of central control, Frederick Taylor. The science of public management is inherently oriented toward constructing large central agencies from whence professional systems of command and control can design and implement solutions to problems by way of standardized rule-making. The value of scientific management has supplanted the value of democratic participation. The entire administrative apparatus of the central government is presently designed on the assumption that expert decision makers are more capable of solving social and economic problems than smaller self-governing arrangements.

If democracy is to be defended, perhaps new rule-making arrangements must be devised to enable citizens to articulate their concerns and aspirations and to mediate their conflicting demands in ways that strengthen, not weaken, their participation and sense of connection. . . . the strengthening of democracy may thus hinge on new forms of federalism which create smaller authority centers for citizens, in greater numbers, and at lower levels of government.

Societies are complex configurations of relationships, and any society as complex and diverse as America's cannot possibly be governed from the top in any meaningful sense. This is particularly true as national self-identity is being slowly weakened by forces above and below the national government: global economics on the one hand, and regional power and local democratic movements on the other.

Governments must keep pace with the economic and technological forces that are shaping society, lest they obstruct progress. What are those forces today? According to futurist Alvin Toffler, they are the pull of power away from the center, political "demassification," and "mosaic democracy."

Although the means for expanding citizen participation exist in the form of decentralization and a rapidly developing electronic technology, the nation's dominant political culture, of both the Left and the Right, fight to keep the focus of debate on centers of authority that are the most remote from the people. Why, in today's world of technology, regional economics, and declining power and legitimacy of central states, does the national government, or even state governments for that matter, continue to monopolize power?

Citizens must be granted greater control over issues and debates, particularly those at the local level. Politics driven from the bottom at the level of community may prove to have a number of advantages for a democracy in search of renewal, providing basic protections are provided for political minorities. For one, local decision making may be more inclined to search for consensus than a national politics of ideological conflict and interest group gridlock.

For another, local participatory democracy may dramatically reduce public cynicism toward a political process that Ross Perot described as "a costly sideshow." And finally, decentralist republicanism might dramatically reduce the cost of government. If resources are not wasted by being passed through multiple levels before arriving at their intended destination, public support for needed public action would be more sustainable.

## AMERICAN CIVIL INSTITUTIONS

### Societal Fragmentation

In *Habits of the Heart,* communitarian sociologist Robert Bellah mapped the destruction of America's civic life and the weakening of vital institutions that sustain human society such as marriage, family, and religion. Bellah singled out individualism as the source of the disintegration of human relationships into ever more tenuous, fragmented, and distrustful patterns.

The 20th century has witnessed the emergence of the self as the main form of reality and a withdrawal by individuals from the public sphere to pursue purely private ends. What Bellah was describing was the stress fractures of American civil society; the weakening of the very minisocieties that provide order and meaning to human beings. For civil society to be repaired, it must tackle the tough issues: issues of race and ethnicity, the underclass, religion, and marriage and family. . . .

### The Role of Religion

Americans remain a deeply and incorrigibly religious people. Yet how religion has or should influence our common life together is a matter of deep confusion and division—between secularists who speak the sterile language of economic entitlement and political rights, and religious movements that refuse to accommodate, in language or methods, democratic pluralism.

For tens of millions of Americans, religion is of paramount importance. Religion concerns the "first things" which inform and shape all other aspects of life. President Clinton publicly acknowledged this when he reaffirmed that "we are a people of faith" and that "religion helps to give our people the character without which a democracy cannot survive."[29]

The work of the spiritual realm entails primarily that of ministering to the conscience and souls of people, and by extension, the soul and conscience of society. A society searching for order plunges into deeper disorder when this harmony is lost. Political scientist David Walsh has said that a civilization is in a state of crisis "not when its order has broken down for one reason or another, but when the attempt to restore the authoritative order of society is itself ineffective and thereby serves only to exacerbate the original problem."[30]

In a free society, the state must be a subsidiary of society—it must be society's servant, not its master. It is equally wrong for either secularists or religious believers to try to master society through the state. The more realistic mission would be to minimize the intrusive powers of the state in society, working to restore the institutions of civil society.

The solution, as presented by William Boxx, . . . is in rediscovering the notion of subsidiarity, developed by John Courtney Murray. Disharmony has been created in America by the state's secularization of life and the marginalization of religious faith in public life. But disharmony can also be created by religious activism that confuses the legitimate need to restore values with illegitimate attempts to place religious authority over government, and by extension, the populace.

Religion, according to historian Christopher Dawson, serves a higher order and must not obscure or trivialize its mission by letting herself be used "as the instrument of secular power and politics." The idea, according to Dawson, that the spiritual life of society should "be ruled and guided by a political party would have appeared to our ancestors a monstrous absurdity."[31]

This struggle between the secular and sacred has always been settled in favor of properly ordered spheres of influence. In *Democracy in America,* Alexis de Tocqueville was struck both by religion's force in American society as well as the proscribed nature of the realm that religion occupied.

Religion, according to Tocqueville, "retained a greater influence over the souls" of people than any country in the world. Religion's role in the society of the early 1800s was one of nourishing the habits of restraint, industry, and tranquility that were thought necessary to maintain republican institutions. Although religion's influence over manners and morals—the "habits of the heart"—was vast, organized religion maintained a studied distance when it came to political parties and public affairs. In America, said Tocqueville, religion "exercises but little influence upon the laws and upon the details of public opinion; but it directs the customs of community, and, by regulating domestic life, it regulates the state.[32] The clergy, in particular, were careful to preserve the unique and honored station they occupied in society. They eschewed all parties," filled no public appointments, and were "excluded by public opinion" (and by law in several states) from serving in legislatures.[33]

Tocqueville's recorded observations, now over 150 years old, serve as a timely reminder of what is at stake for religious belief when boundaries are confused. Tocqueville observed, "The church cannot share the temporal power of the state without being the object of a portion of that animosity which the latter excites."[34]

The aim of religious believers must be to create a good society by voluntary means, not through political movements seeking legislative solutions to moral and spiritual problems. Such a movement would aim to revitalize culture, not just correct a nation's politics, understanding that culture more shapes, than is shaped by politics.

Leading sociologist Os Guinness hopes for a new American renaissance that introduces Americans again to á public philosophy that would include a commitment to a new civil public discourse and the pursuit of justice for people of all faiths as well as no faith. This philosophy . . . . is a shared vision for the common good. It embraces liberty for individuals, tolerance for America's ethnic mosaic, and a well-moderated pluralism. It finds strength in what Americans have in common. A return to a strong public philosophy can heal a land divided by race, class, religion, and ideology.

A genuinely American public philosophy would attempt to advance a shared vision for the common good, producing a truce of sorts between the reimposers and the removers of religion. Those who discount the importance of our common ideals fail to appreciate their role in making one nation out of many.

## PERSPECTIVES ON CIVIL SOCIETY

. . . The questions that this [reading] ponders may take decades to fully answer: Can the political and civic order be renewed without resort to demagoguery, reaction, or worse, bitter and possibly violent conflict? Can the American spirit of progress based upon optimism, hope, and perseverance be reimparted to a generation more alienated and compulsively skeptical than perhaps any other? Can religious faith serve again to fortify a weakened society without resort to a politicized and sectarian public orthodoxy? Can society curb excessive victim status without forgetting that, for too many, being left behind in the world's most prosperous society can be embittering? Can an authentic appeal be made for private individuals to embrace an American creed, and can a uniquely American sense of identity be restored without canceling out ethnic and religious differences? All Americans of good will are invited to contribute to the resolution of these questions.

## NOTES

1. Eleanor Clift, "Interview: I Try to Be Who I Am," *Newsweek,* 28 December 1992, 24.

2. John W. Gardner, *Building Community,* Independent Sector, September 1991, 16.

3. George Weigel, "Christian Conviction and Democratic Etiquette," *First Things* (March 1994): 34.

4. *The Responsive Communitarian Platform: Rights and Responsibilities,* Communitarian Network, 1.

5. Edward H. Crane, "Civil Vs. Political Society," *The Cato Policy Report,* Cato Institute, Spring 1993.

6. Michael Novak, "The Conservative Momentum," Center for the American Experiment, June 1993, 5.

7. Richard Eckersley, "The West's Deepening Cultural Crisis," *The Futurist* (November–December 1993): 10.

8. Peter L. Berger, *A Far Glory: The Quest for Faith in an Age of Continuity* (New York: Free Press, 1992), 45.

9. Gardner, *Building Community,* 5.

10. John A. Howard, "A Sure Compass," Rockford Institute, 1992, 6.

11. Arthur M. Schlesinger, Jr., *The Disuniting of America: Reflections on a Multicultural Society* (New York: W.W. Norton, 1992), 17.

12. Michael Sandel, quoted from Digby C. Anderson, ed., *The Loss of Virtue: Moral Confusion and Social Disorder in Britain and America* [London: Social Affairs Unit, 1992], 86.

13. Robert Bellah et al., *The Good Society* (New York: Knopf, 1991), 15.

14. David G. Green, *Reinventing Civil Society The Rediscovery of Welfare Without Politics* (London: Institute of Economic Affairs, 1993), 3.

15. Peter L. Berger and Richard John Neuhaus, *To Empower People: The Role of Mediating Structures in Public Policy* (Washington, DC: AEI Press, 1977), 4.

16. Ibid.

17. Christopher Dawson, *Beyond Politics* (Freeport, NY: Books for Libraries, 1971), 21.

18. Berger and Neuhaus, *To Empower People,* 6.

19. Robert Royal, "Reinventing the American People," *The American Character* 6 (Fall 1993): 3 [published by the Ethics and Public Policy Center].

20. Ibid., 26.

21. Schlesinger, *The Disuniting of America,* 17.

22. Peter H. Gibbon, "In Search of Heroes," *Newsweek,* 18 January 1993, 9.

23. Jonathan Rauch, "Caesar's Ghost," *Reason Magazine* (May 1993): 55.

24. Ibid., 57.

25. John Courtney Murray, *We Hold These Truths: Catholic Reflections on the American Proposition* (New York: Sheed and Ward, 1960), 11–12.

26. Thomas Hobbes, *Leviathan* [London: Printed for Andrew Crooke, at the Green Dragon in St. Pauls Churchyard], 1651, chaps. 14–15.

27. Philip E. Johnson, "Nihilism and the End of Law," *First Things* (March 1993): 19.

28. A. Lawrence Chickering, *Beyond Left and Right: Breaking the Political Stalemate* (San Francisco: ICS Press, 1993), 159.

29. Quoted from Weigel, "Christian Conviction and Democratic Etiquette," 30.

30. David Walsh, *After Ideology* (San Francisco: Harper, 1990), 18.

31. Dawson, *Beyond Politics,* 26.

32. Alexis de Tocqueville, *Democracy in America,* vol. 1 (New York: Vintage Books, 1945), 315.

33. Ibid., 320.

34. Ibid., 322.

# "BARN RAISING"

## DANIEL KEMMIS

In many instances in which public undertakings or community development initiatives are blocked, there is a latent public consensus that would be more satisfying to most of the participants than what finally emerges. But in fact this consensus rarely sees the light of day. Another way to say this is that in most of these cases there is more common ground, and higher common ground, than the people involved ever succeed in discovering. The common ground is there (just as it was in the stock sale or the trace race), but our prevailing way of doing things blocks us from realizing it. Our failure to realize is twofold: we do not recognize the common ground (a failure to realize its existence), and we do not make it a reality (a failure to realize its potential). This twofold failure leaves our communities poorer than they need to be.

What is it that could block us from realizing common ground? To a certain extent it is a problem of language—of how we speak publicly. This is one of the major problems identified by Robert Bellah and his coauthors in their stimulating examination of American public life, *Habits of the Heart*. In preparation for writing the book, the five authors interviewed over two hundred Americans from various walks of life, attempting to discover how these people thought about or became involved in public life. Of all the themes that emerged, the one that most consistently caught the attention of the authors was the way that people used language which portrayed them as being more isolated, more cut off from the world, than their stories showed them in fact to be. When they talked about what they valued, for example, they would consistently speak as if they had chosen those values entirely on their own, or as if they could choose others at a moment's notice, whereas their stories made it clear that the values were deeply rooted in their backgrounds, their associations, the way they lived. Or they would speak of themselves as being motivated by purely selfish considerations, when it was perfectly obvious that in their family or professional lives they were deeply committed to the common good and derived substantial satisfaction from improving the lives of other people. The consistent appearance of this incongruity became a central concern to the authors. Here are two of their ways of summarizing what they saw in the people they interviewed:

They are responsible and, in many ways, admirable adults. Yet when each of them uses the moral discourse they share, what we call the first language of individualism, they have difficulty articulating the richness of their commitments. In the language they use, their lives sound more isolated and arbitrary than, as we have observed them, they actually are . . . .

We want to make it clear that we are not saying that the people to whom we talked have empty selves. Most of them are serious, engaged, deeply involved in the world. But insofar as they are limited to a language of radical individual autonomy, as many of them are, they cannot think about themselves or others except as arbitrary centers of volition. They cannot express the fullness of being that is actually theirs.[1]

If these findings are even partially true, they would have far-reaching implications for the way we do public business. My own experience with public life persuades me that the findings are quite accurate, and that we live with their implications all the time. The most costly of those implications involves our difficulty in articulating, staking out, and building on common ground. Our story is really very much like that of the blind men and the elephant. Unable to come to a common articulation of what we are touching, we are chronically unable to benefit from its existence. To the extent that our language of individualism keeps us from naming and building upon what we have in common, we are impoverished, not only in language, but in many other ways as well.

I had a sense of this impoverishment the night that I sat in the Missoula City Council chambers and listened to the testimony on the draft comprehensive land use plan. The rural residents spoke passionately of their property rights and of their undying opposition to the urban arrogance which would presume to limit those rights in any way. The city dwellers who supported the plan spoke just as passionately of the quality of life which was so important to them and of the need they felt for some regulations to protect that quality of life against the developments which threatened it. What I did not hear was any sense of how these people's fates were woven together, how the good life that they each wanted depended upon the others being secured in a different but equally good life. It seemed to me likely that if one were to ask most of the people from outside Missoula why they lived where they did, and if they could be persuaded to speak honestly, they would talk not only about living on some particular piece of land in the country but also about living within driving distance of a town like Missoula. Missoula, in other words, is part of the place they had chosen to live—not an accidental, but an integral part. And the same would be true of the Missoulians: part of the quality of their lives depends upon their living surrounded by rural land and rural living. They have a stake in that rural life; they have a stake in its being a good life.

I heard, that evening, almost no expression of that mutual stake in the shape of one another's lives. In this sense, then, it is certainly true that, when people testify at a public hearing, "their lives sound more isolated and arbitrary than . . . they actually are."[2] People in this situation do not speak of what they have in common, or of how the common good might be guarded and enhanced. What they speak of is how a proposed initiative (in this case the land use plan) either enhances or threatens their individual rights. They speak in terms of the ideologies most conducive to their particular rights, and they leave the decision makers to choose between those opposing ideologies. Whichever side the decision makers opt for, the losing parties will either appeal to a higher decision maker or begin building political coalitions to reverse the dangerous trend which they see in this and similar decisions. But no matter how successful their coalition building, they will never, as Thurow says, be strong enough to ensure the adoption of their own initiatives. What they will ensure is the effort to build a majority coalition on the other side. So in most localities on most issues, the political pendulum is pushed back and forth endlessly, but the higher public good which everyone feels must be there never emerges.

In the example of the comprehensive plan, that higher public good can be spoken of, both figuratively and literally, as "common ground." If we try to imagine the kind of thinking which would lead people, in that situation, to work for the "higher common good," we see them acknowledging to one another that they all want to live well on a certain, very definite part of the earth. If the hearing on the comprehensive plan had been a genuine "public hearing," the people from the country would have been able to hear how deeply the city residents are attached to this place, how they consistently pass up higher paying work elsewhere just to be in this place, how they want their children to be able to have the benefits of open space, of small-town neighborliness which means so much to them. And they, in turn, would have heard their rural neighbors speak of how important it was to them that their children have the experience of raising some animals, of chopping wood—some of those simple, subsistence practices which Jefferson found so important. If people could actually hear the ways in which their neighbors' lives and hopes are rooted in this particular part of the earth that they all call home, they might be able to begin figuring out how to go about living well together here. But the oscillation between unrestrained individualism and stifling bureaucracy never seems to come to rest on that question.

This "missing middle" which our public policy seems never to find is in fact the *res*, the "public thing" of the "republic"—the vanishing table which could "gather us together and yet prevent us from falling over each other." It is that higher common ground which we share, yet cannot find and therefore cannot occupy. When the authors of *Habits of the Heart* describe people who "sound more isolated and arbitrary than . . . they actually are," they are seeing Arendt's "seance" in which the gathering force of the "public thing" has been removed. This, they say, is what happens when people rely, as they usually do (as they nearly always do "in public"), upon their "first language of individualism." Yet the authors of *Habits of the Heart* also found, in most cases, that these same people have some access to what they call "second languages": "[I]f the language of the self-reliant individual is the first language of American moral life, the languages of tradition and commitment in communities of memory are 'second languages' that most Americans know as well, and which they use when the language of the radically separate self does not seem adequate."[3]

It is well to take a deep breath in turning from this one way of speaking to the other. It takes a while to get acclimated to an entirely new linguistic (and moral) landscape. "Tradition," "commitment," "memory," "hope"—these are not familiar landmarks in the procedural republic. We are prone to doubt that the same set of people can actually use both of these languages or occupy both of these landscapes. Is it really possible that people who can be "unencumbered selves" at a public hearing can be something quite different in another setting? In fact, it is possible, and indeed quite common. The people at the comprehensive plan hearing were certainly relying there upon their "first language of individualism"—and as a result, there was precious little public hearing going on. But there is reason to believe that those same people have access to "second languages" of "tradition and commitment" that might enable them to do a better job of seeing and articulating what they have in common.

For a start, we might refer again to language like that in the preamble to Montana's constitution. It seems likely that people on both sides of that land use hearing would feel very much the same emotional response to the words, "We the people of Montana, grateful to God for the quiet beauty of our state, the grandeur of its mountains, the vastness of its rolling plains. . . ." Here is common language, describing a relationship of diverse individuals to "common ground." The language is not that of individual rights, but of shared gratitude, echoing of humility and hope. Such language is a start toward the articulation of common ground, but standing by itself, it can readily be dismissed as mere sentimentality.

We move a step further in the right direction by remembering Wallace Stegner's call upon the people of the West (the "native home of hope") to create a "society to match its scenery." Stegner's argument is that this can only be done by relying upon some other language and culture than that of "rugged individualism." In calling for a renewal of the culture of cooperation, he invokes the barn building culture of the not yet forgotten days of the frontier. He thus calls to mind precisely a "language of tradition and commitment" which Westerners as a "community of memory" can still recall. Calling that culture to mind is a natural function of centennial celebrations. We need to spend time in such celebrations, in such remembering. But this should not be merely an exercise in nostalgia. At its best, such recalling can serve the same purpose as language like that of Montana's preamble: it can help to remind us, in an active, creative way, of what we have in common. In describing the "second language of commitment," Bellah speaks of the importance of these common bonds: "We know ourselves as social selves, parents and children, members of a people, inheritors of a history and a culture that we must nurture through memory and hope. . . . In order not to forget that past, a community is involved in retelling its story, its constitutive narrative, and in so doing, it offers examples of the men and women who have embodied and exemplified the meaning of community."[4]

In this spirit, I will tell one brief story about some men and women who have helped me understand what "cooperation" might mean. Most of us could tell different versions of this same story.

By the time I was eight or nine years old, the wind that blew almost ceaselessly across the high plains of eastern Montana had taken its toll on our barn. We planned to tear down the old one and from its remnants build a new barn in the swale of a dry creek bank, high enough to avoid the torrents that roared through every year or two. It never would have occurred to us, in the early 1950s, to tackle this massive job without calling on the neighbors for help.

Since my brother and I were too young to be of much help to the builders, we spent most of the day down among the box elders on the creek bottom playing with the neighbor children. That day stands out clear in my mind, not so much for the image of the new barn rising out of the old, but for the fact that our neighbor, Albert Volbrecht, had brought his children along. We didn't exactly play with Albert's children; we listened to them tell dirty stories that would have made our mother, Lilly, frying chicken up in the house, cry with rage. What fascinated me was the fact that the little Volbrecht girl was the one in the family with the best stock of stories. Her younger brothers revered her, at least on that score, for her prowess.

Though my mother did not know the exact wording of the stories the Volbrecht girl was entertaining us with, she did know the kind of language the child used under other circumstances, and she heartily disapproved. She would have done anything in her power to deny my brother and me that part of our education. But there was nothing she could do about it. The Volbrechts had to be at the barn raising, just as they had to be there when we branded calves. They were neighbors, and that was that. Albert's presence loomed large on the scene no matter the situation. His hat was the biggest in the corral, his voice the loudest, his language the foulest, his intake of beer the most prodigious. His influence was pervasive. I saw my father drink a can of beer once after the last calf was branded. I was astonished to see him do such a thing, and so was my mother. The blame for my father's indiscretion came to rest on Albert. Like his children, Albert was too fond of off-color stories for my mother's taste. The simplest event became colorful, wild, when Albert retold it. My mother accused him of being unable to open his mouth without storying. And Albert, for his part, delighted in watching my mother squirm at his bawdy jokes.

In another time and place, Albert and Lilly would have had nothing to do with one another. But on those Montana plains, life was still harsh enough that they had no choice. Avoiding people you did not like was not an option. Everyone was needed by everyone else in one capacity or another. If Albert and Lilly could have snubbed one another, our barn might not have been built, and neither our calves nor Albert's branded. Lilly and Albert didn't like each other much better at the end of the barn raising than at the beginning. But that day, and many others like it, taught them something important. They learned, whether they liked it or not, a certain tolerance for another slant on the world, another way of going at things that needed doing. They found in themselves an unsuspected capacity to accept one another. This acceptance, I believe, broadened them beyond the boundaries of their own likes and dislikes and made these personal idiosyncrasies seem less important. In addition, they learned that they could count on one another. If Albert said he would be there with a "farmhand" attachment on his tractor to lift the roof into place, he would be there with the "farmhand." If Lilly said she would fry the chicken, she would do it whether she was in the mood that morning or not. And because Albert and Lilly and the rest of our neighbors were able to count on one another, they experienced the satisfaction of accomplishing a big, tough job by working together.

This eastern Montana of my boyhood still echoed of the frontier. From Plymouth Rock on-wards, Americans on the frontier had found themselves united with their neighbors in the face of an often hostile and precarious existence. Over the generations, the lessons of cooperation wove themselves into something that can only be called political education. People who had learned by repeated experience that they could count on each other, and in doing so accomplish difficult and important tasks together, were the people who eventually formed cooperatives to bring electricity to the most remote areas or to market wheat or beef out of those areas. This way of working together was still taken for granted in my childhood. When early in the 1950s the rural electric association lines marched across the hill to our farmstead, bringing us the magic of electricity, I was oblivious to the fact that generations of Alberts and Lillys learning to work together were behind this miracle.

The point here is not nostalgia. We cannot re-create the world of the frontier, even if we thought we wanted to. But there is something to be learned from the subtle but persistent process by which frontier families learned the politics of cooperation. They learned it the way almost anything worthwhile is learned—by practice. Republican theorists have always understood that citizens do not become capable of democratic self-determination by accident. As Bellah points out, republicans from Montesquieu to Jefferson (and we might add the populists) had recognized that the character which is required for participation in face-to-face self-government can only be instilled through repeated experiences of a very specific kind. For these democratic republicans, " . . . the virtuous citizen was one who understood that personal welfare is dependent on the general welfare and could be expected to act accordingly. Forming such character requires the context of practices in which the coincidence of personal concern and the common welfare can be *experienced* [emphasis added]."[5]

From childhood, Albert and Lilly and all of their neighbors were schooled in those experiences. Because of that practical education, they could overcome many of their differences; they could recognize their need for one another and act accordingly. By contrast, the people at the comprehensive plan hearing had gone to a very different school, and they, too, acted accordingly. Their differences seemed insurmountable to them, and they seemed to see little of their mutual need for one another. The political education which had created this pessimism and isolation is exemplified by another brief story.

A group of citizens in a western town recently began making plans to initiate a major annual art and music festival. During the first summer, they wanted to hold a small one-day preview event, both to raise awareness within the community about the larger festival idea and to raise some money for the next year's festival. They settled on the idea of an old-fashioned box social, where people would be asked to bring picnic lunches, which could then be auctioned. The idea gathered momentum quickly and seemed like a nearly certain success until someone pointed out the possibility of a lawsuit. What if someone got sick from one of the lunches and filed suit? With that question, the box social was laid to rest.

How could it be that my parents and their neighbors could have box socials but we cannot? I have tried to imagine Albert suing us because my mother's fried chicken laid him up or because he got hurt in our corral. But it is truly unimaginable. He no more had that option than we had the choice of not inviting him to help with the barn because we disapproved of the way he or his kids told stories. Most of us now do have those options, and as a society, we pursue them with a vengeance. We have as little as possible to do with those whose "life-styles" make us uncomfortable. If we are injured by one of "them" (or even by one of "us"), we will not lightly shrink from a lawsuit. Short of that, we readily and regularly oppose each other at public hearings, avidly pursuing our own interests and protecting our own rights with no sense of responsibility to hear or respond to the legitimate interests of those on the "other side" or to discover common ground. More and more often, the result is deadlock—and then frustration and withdrawal from all things public. Whereas the politics of cooperation gave people a robust sense of their capacity to get big, tough jobs done together, we increasingly come to the gloomy conclusion that "anybody can wreck anything," so there is no purpose in trying anything. We have been practiced in the politics of alienation, separation, and blocked initiatives rather than in any "practices of commitment" which might "give us the strength to get up and do what needs to be done."

Yet one of the lessons of *Habits of the Heart* is that even those who testified on the Missoula comprehensive plan, even those who never got to testify on the Dubois forest plan, do have some experience with "the second languages and practices of commitment." They do not have enough of that experience to change the way they behave at a public hearing, and that is a growing problem for our society. But they do have snatches of such experience, and it is there that the possibility of reform must be sought. Here, picked almost at random, are a few examples of such "practices of commitment":

- Children in a 4–H club are taught to raise and care for animals, preparing for the competition at the county fair.
- Residents of a rural community form and maintain a volunteer fire department.
- Urban residents create a neighborhood watch program for their block.
- Other urban residents form a softball league, carrying their competition through to the fall championships.

As with instances of community deadlock, these examples can also be multiplied almost endlessly. And as with those other examples, these, too, share certain common features. Those features are essential ingredients in any revitalization of public life, either in this region or in the nation itself. But because that connection is far from obvious, it will bear some deliberate looking into.

There are two basic ingredients of practices such as those listed here or the thousands of others which might have been listed instead. It is the combination of these two ingredients which give to these practices the potential for revitalizing public life. The two elements are: (1) a central concern with value, with standards of excellence, with what is good; and (2) a rigorous objectivity. What these practices promise (and what, in fact, nothing else can provide) is the kind of experience which would enable us to identify and build upon common ground. The common ground we need to find is like a high, hidden valley which we know is there but which seems always to remain beyond our reach. This hidden valley may be called common ground because it is a place of shared values. The values are shared because they are objective; they are, in fact, public values. This is what makes this common ground valuable, but it is also what keeps it hidden from us. It is valuable because the reclaiming of a vital, effective form of public life can only happen if we can learn to say words like *value* in the same breath with words like *public* or *objective*.

But this valley of common ground remains hidden because we all inhabit a world in which values are always private, always subjective. Always, that is, except when we are engaged in practices. What barn building and violin playing, softball and steer raising all have fundamentally in common is this: all of them deal with questions of value, with what is good or excellent (a well-built barn; a well-executed double play), but they all do so in an explicitly social setting, wherein purely subjective or individualistic inclinations are flatly irrelevant, if not counterproductive. MacIntyre explains why:

> If, on starting to listen to music, I do not accept my own incapacity to judge correctly, I will never learn to hear, let alone to appreciate, Bartok's last quartets. If, on starting to play baseball, I do not accept that others know better than I when to throw a fast ball and when not, I will never learn to appreciate good pitching let alone to pitch. In the realm of practices the authority of both goods and standards operates in such a way as to rule out all subjectivist and emotivist analyses of judgment.[6]

What MacIntyre says here of baseball or Bartok may also be said of the thousands of examples of practices which people engage in, from raising steers to running a rural fire department. No one can do these things in a "practiced" way while maintaining a purely subjectivist approach to values. People who engage in these kinds of activities experience what it is to operate within a system of shared values, in which it is clearly not enough to say, "Well, those may be your values, but these are mine." Everyone who testifies at a public hearing may act on that occasion as if all values are subjective and may therefore contribute to the difficulty we have in acting upon shared values. But for most of those people, there is at least one part of their lives where they act, think, and talk very differently. Whether they are cross-country skiing or raising prize irises, they come into relationship with other people in a very particular way. "Every practice," according to MacIntyre, "requires a certain kind of relationship between those who participate in it."[7] What that relationship instills, over time, are precisely the "civic virtues"—those habits which would be necessary if people were ever to relate to each other in a truly public way. Here is how MacIntyre describes how even our homeliest practices gradually instill these civic virtues:

> We have to learn to recognize what is due to whom; we have to be prepared to take whatever self-endangering risks are demanded along the way; and we have to listen carefully to what we are told about our own inadequacies and to reply with the same carefulness for the facts. In other words we have to accept as necessary components of any practice with internal goods and standards of excellence the virtues of justice, courage and honesty.[8]

What Hannah Arendt calls the "weirdness" of our modern situation may be reduced to this: that in what we call the "public" realm, all of these virtues which might in fact enable us to be public are suddenly overshadowed. The "second language of commitment," which so many public hearing contestants speak in their softball leagues or their PTA meetings, becomes suddenly silent when these people think they are in a public setting. Instead, in that setting, they speak their "first language of individualism," with consequences which are all too familiar. The person who grew up knowing that she could not arbitrarily decide what constitutes a prize-winning steer or a good time to bunt now accepts as utterly natural the idea that what she considers a good community is "her value" and what her opponent considers a good community is "his value." What this does to the tune of "public" discourse is predictable: "From our rival conclusions we can argue back to our rival premises, but when we do arrive at our premises argument ceases and the invocation of one premise against another becomes a matter of pure assertion and counter-assertion. Hence perhaps the slightly shrill tone of so much moral debate."[9]

At the root of this difficulty, MacIntyre discovers the same feature which struck the authors of *Habits of the Heart* so forcibly: the feeling on the part of most people that, in the end, their positions (and certainly the positions of their opponents) are a result, not of reason, but of individual inclination.

> [I]f we possess no unassailable criteria, no set of compelling reasons by means of which we may convince our opponents, it follows that in the process of making up our own minds we can have made no appeal to such criteria or such reasons. If I lack any good reasons to invoke against you, it must seem that I lack any good reasons. Hence it seems that underlying my own position there must be some non-rational decision to adopt that position. Corresponding to the interminability of public argument there is at least the appearance of a disquieting private arbitrariness. It is small wonder if we become defensive and therefore shrill.[10]

We are faced, then, with a considerable paradox. While many people do receive training in civic virtues, and are therefore capable of at least a halting fluency in the "second language of commitment," the place that they are least likely to use that language is in what we call public settings. Our public discourse is couched almost entirely in the framework which MacIntyre identifies as the dichotomy of regulated versus unregulated individuality. If people think of public choices only in these terms, it is not surprising that they use in any public setting the "first language of individualism." This, then, is where people "have difficulty articulating the richness of their commitments"; this is where "their lives sound more isolated and arbitrary than . . . they actually are"; here, where the capacity to identify shared values is most acutely needed, it is most consistently lacking.[11] So what can be done about this deadly paradox? If it is true that people attain civic virtues through practices, and if it is true that many people gain such education outside the public arena, the obvious question is: "What can be done to establish practices which would teach people to act and speak in a truly public way in public?"

There is no simple answer to that question. But one part of the answer may emerge from understanding how important to practices is the concrete, the specific, the tangible. It is precisely that element of concreteness which gives to practices their capacity to present values as something objective, and therefore as something public. Again, we need to recall Arendt's table—that actual, physical thing, the rest which makes the public (the common unity or community) possible at all. Lawrence Haworth has perhaps best understood the essential connection between the concepts of community and objectivity: "In any genuine community there are shared values: the members are united through the fact that they fix on some object as preeminently valuable. And there is a joint effort, involving all members of the community, by which they give overt expression to their mutual regard for that object."[12]

In the case of my parents and their neighbors, this matter of objectivity may be viewed on several levels. The barn itself was an "object" which, being a straightforward matter of life and death, seems to qualify as being "preeminently valuable." The barn was as real, as objective, as anything could be, but it only acquired its urgency within the context of the broader and even more compelling objectivity of the land and the weather to which it was a response. However Albert and Lilly may have differed in some of their personal values, they differed not at all in their experience of winter on the high plains. For both of them alike, the prairie winter was cold and deadly, and it absolutely required a good barn.

Strangely enough, that objective requirement of a good barn meant that they were not free to treat their values as being purely subjective. In some things they could afford to be subjective, to be sure. Albert could value beer and salty language in a way that Lilly never would. But when it came to values like reliability, perseverance, or even a certain level of conviviality, they found themselves dealing in something much more objective than we generally think of "values" as being. In fact, those people could no more do without those values than they could do without their barns, simply because they could not get the barns built without the values. The shaping of their values was as much a communal response to their place as was the building of their barns.

The kinds of values which might form the basis for a genuinely public life, then, arise out of a context which is concrete in at least two ways. It is concrete in the actual things or events—the barns, the barn dances—which the practices of cooperation produce. But it is also concrete in the actual, specific places within which those practices and that cooperation take place. Clearly, the practices which shaped the behavior and the character of frontier families did not appear out of thin air; they grew out of the one thing those people had most fundamentally in common: the effort to survive in hard country. And when the effort to survive comes to rely upon shared and repeated practices like barn raising, survival itself is transformed; it becomes inhabitation. To inhabit a place is to dwell there in a practiced way, in a way which relies upon certain regular, trusted habits of behavior.

Our prevailing, individualistic frame of mind has led us to forget this root sense of the concept of "inhabitation." We take it for granted that the way we live in a place is a matter of individual choice (more or less constrained by bureaucratic regulations). We have largely lost the sense that our capacity to live well in a place might depend upon our ability to relate to neighbors (especially neighbors with a different life-style) on the basis of shared habits of behavior. Our loss of this sense of inhabitation is exactly parallel to our loss of the "republican" sense of what it is to be public.

In fact, no real public life is possible except among people who are engaged in the project of inhabiting a place. If there are not habituated patterns of work, play, grieving, and celebration designed to enable people to live well in a place, then those people will have at best a limited capacity for being public with one another. Conversely, where such inhabitory practices are being nurtured, the foundation for public life is also being created or maintained. This suggests a fairly intimate connection between two potent strains of contemporary American thought. One is the revival of interest in civic republicanism. The other appears frequently under the title of "bioregionalism." That word raises issues of definition which need not detain us here. (I mean specifically the challenge of defining any particular bioregion with lines on a map.) What is of particular interest in this context is the tendency of bioregionalists to identify their work by the word *re-inhabitation*. In a talk with that title, Gary Snyder evokes the connection between "coming into the country" and the habituated ways which make it possible to stay there: "Countless local ecosystem habitation styles emerged. People developed specific ways to be in each of those niches: plant knowledge, boats, dogs, traps, nets, fishing—the smaller animals, and smaller tools." These "habitation styles" carried with them precisely the element of objectivity which MacIntyre and Haworth emphasize. Habitation, in other words, implies right and wrong ways of doing things: "Doing things right means living as though your grandchildren would also be alive, in this land, carrying on the work we're doing right now, with deepening delight."[13]

In this talk, as elsewhere, Snyder acknowledges his debt to Wendell Berry's teachings about practiced ways of living in places. Berry makes clear to us why the concept of inhabitation is broader and deeper than "environmentalism":

> The concept of country, homeland, dwelling place becomes simplified as "the environment"—that is, what surrounds us. Once we see our place, our part of the world, as *surrounding* us, we have already made a profound division between it and ourselves. We have given up the understanding—dropped it out of our language and so out of our thought—that we and our country create one another, depend on one another, are literally part of one another; that our land passes in and out of our bodies just as our bodies pass in and out of our land; that as we and our land are part of one another, so all who are living as neighbors here, human and plant and animal, are part of one another, and so cannot possibly flourish alone; that, therefore, our culture must be our response to our place, our culture and our place are images of each other and inseparable from each other, and so neither can be better than the other.[14]

Berry and Snyder present some of the best thinking and writing about this intimate relationship of place and culture, including, crucially, the awareness of how places, by developing practices, create culture. The civic republicans, in a sense, take up where these writers leave off. That is, they recognize the crucial role of practices, not only in the development of culture, but also in the revitalization of public life. Here is how Bellah speaks of what he calls "practices of commitment":

> People growing up in communities of memory not only hear the stories that tell how the community came to be, what its hopes and fears are, and how its ideals are exemplified in

outstanding men and women; they also participate in the practices—ritual, acesthetic, ethical—that define the community as a way of life. We call these "practices of commitment" for they define the patterns of loyalty and obligation that keep) the community alive.[15]

There is considerable room for more mutual reinforcement of these two strains of understanding. The political philosophy of the bioregionalists tends to be vague, uncertain, often more than a little precious and utopian. A more solid, and therefore more confident understanding of how place-centered practices could transform public life would do much to make re-inhabitory politics more credible. The civic republicans are developing very valuable insights into this potentially transformative power of homely practices; what they tend to underemphasize is precisely what the bioregionalists understand so well: the essential role of place in developing those practices.

It is in what Bellah calls "communities of memory" that these "second languages and practices of commitment" have been most carefully preserved. Because of this, it has seemed appropriate to take the occasion of centennial (and bicentennial) celebrations to help us recall how cooperation could once have become such an important part of our culture. But of course there comes a time for turning from what was to what may be. If public life needs to be revitalized, if its renewal depends upon more conscious and more confident ways of drawing upon the capacity of practices to make values objective and public, if those practices acquire that power from the efforts of unlike people to live well in specific places, then we need to think about specific places, and the real people who now live in them, and try to imagine ways in which their efforts to live there might become more practiced, more inhabitory, and therefore more public.

There are two arenas within which this move toward a more inhabitory and more public life must occur if it is to sustain itself. One is the arena of economics; the other, that of politics. . . .

## NOTES

1. Robert N. Bellah et al., *Habits of the Heart* [Berkeley: University of California Press, 1985], 20–21, 81.

2. Ibid., 21.

3. Ibid., 154.

4. Ibid., 138, 153.

5. Ibid., 254. Emphasis added.

6. Alasdair MacIntyre, *After Virtue* [Notre Dame, IN: University of Notre Dame Press, 1984], 190.

7. Ibid., 191.

8. Ibid.

9. Ibid., 8.

10. Ibid.

11. Bellah et al., *Habits of the Heart*, 20–21.

12. Lawrence Haworth, *The Good City* [Bloomington: Indiana University Press, 1963], 86.

13. From remarks given at the "Reinhabitation Conference" at North San Juan School, held under the auspices of the California Council on the Humanities, August, 1976.

14. Wendell Berry, *The Unsettling of America* [New York: Avon Books, 1977], 22.

15. Bellah et al., *Habits of the Heart*, 154.

# PART IV

# SOCIAL EQUITY AND
# ECONOMIC EFFICIENCY

Efficiency, completing tasks with the least possible expenditure of resources, is a central concern of public administration, as important during the early years of the nation as it is to public service practitioners today. It seems there is always much more to be done in the public sector than there are resources available. Public administration is an applied field, so it is natural for people to wish to make each unit of expenditure do as much work as it can. However, there is more to public administration than efficiency, including effectiveness (accomplishing what people want to have done), accountability, and democratic process. Public-sector action is the result of a process of citizen discussion, identification of problems and issues, creating policies or laws, and organizing people to implement programs and services. All this occurs in a public setting, so it carries expectations of openness, citizen access, and accountability to those paying for the result.

The public setting of discussion and decision making on a particular issue occurs within the larger context of American society. Citizen preferences about the size and scope of government and what issues are appropriate for public-sector action change over time, in each neighborhood, community, region, state, and the nation. During the Great Depression of the 1930s and World War II in the first half of the 1940s, the national government grew dramatically as the nation coped with serious problems and threats. Following World War II, at the national level the 1950s were a time of preference for limited government. The civilian part of the national government grew in the 1960s and 1970s to meet challenges of race, poverty, and the environment, and the military grew in the late 1960s and early 1970s in response to the Vietnam War, shrinking again in the late 1970s. The trend toward growth of the civilian national government was reversed in the 1980s (though there was a significant military buildup as an ideological response to the Soviet Union), a time of increasing inequalities in wealth, and reaction against a large role for government in society. Economic growth was rapid in the 1990s; the role of the national government in society followed a middle course between activism and withdrawal, and government focused on efficient management practices.

These fluctuations in national economics, politics, and attitudes about the role of government were related to, though they did not always parallel, events at the state and local levels. Local government grew significantly in the last half of the twentieth century as many people and businesses moved out of central cities, often to suburbs. This movement was accompanied by construction of highways, infrastructure, commercial and industrial structures, new and larger homes, and corresponding growth in government organizations.

Among the effects of this growth and shift in population were urban decay and greater separation of people by race, as whites left the inner city to blacks and others who could not afford to
leave, and the tax base of cities eroded. Those who moved to the suburbs sought the American
dream; many women stayed at home to take care of children while their husbands drove to the
city to work. The dream was not always what people expected, as the large, inefficient cars men
drove to work polluted the environment, women had few career choices and often were relatively
isolated in their suburban dwellings, and the social problems inherent in an unequal, racially
segregated society were soon to contribute to large-scale disruption of the status quo.

The immediate postwar period also brought new expectations about the standard of living
in America, the dominant world economic power. Never before had so many Americans
expected, almost as a matter of right, to have large houses, autos, and many consumer goods,
and that they would be better educated and wealthier than their parents. In later decades, as
the economy sometimes faltered and these expectations were not always met, people thought
something was fundamentally wrong and should be fixed. It did not occur to them that much
of the rest of the world did not (and does not) live this way, that Americans used a relatively
large proportion of resources consumed worldwide to support this standard of living, and
that it was a recent phenomenon.

The readings in Part IV contrast two very different responses in American public administration to conditions in society, emphasizing the conceptual bases of each and offering critique of
the currently dominant paradigm in public administration. Today, economic inequality, the problems of cities and immigrants, and damage to the physical environment continue in the United
States, but public administration as a field seems to some to be disengaged from them. The purpose here is to challenge the reader to reflect on the responsibility of public administration practitioners and academicians to help citizens and elected representatives identify and deal with
challenging social and environmental conditions.

## PUBLIC ADMINISTRATION TAKES AN OUTWARD-LOOKING
## APPROACH TO SOCIAL CONDITIONS . . .

As unrest grew in the 1960s over conditions in the cities, environmental degradation, and the
Vietnam War, many people came to believe that government at all levels was controlled by unresponsive politicians representing the wealthy elite and the status quo. In the first reading, H.
George Frederickson describes an approach advocated by public administration scholars in response to this situation, *New Public Administration*. It emphasized *social equity*, the idea that
public-service practitioners should take responsibility for the impacts on people of public organizations, especially the disadvantaged and those without much power in the political system.

Taking responsibility for conditions in society means becoming an advocate for particular
policies, entering the discussions of citizens and elected officials about what problems should
be addressed by the public sector and how it should be done. For public employees, this is a
departure from the traditional model of bureaucratic neutrality portrayed by the *politics-administration dichotomy*. It may well have been that public employees were always involved to
some extent in influencing policy, but New Public Administration (NPA) went further by suggesting this as a primary function of public practitioners. If asked, many Americans would
likely say this strains the boundaries of acceptable behavior by non-elected public employees.
The effects of NPA were likely felt more in academia than in the practice of public administra-

tion. However, NPA may have raised practitioner awareness about the relationship between governmental action and social equity and encouraged some practitioners to take into account service impacts on disadvantaged people.

## ... THEN TURNS INWARD TOWARD MANAGERIAL EFFICIENCY

By the 1990s, more than twenty years after the beginning of NPA, the social setting had changed considerably. Belief in government as a means to solve collective problems had faded following Great Society social welfare initiatives begun in the 1960s, the Vietnam War of the 1960s and 1970s, and the Watergate scandal of the 1970s. In the 1980s, in part due to the business-oriented, antigovernment rhetoric of the Reagan administration, citizen perceptions of government nationally had changed from thinking of it as a viable problem-solving mechanism to regarding government itself as a significant problem. There was also a shift in the conceptual basis for analyzing the role of government in society. Instead of societal values such as democracy, fair treatment of citizens, or social betterment, the focus narrowed to economic efficiency, making government smaller so it did not function as a drag on the economy by excessively regulating business and taxing heavily.

The earlier values exemplified by NPA did not entirely disappear, but during the 1990s they became secondary to economic efficiency, to "running government like a business." This meant, at least in concept, keeping public employees away from policymaking and encouraging them to find innovative ways to accomplish public purposes with fewer resources. The public sector adopted some of the language and practices of the market, such as *customer service*, *privatization*, and *entrepreneurial* management. The American public sector was a latecomer to this wave of efficiency crowding out other values, which had been under way in several other nations for some time.

Though there have historically been reform efforts at all levels of government, the Reagan administration in the 1980s began the recent national discussion of efficiency. It was in the 1990s that governmental structures and practices became significantly involved in implementation of the emphasis on economic efficiency. The most visible evidence of this took the form of the Clinton administration's National Performance Review, but state and local governments were active as well (the market-based concepts used by the Clinton administration were first applied in local government). Privatization and contracting out services became more common, quantitative performance measurement often replaced other means of evaluating public services, and a new view of *citizens as customers* competed with the earlier view of *citizens as owners* of government and participants in creating public policy.

The general term currently used to identify the body of governmental practices based on economics is *New Public Management* (NPM). Use of the words "new public" makes it easy to confuse New Public Administration and New Public Management, but the two bodies of thought are quite different. They emerged under different conditions, NPA during a time of social change and NPM in a time of focus on economics and profiting from the private market. Though there are similarities, NPA and NPM represent opposites on a continuum of values in public administration. NPA had a "macro" orientation, with concern about conditions in the broader society, especially substantive equality and social justice. NPM has a "micro" orientation, concentrating on delivering services at the lowest per-unit cost. A particularly important difference between NPA and NPM is that the language and substance of NPM has become integrated into public administration practice.

It was intended that citizens and public-service practitioners in NPA work closely together as active participants in the process of deciding what government should do and how it should do it. In concept, citizens in NPM are separate from government in the same way customers are separate from business firms. They are the people who consume services, rather than the people who gather to decide what should be done. The expectation of public-service practitioners in NPM is that they carry out policy decisions from elected officials and survey citizen opinion to fine-tune the techniques used in delivering services. This contrasts with the direct involvement in NPA of public practitioners, with citizens, in policy creation and implementation.

New Public Management fits current public skepticism about the role of government in American society and the lack of perceived need for significant change in social conditions. Its apparent separation of practitioners from the policymaking process may be in keeping with traditional American expectations about the subordinate role of nonelected public employees. The word "apparent" is used because it has been recognized for some time that public employees are involved in creating and implementing public policy. Because of the size and complexity of public-service programs and organizations, it may be unrealistic to expect the people with knowledge of practice, career public-service practitioners, to be removed from shaping the nature of government and its relation to society. Citizens and elected officials continue to need the expertise of career public professionals, not just in efficiently carrying out directives, but in interpreting social needs and tailoring public action to meet them.

New Public Management is one in a long series of efforts, extending back at least to the late nineteenth century, to change government, making it more accountable and efficient. These efforts are often called *reforms*; they sometimes move quickly into the field, they often come from innovations in the private sector, some have significant impacts while others have little, and they vanish as people grow tired of them and move on to the next wave of "new" ideas (in quotes because they are often old ideas dressed up to appear new to those without knowledge of the history of management practice). However, reform movements can be more than passing fads, as portions of them sometimes remain engrained in practice. For example, few organizations use pure versions of reforms from the past such as Performance Planning Budgeting System (PPBS), Zero-Based Budgeting (ZBB), or Management By Objectives (MBO), but significant aspects of these approaches to management can be found in contemporary practice. For the study and practice of public administration, there is little point in characterizing such reforms as bad or good. They have a variety of effects, and the question is whether and how they fit with the values and wishes of citizens, elected representatives, and practitioners.

In the second reading, Richard Box analyzes the intellectual underpinnings of NPM, assesses its impact on the size and scope of government, and discusses possible approaches practitioners can take to defining their role in the public policy process. According to John Kirlin, in the third reading, public administration is turning away from the "big questions" of its role in society, concentrating instead on the economic efficiency of management and service delivery. Few people would argue that efficiency, in itself, is a bad thing—managers in public administration have always known the importance of efficiency. However, Box and Kirlin suggest that the contemporary focus on "running government like a business" seems to have distracted people in public administration from asking what role public service practitioners can, and should, play in dealing with larger issues in society.

# Reading 4.1

# "INTRODUCTION"

## H. George Frederickson

> The problem is: how can we make government competent and authoritative
> without destroying the values of democratic participation and responsibility?
> —*Don K. Price, "1984 and Beyond: Social Engineering or Political Values"*

The national bicentennial in 1976 marked two important birthdays for public administration. It was the ninetieth anniversary of the appearance of the first fully developed essay on what was considered a "new" or at least a separately identified field—public administration. In that essay, the young political scientist Woodrow Wilson wrote the now famous words, "administration lies outside the proper sphere of politics. Administrative questions are not political questions; although politics sets the tasks for administration, it should not be suffered to manipulate its offices."[1]

And it was exactly fifty years since the publication of Leonard White's text, *Introduction to the Study of Public Administration,* the first in the field.[2] White's book was, for his time, an advanced and sophisticated attempt to marry the science of government and the science of administration. Whereas Wilson had argued that public administration is "a field of business" and should be separate from "politics," White forty years later countered that public administration can be effective only if it constitutes an integration of the theory of government and the theory of administration.

As fields or professions go, public administration is young. Its early impetus was very much connected with civil service reform, the city manager movement, the "good government" movement, and the professionalization of the administrative apparatus of government. It was in this era that "principles of administration" were developed and the first academic programs in the field were established at American universities. This was a heady era, during which the United States civil service was developed, an innovation adopted in many American states and municipalities. Formal systems of budgeting and purchasing were adopted, and other aspects of the science of management were applied to government affairs. Many of the early leaders in this reform movement also played out important political roles, most notably Theodore Roosevelt and Woodrow Wilson. Public administration was new, a response to a rapidly changing government.

The second "era" in public administration could be said to have begun with the Depression and the New Deal, followed by World War II. This era was characterized by the remarkably rapid growth of the government, particularly at the national level, the development of major American social programs, and ultimately the development of a huge defense program. At this time it became apparent that a large and centralized government can accomplish heroic tasks. Patterns were being developed and attitudes framed for the conduct of American government and the practices of public administration for the coming twenty years. This era also produced most of the major American scholars in public administration who were to dominate the scene from the 1940s into the 1970s.

The period that followed was characterized by rapid growth in the public service and by extensive suburbanization and urbanization. But it was also a period of great questioning of the purposes and premises of public administration. A broad variety of social programs and services were developed, a cold war machine was maintained, and the public service continued both to grow and to professionalize. It seemed as if such expansion could go on endlessly. But by the mid-1960s several crises were developing simultaneously. In many ways, these crises seemed in part to result from the excesses of an earlier time. In other ways, they seemed to be an expression of old and unanswered problems built into our society and our system of government. The urban crisis resulted from relentless suburbanization—governmentally supported. The racial crisis is closely connected, resulting in part from the serious ghettoization of American minorities in the central sections of our great cities. As the central cities have deteriorated, so have their public services. We continue to have unacceptable levels of unemployment, especially among minorities. And our welfare system is badly overloaded. The rapid depletion of our fuel resources results in an energy crisis, which comes hard on the heels of the environmental crisis. And, of course, there is health care, transportation, and on and on. All of these crises have affected public administration.

Three particular events or activities occurred between the mid-1960s and 1970s that indelibly marked the society and the government and, hence, public administration: the war in Vietnam, the urban riots and continued racial strife, and Watergate. These crises and events resulted in new government programs and changed ways of thinking about and practicing public administration.

Frederick C. Mosher and John C. Honey studied the characteristics and composition of the public service in the mid-1960s.[3] Their basic finding was that most public servants feel little or no identity with the field of public administration. Few have ever had a course and fewer still hold a degree in the subject. Public administration at the time seemed to have a rather narrow definition of its purposes, centering primarily on budgeting, personnel, and organization and management problems. Most public servants, it was found, identify with some or another professional field, such as education, community planning, law, public health, or engineering. Even many of those who would be expected to identify with public administration are more particularly interested in some subset of the field, such as finance, personnel, policy analysis, and the like. There was very little policy emphasis in public administration—very little discussion of defense policy, environmental policy, economic policy, urban policy. There was, at the time, much talk of public administration as everyone's "second profession." Education for public administration in the mid-1960s hardly sparkled. The early furor of the reformers had died. The American Society for Public Administration [ASPA] was beginning to struggle.

By the late 1970s, public administration had changed, both in its practice and its teaching.

There are many indicators: the Intergovernmental Personnel Act; Title IX of the Higher Education Act; the Federal Executive Institute and the Federal Executive Seminars; the remarkable growth and vigor of education for public service; the President's Management Intern Program; the Harry S. Truman Foundation; the size and quality of ASPA; the development of the Consortium on Education for the Public Service; several HUD [Housing and Urban Development] grants to public administration-related activities; a much heavier policy emphasis; a renewed concern for ethics and morality in government service; and the continued professionalization of the public service coupled with refinement of management methods at all levels of government.

Public administration is both changed and new. What, then, can be said of this field, or occupation, or profession, that is new? How can it be described? To what extent is it the culmination of the thinking and practices of those who have built the public service, and to what extent is modern public administration different than predicted or expected? Is it a creature as responsive to its times as it is to its lineage? The purpose of this book is to sketch the outlines of contemporary public administration and to set out my "best guesses" as to the likely characteristics and behavior of our field over the near-term future (say, ten to twenty years).

## WHAT IS NEW PUBLIC ADMINISTRATION?

To affix the label "new" to anything is risky business. The risk is doubled when newness is attributed to ideas, thoughts, concepts, paradigms, theories. Those who claim new thinking tend to regard previous thought as old or jejune or both. In response, the authors of previous thought are defensive and inclined to suggest that aside from having packaged earlier thinking in a new vocabulary there is little that is really new in so-called new thinking. Accept, therefore, this caveat: Parts of new public administration would be recognized by Plato, Hobbes, Machiavelli, Hamilton, and Jefferson as well as by many modern behavioral theorists. The newness is in the way the fabric is woven, not necessarily in the threads that are used. And the newness is in arguments as to the proper use of the fabric—however threadbare.

The threads of the public administration fabric are well known. Herbert Kaufman describes them simply as the pursuit of these basic values: representativeness, politically neutral competence, and executive leadership.[4] In different times, one or the other of these values receives the greatest emphasis. Representativeness was preeminent in the Jacksonian era. The eventual reaction was the reform movement emphasizing neutral competence and executive leadership. Now we are witnessing a revolt against these values accompanied by a search for new modes of representativeness.

Others have argued that changes in public administration resemble a zero-sum game between administrative efficiency and political responsiveness. Any increase in efficiency results a priori in a decrease in responsiveness. We are simply entering a period during which political responsiveness is to be purchased at a cost in administrative efficiency.

Both the trichotomous and dichotomous value models of public administration just described are correct as gross generalizations. But they suffer the weakness of gross generalizations: they fail to account for the wide, often rich, and sometimes subtle variation that rests within. Moreover, the generalization does not explain those parts of public administration that are beyond its sweep. Describing new public administration in some detail is a means by which these generalizations can be given substance.

**Social Equity**

Educators have as their basic objective, and most convenient rationale, expanding and transmitting knowledge. The police are enforcing the law. Public health agencies lengthen life by fighting disease. Firemen, sanitation men, welfare workers, diplomats, the military, and so forth, all are employed by public agencies. Each specialization or profession has its own substantive set of objectives and therefore its rationale.

What, then, is public administration? What are its objectives and its rationale?

The classic answer had always been the efficient, economical, and coordinated management of the services listed above. The focus has been on top-level management (city management as an example) or the basic auxiliary staff services (budgeting, organization and management, systems analysis, planning, personnel, purchasing). The rationale for public administration is almost always better (more efficient or economical) management. New public administration adds social equity to the classic objectives and rationale. Conventional or classic public administration seeks to answer either of these questions: (1) How can we offer more or better services with available resources (efficiency)? or (2) How can we maintain our level of services while spending less money (economy)? New public administration adds this question: Does this service enhance social equity?

Social equity is a phrase that comprehends an array of value preferences, organizational design preferences, and management style preferences. Social equity emphasizes equality in government services. Social equity emphasizes responsibility for decisions and program implementation for public managers. Social equity emphasizes change in public management. Social equity emphasizes responsiveness to the needs of citizens rather than the needs of public organizations. Social equity emphasizes an approach to the study of and education for public administration that is interdisciplinary, applied, problem solving in character, and sound theoretically.

**Inequality**

One of the basic concerns of new public administration is the equitable treatment of citizens. Social equity works from these value premises. Pluralistic government systematically discriminates in favor of established, stable bureaucracies and their specialized minority clientele (the Department of Agriculture and large farmers as an example) and against those minorities (farm laborers, both migrant and permanent, as an example) who lack political and economic resources. The continuation of widespread unemployment, poverty, disease, ignorance, and hopelessness in an era of economic growth is the result. This condition is morally reprehensible and if left unchanged constitutes a fundamental, if long-range, threat to the viability of this or any political system. Continued deprivation amid plenty breeds widespread militancy. Militancy is followed by repression, which is followed by greater militancy, and so forth. A public administration that fails to work for changes to try to redress the deprivation of minorities will likely eventually be used to repress those minorities.

For a variety of reasons—probably the most important being committee legislatures, seniority legislatures, entrenched bureaucracies, nondemocratized political-party procedures, inequitable revenue-raising capacity in the lesser governments of the federal system—the procedures of representative democracy presently operate in a way that either fails or only very gradually attempts to reverse systematic discrimination against disadvantaged minorities. Social equity, then, includes activities designed to enhance the political power and economic well-being of these minorities.

## Value-Free Public Administration?

A fundamental commitment to social equity means that new public administration attempts to come to grips with Dwight Waldo's contention that the field has never satisfactorily accommodated the theoretical implications of involvement in "politics" and policy making.[5] The policy-administration dichotomy lacks an empirical warrant, for it is abundantly clear that administrators both execute and make policy. The policy-administration continuum is more accurate empirically but simply begs the theoretical question. New public administration attempts to answer it in this way: Administrators are not neutral. They should be committed to both good management and social equity as values, things to be achieved, or rationales.

## Change

A fundamental commitment to social equity means that new public administration is anxiously engaged in change. Simply put, new public administration seeks to change those policies and structures that systematically inhibit social equity. This is not seeking change for change's sake nor is it advocating alterations in the relative roles of administrators, executives, legislators, or the courts in our basic constitutional forms. Educators, agriculturists, police, and the like can work for changes that enhance their objectives and resist those that threaten those objectives, all within the framework of our governmental system. New public administration works in the same way to seek the changes that would enhance its objectives—good management, efficiency, economy, and social equity.

A commitment to social equity not only involves the pursuit of change but attempts to find organizational and political forms that exhibit a capacity for continued flexibility or routinized change. Traditional bureaucracy has a demonstrated capacity for stability, indeed, ultrastability.[6] New public administration, in its search for changeable structures, tends therefore to experiment with or advocate modified bureaucratic organizational forms. Decentralization, devolution, termination, projects, contracts, evaluation, organization development, responsibility expansion, confrontation, and client involvement are all essentially counterbureaucratic notions that characterize new public administration. These concepts are designed to enhance the potential for change in the bureaucracy and to further policy changes that increase possibilities for social equity.

Other organizational tools such as programming-planning-budgeting [PBB] systems, policy analysis, productivity measurement, zero-base budgeting, and reorganization, can be seen as enhancing change in the direction of social equity. They are almost always presented in terms of good management as a basic strategy, because it is unwise to advocate change frontally.[7] In point of fact, however, these tools can be used as basic devices for change. Both policy analysis and productivity measurement are fundamental to determining the quality and distribution of public costs and benefits. PPB and zero-base budgeting are useful tools for challenging the present patterns of public expenditures and planning public programs with some promise of accomplishing specified objectives. Reorganization is a basic tool for realigning organizational skills to best meet public needs. All three of these notions have only a surface neutrality or good-management character. Under the surface they are devices by which administrators and executives try to bring about change. It is no wonder they are so widely favored in public administration circles. And it should not be surprising that economists and political scientists in the "pluralist" camp regard devices such as PPB as fundamentally threatening to their conception of democratic government.[8] Although they are more subtle in terms of change, PPB, productivity measurement, and policy analysis belong to the same genre as more frontal change techniques such as organizational development, projects, contracts, decentralization, and the like. All enhance change, and change is basic to new public administration.

New public administration's commitment to social equity implies a strong administrative or executive government—what Alexander Hamilton called "energy in the executive." The policy-making powers of the administrative parts of government are increasingly recognized. In addition, a fundamentally new form of political access and representativeness is now occurring in the administration of government, and this access and representativeness may be as critical to major policy decisions as is legislative access or representativeness. New public administration seeks not only to carry out legislative mandates as efficiently and economically as possible, but both to influence and to execute policies that more generally improve the quality of life for all. Forthright policy advocacy on the part of the public servant is essential if administrative agencies are basic policy battlefields. Policy advocacy is as old as management. Where is the department head or bureau chief who does not try to improve the department budget, salary, facilities, benefits? And certainly no one would want a police chief, a school superintendent, or a secretary of defense who did not believe the function of that agency to be anything less than vital to the well-being of the polity. New public administration emphasizes the social purposes of the agency rather than well-being of the agency—recognizing, however, that both are important.

Classic public administration emphasizes developing and strengthening institutions that have been designed to deal with social problems. The public administration focus, however, has tended to drift from the problem to the institution. New public administration attempts to refocus on the problem and to consider alternative possible institutional approaches to confronting problems. The intractable character of such public problems as urban poverty, unemployment, and health care lead public administrators seriously to question the investment of ever more money and manpower in institutions that seem only to worsen the problems. They seek, therefore, either to modify these institutions or to develop new and more easily changed ones designed to achieve more proximate solutions. New public administration is concerned less with the Defense Department than with defense, less with civil service commissions than with the manpower needs of administrative agencies on the one hand and the employment needs of the society on the other, less with building institutions and more with designing alternate means of solving public problems. These alternatives will no doubt have some recognizable organizational characteristics, and they will need to be built and maintained, but they will seek to avoid becoming entrenched, nonresponsible bureaucracies that become greater public problems than the social situations they were originally designed to improve.

The movement from an emphasis on institution building and maintenance to an emphasis on social anomalies has an important analogue in the study of public administration. The last generation of students of public administration generally accept both Herbert Simon's logical positivism and his call for an empirically based organization theory. They focus on generic concepts such as decision, role, and group theory to develop a generalizable body of organization theory. The search is for commonalities of behavior in all organizational settings.[9] The organization and the people within it are the empirical referent. The product is usually description, not prescription, and if it is prescription it prescribes how to manage the organization better internally. The subject matter is first organization and second the type of organization—private, public, or voluntary. The two main bodies of theory emerging from this generation of work are decision theory and human relations theory. Both are regarded as behavioral and positivist. Both are at least as heavily influenced by sociology, social psychology, and economics as they are by political science.[10]

New public administration advocates could be best described as "postbehavioralists." Unlike his progenitor, the postbehavioralist emphasizes the public part of public administration. The postbehavioralist accepts the importance of understanding as scientifically as possible how and why organizations behave as they do, but he tends to be rather more interested in the impact of that organization on its clientele and vice versa. He or she is not antipositivist or antiscientific although probably less than sanguine about the applicability of the natural science model to social phenomena. He or she is not likely to use behavioralism as a rationale for simply trying to describe how public organizations behave. Nor is he or she inclined to use behavioralism as a façade for so-called neutrality, being more than a little skeptical of the objectivity of those who claim to be doing science. The postbehavioralist attempts to use scientific skills to aid analysis, experimentation, and evaluation of alternative policies and administrative modes. In sum, then, the postbehavioralist is less "generic" and more "public" than his forebear, less "descriptive" and more "prescriptive," less "institution oriented" and more "client-impact oriented," less "neutral" and more "normative," and, it is hoped, no less scientific.

## Participation

New public administration's commitment to responsiveness and social equity implies participation. Some advocates of new public administration emphasize internal participation. The positive effects of the ability of public servants to be involved in and influence the policies that govern their work has been empirically demonstrated. Open and fully participative decision processes have long been canons of good management practice. Other advocates of new public administration emphasize citizen participation in the policy-making process. This view is most commonly found among those who practice or teach local government and work with the so-called street-level bureaucracies. Thus, citizen participation, neighborhood control, decentralization, and democratic work environments are standard themes in new public administration.

This has been a brief and admittedly surface description of new public administration from the perspective of one analyst. If the description is even partially accurate, it is patently clear that fundamental changes are occurring in public administration that have salient implications for both its study and practice as well as for the general conduct of government.

## NOTES

1. Woodrow Wilson, "The Study of Administration," *Political Science Quarterly* 2(1) (June 1887), as reprinted in 56 (December 1941): 493–509.

2. Leonard D. White, *Introduction to the Study of Public Administration* (New York: Macmillan, 1926).

3. John C. Honey, "A Report: Higher Education for Public Service," *Public Administration*

*Review* 27 (November 1967): 294–321; Frederick C. Mosher, *Democracy and the Public Service* (New York: Oxford University Press, 1968).

4. Herbert Kaufman, "Administrative Decentralization and Political Power," *Public Administration Review* 29 (January–February 1969): 3–15.

5. Dwight Waldo, "Scope of the Theory of Public Administration," in *Theory and Practice of Public Administration: Scope, Objectives and Methods*, ed. James C. Charlesworth (Philadelphia: American Academy of Political and Social Science, 1968), 1–26.

6. Ibid.

7. Anthony Downs, *Inside Bureaucracy* (Boston: Little, Brown, 1967).

8. See especially Charles L. Schultze, *The Politics and Economics of Public Spending* (Washington, DC: The Brookings Institution, 1969); and Aaron Wildavsky, *The Politics of the Budgetary Process* (Boston: Little, Brown, 1974).

9. Charles Lindblom, *The Intelligence of Democracy* (New York: Free Press, 1966).

10. See especially James March and Herbert Simon, *Organizations* (New York: Wiley, 1958).

# "RUNNING GOVERNMENT LIKE A BUSINESS: IMPLICATIONS FOR PUBLIC ADMINISTRATION THEORY AND PRACTICE"

## RICHARD C. BOX

*The public sector faces increasing demands to run government like a business, imparting private-sector concepts such as entrepreneurism, privatization, treating the citizen like a "customer," and management techniques derived from the production process. The idea that government should mimic the market is not new in American public administration, but the current situation is particularly intense. The new public management seeks to emphasize efficient, instrumental implementation of policies, removing substantive policy questions from the administrative realm. This revival of the politics-administration dichotomy threatens core public-sector values of citizen self-governance and the administrator as servant of the public interest. The article examines the political culture that encourages expansion of market-like practices in the American public sector, explores the issues of the purpose and scope of government and the role of the public-service practitioner, and offers a framework for the study and practice of public administration based on citizenship and public service.*

Increasingly, public administration practitioners and academicians are faced with demands from politicians and citizens that government should be operated like a business. By this, they mean that it should be cost efficient, as small as possible in relation to its tasks, competitive, entrepreneurial, and dedicated to "pleasing the customer." But, despite the considerable success of market-like reforms in increasing the efficiency of governmental bureaucracies, there remains a sense that something is wrong. For people who are concerned about the quality of public service and attention to issues of social injustice, fairness in governmental action, environmental protection, and so on, something about running government like a business does not feel right. It seems to degrade the commitment to public service, reducing it to technical-instrumental market functions not unlike the manufacture and marketing of a consumer product. Gone is the image of citizens determining public policy and its implementation to shape a better future because customers do not actively participate in governance but wait passively to respond to an "agenda set by others" (Schachter 1997, 65).

From *American Review of Public Administration* 29, no. 1 (March 1999): 19–43. Copyright © 1999 Sage Publications, Inc. Reprinted with permission.

The idea that the public sector should conduct its affairs in a businesslike way is not new in the United States. Though there are enduring classical republican elements in American political thought that emphasize citizens working together for the good of the community, the American public sector exists within a context of market capitalism and classical liberalism, The values of this context include limited and efficient government in combination with individual liberty and political competition. Relatively little attention is given to problems associated with the workings of the market, such as economic inequality or reduced opportunities for collective citizen decision making through discourse.

A strong governmental apparatus can operate to set the parameters of market activity and its impact on the lives of citizens, but in the United States big government must exert control without seeming to be like a centralized European-style state. Although they wanted a stronger government than that provided by the Articles of Confederation, the founders of the United States intended to avoid forming a state apparatus with a purpose and values of its own and a mandate to shape the broader society. This initial "statelessness" (Stillman 1991) is manifest in contemporary public administration debates over the issue of legitimacy (*Public Administration Review* 1993). The concept of statelessness can be overdrawn, as Americans built an extensive government to meet the challenges of the years 1877–1920, including "the emergence of a nationally based market" and "the growth of trusts and oligopolies with national orientations and national economic power" (Skowronek 1982, 11). Because of this institution-building effort, contemporary American government has a significant interactive relationship with the private economy, but it retains from the founding era the cultural expectation of minimal interference in the private sector. This expectation forms a political-cultural context in which the values of the private sector are primary and the values of collective citizen deliberation and the public interest are secondary. Even in this setting, there historically has been recognition of a unique and different role for the public sector, however difficult to define. This was true in the founding era, in the era of Jacksonian democracy, in the reform era, and through several decades of the post–World War II era.

Today, even those elusive public-private differences are fading as the public sector is increasingly penetrated by the metaphor of the market, of "running government like a business." The expansion of such thinking in the public sector has important implications for theory and practice. This article examines the nature of the political culture that encourages market-based practices in the American public sector, explores the issues of the size and scope of government and the role of the public-service practitioner, and offers a framework for the study and practice of public administration in this economistic environment that is based on citizenship and public service. The reading is not about the specifics of any particular reform effort, and the intent is not to bemoan the condition of the public sector. Rather, the reading suggests constructive ways that public-service practitioners and academicians can approach these issues in their work, seeking to preserve and enhance the essence of public service within the market context.

## PUBLIC ADMINISTRATION'S RESPONSE TO MARKET PRESSURES

Expansion of market concepts in the public sector is taking place at the end of the 20th century thrust to build administrative systems that address the problems of a growing urban–industrial nation. That thrust produced a public sector that appears today to be large, cumbersome, wasteful, and beyond citizen control (King & Stivers 1998b, 11), isolated from and out of touch with the rest of society (Peters & Pierre 1998, 228–229). Large government requires that a few elected people represent the wishes of the masses, and representative democracy has grown so remote from the everyday lives of people that it no longer bears a clear relationship to common experience (Hummel & Stivers 1998). Many citizens are so alienated from the concept of self-governance that they think of government as something separate, not a reflection of their own will, though some others would like to participate directly in re-creating the machinery of government to allow for genuine self-governance. As a potential remedy, many politicians and citizens believe that government should be run more like businesses, becoming trim and lean, exhibiting competitive behaviors, and giving greater attention to the needs of "customers."

Evidence of the expansion of market concepts in the public sector may be found in the literature and practice of public administration in an emphasis on a constellation of cost-cutting and production management concepts taken from the private sector, currently drawn together as new public management. These concepts include, among others, privatization, downsizing, rightsizing, entrepreneurism, reinvention, enterprise operations, quality management, and customer service. New public management seeks to separate politics (in the sense of decision making by the people or their representatives) from administration, allowing (or making) managers to manage according to cost-benefit economic rationality, largely free from "day-to-day democratic oversight" (Cohn 1997). Such a separation resembles the old politics-administration dichotomy and Herbert Simon's description of administrative decisions that are largely "factual" (Simon 1997). In this reformed management setting, the public-private distinction is "essentially obsolete," and management is generic across sectors (Peters & Pierre 1998, 229).

This desire to separate the activities of politics (deciding about public policies) and administration (implementing them) is part of a redefinition of the function of government based on "a new elite consensus on the role of the state in society. A substantial public sector is to be maintained, but its purposes and operating values are considerably different from that which was characteristic of the social welfare state. The goal is no longer to protect society from the market's demands but to protect the market from society's demands" (Cohn 1997, 586). This is both an American and international phenomenon (Cheung 1997; Cohn 1997; Hood 1996; Kettl 1997; Lan & Rosenbloom 1992), and it may be seen as evidence of a new equilibrium in relations between economic classes. We may no longer find useful the "stale discourse of class warfare" (Barber 1998, 8), but expansion of economistic concepts in the public sector could reflect the reality that "big government has always been an ally of the little guy, and downsizing it has generally been a recipe for upgrading the power of private-sector monopolies. Schoolroom bullies are forever questioning the legitimacy of hallway monitors" (Barber 1998, 5).

At the level of governmental operation, the question is the extent to which the functions of government should be modeled after the private sector. This gets to the heart of the matter for public-service practitioners, who want to know what is expected from public agencies, how they should relate to citizens (their customers, to use the language of the market), and what is the proper source of policy direction—professional interpretation of the public interest, decisions by elected officials, or the desires of citizens. For over a century, public administration practitioners and academicians have debated the normative role of practitioners, with opinions ranging from neutral implementers of policies determined by others to practitioners as active participants in the policy-making process (Kass & Catron 1990; McSwite 1997). Despite the intent of new public management, market concepts are not likely to remove the practitioner from policy making because government is so complex that citizens and elected representatives cannot govern alone.

Instead, running government like a business means that public managers increasingly regard the public as customers to be served rather than as citizens who govern themselves through collective discourse processes. They keep the public at a distance by conducting surveys and focus groups to identify existing opinions rather than engaging citizens face-to–face in exchanges of information, ideas, and values that result in informed governance. As elected officials withdraw from direct and frequent involvement in administration, the balance of control shifts toward professionals (Cope 1997). With citizens excluded from collective governance and elected officials withdrawn from the daily world of policy implementation, the question becomes, "Who then is accountable?" (Peters & Pierre 1998, 228).

An important task before public administration theorists is to describe the impact of governing and managing by market theory and practice on public service at all levels of government and to explore how theorists and practitioners can respond. Is a complete transformation of the public sector, mimicking the private sector, the answer in the face of pervasive public preference for use of market-like management practices and the apparent reduction of many processes and interactions to cost-benefit calculations? Is this really a problem, or are we approaching the old ideal of pure businesslike efficiency by walling off "unrelated" matters of politics and preferences from public management, squeezing out of professional practice substantive consideration of whether what we do efficiently, instrumentally, is the right thing to do?

Furthermore, in this market-like environment, is it possible to identify aspects of public service that are in some way fundamental to our notions of a good political culture, aspects that can coexist with market concepts of structure and function? At some point, theorists, and more so practitioners, may find it makes sense to worry less about the apparently unstoppable expansion of the market in the public sector and search for constructive ways to respond to it. For practitioners, one way to do this is to simply comply, mastering the expected economic techniques and carrying out policies as given without taking part in their formulation or questioning them. This response fits well with the traditional split between politics and administration, emulating the model of the neutral bureaucrat. For the academician, this approach means confining research to technical matters of management, such as pay-for-performance plans, budgeting systems, or information technology, thus avoiding critical analysis of the effects of market concepts on the public sphere.

A second way to respond to the expansion of market concepts in the public sector would be to protest vigorously in hopes that someone, someday, will listen, or at least that if the pendulum swings in the other direction in the future, we will be well positioned to say "we told you so." This could be a risky strategy for practitioners in the work world and it could position academicians as useful critics of current practice or place them so far outside the mainstream as to be ignored. A third path in responding to the current situation would be to hope for moderation of the impacts of market concepts in the future, but for today, to seek reasonable ways to adapt public service values to the dominant economic paradigm. This is the path outlined in the final section of the reading.

## THE SOCIAL AND POLITICAL CONTEXT

We can gain a broader perspective by considering the nature of the society that surrounds public administration, the society that creates, supports, and demands services from the public/governmental sector. A description of the nature of society may seem somewhat removed from daily administrative affairs, but of necessity public administration is a reflection of societal values. It is impractical, maybe irresponsible, to operate inside public organizations as if the demands of society do not matter. If we do, sooner or later the external environment will catch up to us, making painful demands for accountability and change that might have been foreseen and dealt with in less traumatic ways.

We have known for some time that the modern market economy would have serious impacts on society and in particular on democracy. In 1906, Max Weber asserted that

> · it is utterly ridiculous to see any connection between the high capitalism of today—as it is now being imported into Russia and as it exists in America—with democracy or freedom in any sense of these words. . . . The question is: how are freedom and democracy in the long run at all possible under the domination of highly developed capitalism? (in Gerth & Mills 1958, 71)

In 1931, John Dewey expressed concern for the future of democratic governance:

> The dominant issue is whether the people of the United States are to control our government, federal, state, and municipal, and to use it in behalf of the peace and welfare of society or whether control is to go on passing into the hands of small powerful economic groups who use all the machinery of administration and legislation to serve their own ends. (in Campbell 1996, 178–179)

Ramos (1984) took for granted the "intrusion of the market system upon human existence," with its accompanying emphasis on instrumental rationality that advances the goals of the market, rather than substantive rationality that offers the individual an opportunity to achieve "truly self-gratifying interpersonal relationships" through reason, the activity of the human psyche (23).

Scott and Hart (1979) documented the transition from a society based on the value of the individual in the preindustrial era to one of organizational values in the 20th century. Now, in an age in which we are replaceable parts of large systems, we think with nostalgic fondness of a time when each person was an integral part of a local community. We spend part of our leisure time watching movies or television shows that glorify the heroic loner, but most of us in real life fulfill our destiny as small productive parts of larger systems.

In the United States, the nation's founders created a governmental structure that allowed limited popular participation in national political life while emphasizing order and stability. In so doing, they established a semidemocratic form in which "the 'people' was no longer being defined, like the Athenian demos, as an active citizen community but as a disaggregated collection of private individuals whose public aspect was represented by a distant central state" (Wood 1996, 219). The focus was on individual rights and protection from the power of government, as contrasted with the classical republican ideal of citizen self-governance.

With the rise of capitalism in the 19th century, it became possible to combine democracy and capitalism by clearly separating the economic and political spheres. Thus, citizens maintained their formal public-sector liberal equality in relation to rights, voting, and the law, whereas private-sector inequalities of wealth and power generated by capitalism were largely off-limits to collective, public action. These were the conditions of creation of modern liberal democracy (Adams, Bowerman, Dolbeare, & Stivers 1990; Wood 1996, 234), a "Lockean accommodation" that "reconciled representative government with capitalism by disenfranchising the group most likely to contest the hegemony of wealth—the working class itself" (Bowles & Gintis 1986, 42). It is semidemocratic in that the mass of people participate in a limited and marginal way in collective decision making. In the balance between the public and economic spheres, the public sector is allowed to trim off the rough edges of economic excess in relation to treatment of workers, consumers, and the physical environment, in exchange for keeping public-sector interference with the inequalities of the economic sector to a minimum. Thus, as the market has "insinuated itself into the domains of sentiment, life-style and psyche," it has "bound the state with subtle threads of economic dependency" (Bowles & Gintis 1986, 34).

There do not at present seem to be any viable alternatives to this semidemocratic capitalist model (Dryzek 1996). On a global level, there were competing models for much of the 20th century, but now those models have largely vanished. There are a few socialist enclaves remaining, and a number of relatively undeveloped countries with authoritarian regimes, now being pressured by the public and private institutions of developed nations to change their economic systems to conform with the semidemocratic capitalist model. Over time, it may be discovered that this model is not optimal for all nations, that it works best for mature, stable societies with institutions that can support it. It may not work well for a range of nations with cultural and political histories very different from those of developed Western societies, nations in which the semidemocratic capitalist model can lead to hardship and social unrest (Kaplan 1997).

Today, even the possibility of alternative systems seems to be disappearing, and the market metaphor is dominant. We are apparently in the midst of postmodern conditions characterized by *thick* interpretations of reality at the micro, local level where people can interact directly and form coherent mutual interpretations of values and identity, and *thin* reality at the macro level of broad classes of people, regions, and nations. This results in a profusion of difference, an "assertion of the random nonpattern and the unassimilable anomaly" (Fox & Miller 1995, 45) that shifts and changes constantly. In such an environment, postmodernists believe it is difficult if not impossible to identify grand themes of common belief or interest across large groups of people or geographic areas.

In the midst of this apparent fragmentation of meaning, the daily mechanics and values of the market permeate social, political, and economic life. Families are pressed by economic circumstances to alter their expectations about work, retirement, child rearing, and care of the elderly. Workers are forced to abandon the certainties of lifetime employment, instead constantly keeping an eye on the job market and the best opportunities to increase earning power. Private, public, and nonprofit organizations must constantly adapt to their rapidly changing environments. In the public sphere, it becomes harder and harder to generate large-scale communities of shared interest through direct discourse and personal action, even at a time when people yearn for a return to a sense of community and personal efficacy (Bellah, Madsen, Sullivan, Swidler & Tipton 1985; Box 1998; Eberly 1994). So, we live in postmodern times characterized by large-scale fragmentation of values and intensification of interest in local action, yet we are surrounded by the seemingly universal, global phenomenon of market mechanisms. Within this universal phenomenon, there appears to be general agreement that people are competitive self-maximizers out to lobby legislatures for their benefit at the expense of others, get the largest quantity of consumer goods their resources will command, climb over the backs of colleagues for career advancement, compete at the community level to draw the best companies to their town at the expense of other towns, and so on. Furthermore, this view seems to reflect not only a description of what we are but also a normative vision of what we should be, in a "celebration of wealth that now threatens to drown all competing values" (Lasch 1996, 22). Times change, and if this competitive, consumerist life pattern affects the world's physical environment and social stratification in ways that clearly threaten individuals, the pendulum of public opinion and political action may shift, as it did during parts of the 1960s and 1970s. But for now, "more is more" instead of "less is more," and the market is our guide.

In such circumstances, it is hardly surprising that the language and methods of the market have made significant inroads into public-sector thought and practice over the past two decades. However, American public administration has not been a pure entity removed from the influence of the market during any period in its development. At all levels of government throughout the nation's history, there has been evidence of market-like behavior, such as 19th-century spoils politics that affected policy implementation at the national level and the local-level graft and machine politics that inspired urban reformers to take action. The progressive-era reaction to these perceived abuses was to separate politics and administration at least to the extent that administration would be more businesslike and scientific. Ironically, this meant that reformers wanted to use the management methods of the market to reduce the extent of market-like behavior in the public sector.

In the 20th century, with the rise of the administrative state, public professionals became more prominent in the formulation as well as implementation of public policy. As the overall scope of government expanded dramatically, the internal management of government retained an expectation of efficiency in the midst of a sense of broader public purpose. There were repeated examinations of management of the national government that advocated application of scientific, businesslike methods to improve efficiency (Arnold 1995). At the local level, the council-manager plan was built on a corporate structural model with the expectation that it would produce efficiency and effectiveness.

Beginning in the 1950s, some economists turned their attention to the public sector, applying their assumptions about individual and collective behavior to the public sphere. By focusing on the individual as the unit of analysis, assuming that individuals seek to maximize their personal preferences in the "political market" as they do in the private sector, and treating the behavior of citizens, elected officials, political appointees, and public professionals as examples of self-seeking regard for their own interests, the public choice scholars discovered a public world very different from that of public administration scholars (Johnson 1991). Where traditional theorists found people searching for a better society and the public interest, economic theorists found the public sector operating like an alternative form of market. In this view, traditional bureaucratic, hierarchical government is not a means to social betterment, but a mechanism that distorts private economic behavior, reduces individual freedom, and makes the economy less efficient. The way to reduce these negative effects, according to economists, is to decentralize government, make it smaller, and introduce market-like concepts such as fees and user charges, vouchers, and systems to monitor employee performance, such as merit pay plans (Jennings 1991, 115–116; Ostrom 1991).

At the conceptual level of the size and scope of government, the economic view is that, when it will be to their benefit, individuals, groups, politicians, and bureaucrats seek to maximize their gains in the public market by competing with others for the benefits offered by collective action (Downs 1957; Niskanen 1971; Olson 1965). Corporations seek to make it harder for potential competitors to enter the market, associations seek tax breaks others cannot have, politicians fight for the power and money of office, and public-sector bureaucrats want their agencies to grow so that their status and freedom to act are increased. Governmental action coerces individuals in society into behaving in a manner consistent with majority will, whether or not they agree with it, and those who stand to benefit the most from governmental action will spend the time and resources to influence public policy decisions. A basic assumption of economics is that free and uncoerced individual choice should be maximized and that, where a clear need for collective action in the public interest is lacking, citizens should be allowed to act alone or in voluntary cooperation without governmental coercion (Schmidtz 1991, chap. 7). But government grows larger and more powerful as people use it to gain advantage over one another. This rent-seeking behavior joins the economic inefficiency of government-as-monopoly-provider-of-services as an argument for smaller and more limited government.

At the conceptual level of public agencies and employees, economic rationality has had a significant impact on scholarly thinking about behavior in organizations. Niskanen's argument that public bureaucrats will seek to increase the size of their agency's budgets (1971; or as modified in 1991, the size of the discretionary budget) was an effort to reconceptualize the behavior of public employees, moving away from models of control by legislatures or a sense of duty to the public interest, to a model of the public professional as a self-interested maximizer of competitive position in the bureaucratic world. Niskanen's ultimate purpose was to shift attention from the attributes of bureaucrats to the characteristics of public agencies, especially the structural features that provide incentives for people to behave in certain ways (Niskanen 1991, 28).

In the past few decades, economically oriented examination of public organizations has been a growth industry, adopting a variety of complex and interesting approaches. Summarizing this work in 1984 in an article on "The New Economics of Organization," Moe noted that it is "perhaps best characterized by three elements: a contractual perspective on organizational relationships, a focus on hierarchical control, and formal analysis via principal–agent models" (739). Among approaches to organizations that fall within the new economics rubric, principal-agent theory is especially applicable in the public sector, where the relationships between citizens (principals) and politicians (their agents) and between politicians (principals in this case) and bureaucrats (their agents) are a constant source of fascination. Agency theory deals with questions that arise because "the desires or goals of the principal and agent conflict and it is difficult or expensive for the principal to verify what the agent is actually doing" (Eisenhardt 1989, 58). It is assumed that agents will naturally do less work than principals want done or fail to do work in the way principals want it done. This is the problem of shirking, and principals meet it by seeking information on the activities of agents; this monitoring is time-consuming and costly. Thus, it behooves the principal (e.g., boss, superior, capitalist, or politician) to seek a wage rate that will motivate agents (employees, subordinates) and a level of monitoring that is not too costly but convinces agents that the risk of being caught shirking is substantial (Bowles & Gintis 1986, 77–78).

There are important implications of this line of thought for behavior in public organizations. The economics-based management tools being applied in the public sector are grounded in the economist's assumptions that employees will shirk and that monitoring is essential (though the assumption of economically rational behavior has been under attack for some time; see Anderson & Crawford 1998). The focus is on explicitly specifying performance, through mechanisms such as clearly articulated contracts and/or pay-for–performance systems. In New Zealand, for example, many public agencies have been changed so that "top managers are hired by contract, rewarded according to their performance, and can be sacked if their work does not measure up" (Kettl 1997, 448). Although this is more extreme than typical implementation of principal-agent concepts in the United States, such thinking can be found in the emphasis on various techniques to measure and reward performance and outcomes, as well as movement away from rigid civil-service systems. Although the elegantly simple structure of principal-agent theory is becoming more cumbersome and problematic with the accumulation of empirical data on its application (Waterman & Meier 1998), it remains a powerful tool in the hands of contemporary governmental reformers.

It was in the 1980s, amid the antigovernment ideology of the Reagan administration and a wave of public sentiment for shrinking the public sector, that market-like concepts broke through the weak wall of separation between the values of the market and the values of public management. Trickle-down, supply-side economics and public choice economics pointed the way to prosperity through smaller government, and it was thought that bureaucratic waste could be eliminated through contracting out and becoming entrepreneurial, and soon the entire public sector would, supposedly, be as efficient as the private sector was assumed to be. The negative aspects of treating public purposes as if they were private became apparent through events such as the savings-and-loan crisis at the national level and reevaluation of the tenets of "reinvention" at the local level (Gurwitt 1994), but the transformational impact of this period cannot be denied. This is reflected in the writing of a deputy project director for the Clinton administration's new-public-management-inspired National Performance Review. Although noting that "there is no single intellectual source for the reinventing government movement," he says that it "evolved during the past 10 to 15 years . . . based, in part, on the pioneering intellectual work of public choice theoreticians such as Mancur Olson, E.S. Savas, Gordon Tullock, and William Niskanen" (Kamensky 1996, 248).

## THE SIZE AND SCOPE OF GOVERNMENT

As Weintraub (1997) has pointed out, there are several meanings of the distinction between public and private, including the following: the liberal-economistic model, based on neoclassical economics, which regards the public-private distinction the same as that between state administration and the market economy; the civic perspective, which views the public realm as separate from both the market and the administrative state; and other perspectives, including feminism, that examine distinctions between public and private as involving the spheres of sociability and family and household (16–17). Here, we take a viewpoint looking outward from inside the administrative state to examine penetration of the market metaphor into public administration, so we are concerned with the liberal-economistic model that is "dominant in most 'public policy' analysis" (16).

Given the development of the political environment described above, it is not surprising that Americans have always been searching for an acceptable balance between what is private and what is public. In the early to mid-19th century, there were debates over the national government's role in banking and funding internal improvements such as telegraph transmission and canals for water transportation. For the most part, the trend was toward resisting expansion of the national role amid prevailing public opinion hostile to action by the national government (White 1954, 437–481). Efforts to expand the role of government were more successful in the late 19th century and into the 20th century, as the regulatory and welfare state was built in response to the changing character of national economic life. Then, beginning with the Reagan administration in the 1980s, the growth of the national government was again brought into question. At the local level, the scope of governmental activity grew steadily, accelerating after World War II as the population expanded and suburbia was built.

Along with questioning the size and scope of government in the 1980s, there was a revival of interest in localism and limited government. If it seems to the individual that the national or state government is too distant, too big, and so dominated by entrenched interest groups that he or she cannot have much effect on public policy, it is natural to turn attention toward a locus of action small enough to offer the possibility of quick and satisfying results. As president, Ronald Reagan encouraged this sentiment as he sought to dismantle the welfare state and return its functions to states, localities, and private and nonprofit-sector organizations. The communitarian movement emphasized nongovernmental action and citizen duties as well as rights. These ideas were given additional thrust by the withdrawal of the national government from many domestic initiatives and the phenomenon of tight resources at all levels of government.

In the midst of the 1980s milieu of negativity toward government, with its bureaucrat bashing and belief that government is the problem rather than the solution, economic thinking about the role of government in society blossomed and became part of the ordinary vocabulary of normative debate. By the 1990s, the idea that government needed to be smaller and more efficient had become accepted as common wisdom (though the reality was different at the state and local levels, as government continued to expand; see Walters 1998). Of those people who spend time thinking about the size and scope of government and its role in society, many have come to hold the view that government at all levels is too big and it would be wise to spin off functions from the national government to the states and localities, or from government to the private and nonprofit sectors.

The economist's conceptual scheme for determining what is public and what is private has become standard fare for students of public affairs and underlies much of the public discourse about the role of government (Mikesell 1995, 1–6). Thus, we distinguish between public goods, such as national defense, and private goods, such as household appliances or a hamburger. Public goods would not ordinarily be offered by the private sector acting on the incentive of making a profit because people cannot be excluded from using it and so have no incentive to pay for it (the market failure to provide a good). If it is provided at all, it is available to everyone, and one person's use of a public good does not exhaust its usefulness to others (because experience the benefits of being defended from foreign aggression does not mean that you cannot experience them, too). Because people could experience the benefits of public goods without paying for them (the free rider problem), government coercively forces members of the public to pay taxes or face financial penalties or imprisonment.

In the real world, things are not so simple, so there are modifications and exceptions to the concept of pure public goods and pure private goods. Toll goods are services that many people use but from which people may be excluded (such as swimming pools open for public use or expressways that charge fees), and common-pool resources are goods that can be exhausted as many people use them but for which exclusion is difficult (notably natural resources such as fisheries). The distinctions between types of goods are often fuzzy, and public-sector decision makers use criteria of public demand and political action to choose which services to offer rather than ideal conceptualizations of types of goods. Thus, government becomes involved in providing a variety of services that might appear to belong in the private or nonprofit sector. In addition, government regulates the activities of private actors as they work with toll, common-pool, or private goods, attempting to control negative effects (externalities) of private economic activity on people or the environment.

To add to the complexity, a distinction is made between the provision of public services and their production. Provision is the fundamental question of whether or not government will cause a service, or good, to be offered. It is a policy question to be decided by the people or their representatives. If the answer is negative ("No, we don't want to provide garbage pickup service"), then either private or nonprofit organizations will provide the service or no one will. If, using the example of garbage pickup, no one provides the service (a market failure), there will likely be a discussion of the public health implications and a revisiting of the negative provision decision. Production is a separate issue. If the provision decision is positive, then the question remains how to actually deliver the service, how to produce it. Osborne and Gaebler argued in their book *Reinventing Government* (1993) that government often does a better job of governance, or *steering* (making policy decisions) than of delivering services, or *rowing* (see chap. 1). Osborne and Gaebler included in the steering-rowing distinction governmental decision making about contracting out services and a governmental role in serving as catalyst for private and nonprofit initiatives such as downtown renewal or building sports facilities.

In an appendix to their book, Osborne and Gaebler built on the work of Savas (1987) to offer decision-making criteria for choosing public, private, or nonprofit action, such as stability, regulation, and enforcement of equity (public-sector strong points), expertise and willingness to take risks (private-sector attributes), and compassion and promotion of community (nonprofit-sector attributes). Vincent Ostrom (1977; 1991; 1994) has written extensively about institutional structures, intergovernmental arrangements, and building institutional capacity that helps people to govern themselves. Using a combination of public choice theory and a historical analysis of American government, Ostrom emphasizes the benefits of a multifaceted, polycentric system of governmental organizations and their private and nonprofit partners, organized to fit the services they offer so that the result is the best possible blend of efficiency with responsiveness to the public (e.g., community police patrol may be more efficiently and effectively organized on a small local scale, whereas police communication systems, detention facilities, and crime laboratories may be handled better through large-scale organization [1977, 1518–1520]). These ideas appear to be helpful, although the circumstances surrounding specific decisions are often complex and uncertain, making application of decision-making criteria difficult.

In the end, it appears that determinations about which goods are provided by the public sector are made according to the rule, "whatever people want and will pay for," and production arrangements are a matter of trial and error according to political preferences. At the national level, our attitudes about the scope and size of government change periodically as we face new challenges and social and economic conditions. In response to evidence of widespread poverty, hunger, and injustice, we mount a campaign to redistribute incomes, taxing the middle class and wealthy to help the poor. In good economic times, politicians find ways to give tax breaks and programmatic "goodies" to the middle class to secure their votes in the next election.

At the local level, the choices made about which services to offer vary significantly from community to community. Some places confine themselves to providing the basics of public safety, streets, and sewer and water. Others provide a wide variety of services, for example, public pools, bicycle trails, recreation programs, public hospitals, downtown redevelopment programs, public-private partnerships to encourage economic development, and so on. In a community in which the author of this article lived for 8 years, the city utilities department discussed offering repair of home appliances to residential customers to make the city more competitive when private-sector firms enter the deregulating market for electric service. As one might imagine, people in the appliance repair business were not pleased with this potential entrepreneurial endeavor, viewing it as public expansion into an area thought of by most as a private-sector activity.

The complexities of intergovernmental relations, deciding whether services should be offered by small local units of government, larger ones, or regional agencies, which services will be provided by government and which will be produced by government or by the nonprofit or private sectors, and so on, are matters often resolved incrementally. To some extent the theory of public goods may influence such decisions, but likely it serves in large part to describe and critique what has taken place after the fact. For some time, there has been disagreement in the public administration community about the size and scope of government and the application of market thinking, with its elements of maximum individual choice, decentralization, and privatization. These issues are unlikely to be resolved anytime soon, if ever (see, e.g., Golembiewski 1997; Kettl 1997; Lyons & Lowery 1989; Phares 1989; Ross & Levine 1996, chap. 11; Stillman 1991, 176–185; Waldo 1981, 97).

## MARKET VALUES AND THE NATURE OF PUBLIC SERVICE

Given the importation of private-sector management techniques into the public sector in the past two decades, many public administrators are expected to be entrepreneurial, offer great customer service, and practice the latest management techniques inside the agency (total quality management, pay for performance, and so on). On the surface, it appears that such techniques would make the public service much more efficient, with results that would please citizens (they get better service), elected officials (they get credit for public agency efficiency), and career public-service practitioners (they get more approval and respect from citizens and elected people). And indeed, anecdotal evidence as well as scholarly research indicates that market-based reforms have produced some desired changes in the way government operates in the United States, as well as significant changes in several other nations (Kettl 1997).

There are, however, potential problems with making the public service more businesslike because there is a difference in the operating norms of private- and public-sector organizations. Terry (1993) described entrepreneurial values as including "autonomy, a personal vision of the future, secrecy, and risk-taking" (393), along with "domination and coercion, a preference for revolutionary change (regardless of the circumstances), and a disrespect for tradition" (394). According to Terry, these values are at odds with values of "democratic politics and administration," such as "accountability, citizen participation, open policy-making processes, and 'stewardship' behavior" (393). In a response to Terry, Bellone and Goerl (1993) espoused "civic-regarding entrepreneurship," which offers a community-minded model of administration that is accountable to the public. Terry, however, is not sure entrepreneurship can easily be combined with public service. His overriding concern is that "public entrepreneurs of the neomanagerialist persuasion are oblivious to other values highly prized in the U.S. constitutional democracy. Values such as fairness, justice, representation, or participation are not on the radar screen" (Terry 1998, 198).

The contemporary emphasis on entrepreneurship makes this debate appear new. However, it is to some extent a repackaging of the old politics-administration question that has been in play since the late 19th century and was highlighted by the Friedrich-Finer argument in the early 1940s over the role of the administrator as relatively independent, expert actor, or tightly constrained agent of political officials (for a description of this argument, see McSwite, 1997, 29–52). The repackaging is occasioned by renewed pressure to manage like a business as economic concepts permeate the thinking of policy makers and implementers. Although administrators are pushed to use entrepreneurial and scientific techniques to please the "customer's" assumed desire for businesslike government, they paradoxically become less accountable to the public, whose members lose some control over administration. For example, administrators may use expenditure-control budgets, which allow flexibility to spend as the professional sees fit, and to save money to carry over for discretionary spending later to avoid direct budgetary control by politicians. The assumptions are that this flexibility will make for more nimble response to changing conditions, give managers an incentive to be frugal, and remove political motivations from what ought to be expert decisions. Using Bellone and Goerl's reasoning, the good public administrator will, in exercising this greater degree of discretionary space, take into account the wishes of citizens in making choices. This logic reopens the question of representative democracy and agency; that is, do public administrators answer to the public, or to their elected representatives (Box 1998; Fox & Miller 1995; Kelly 1998)?

The argument against such flexibility in budgeting is that clear and detailed line-item budgets were created to avoid problems of financial abuse and to ensure that money is spent as citizens or their elected representatives decide it should be spent. Saving up money means that it has not been spent as intended by representatives but instead will be spent as nonelected administrators decide—thus, a question of accountability. Supposedly scientific techniques administered by experts, such as cost-benefit analysis, reengineering, and quality control, may lead to more precise and economically efficient service delivery, but they may also crowd out competing citizen preferences for public policy and service delivery, preferences that can be shown by experts to be inefficient or impractical. This sort of result is often seen in disputes over such relatively minor issues as whether to preserve a historic building or remove esthetically pleasing landscaped medians in major streets to make traffic flow smoothly.

In these and similar instances, there is a conflict between the idea of public management as efficient, businesslike, and scientific, and public management as responsive to these and to other public values as well. Jennings (1991) offered three approaches to public administration that capture the essence of this conflict. The bureaucratic approach "takes efficiency and equal treatment of citizens as its primary values"; the pluralism approach "emphasizes responsiveness to multiple interests," and the market approach "takes efficiency as its prime value," differing from the bureaucratic approach in emphasizing diversity of product and maximum consumer choice (122). It may be argued that public administration theorists have in the past tended to prefer one of the first two approaches. Those who favor greater status and discretion for public administrators lean toward the bureaucratic approach, and those who favor greater citizen discourse and self-governance lean toward the pluralist approach. But today, most agree that some measure of market-like matching of public services to consumer preferences, along with efficient and technically competent management, is inevitable if not desirable. The question is how much, in what ways, and whether there are aspects of public service that should not be governed or managed from a market perspective.

The problem in seeking a reasonable balance between approaches in the face of demands to run government like a business is that operating with private-sector entrepreneurial techniques in the public sphere can subvert values of openness, fairness, and public propriety. In such cases, the public-service practitioner may take on the appearance of an independent actor separated from the public, concerned less about the public interest (however defined) and more about making money and maximizing individual power and freedom to act without review. This decreased accountability carries the possibility of unexpected program outcomes, uneven treatment of citizens, and behaviors that have not generally been thought of as consistent with public service. Again, using the example of city utilities in a community in which the author lived, this time in relation to the question of openness, the utilities department attempted to deny public access to many of its documents on the premise that it must operate secretly to level the playing field with its private-sector counterparts. In relation to fairness and a sense of public propriety, the local publicly owned hospital in the same community is semiautonomous, competes aggressively and successfully for market share with the other hospital in town, advertises its high-tech services widely, makes a sizable surplus, and pays its top executive approximately $300,000 per year, including bonuses based on how much the hospital makes. To some people these may not seem like appropriate behaviors for the public sector.

Movement toward a market model thus may result in loss of citizen self-determination in the creation of public policy and the operation of public organizations. Today, most people recognize that a general return to the participatory democracy of an earlier time and simpler society is impossible. However, in this post-progressive era, many are working to rebuild citizen capacity for self-governance through discourse and active citizenship (Barber 1984; Box 1998; Eberly 1994; King & Stivers 1998a). Not everyone will take part in such efforts, but, as Fox and Miller (1995) put it, having "some-talk" is better than having "few-talk" (129–159). The goal is to move beyond the typical model of citizen participation that is "not designed primarily for citizens but for agencies" (Timney 1998, 98), in which administrators use citizen involvement processes for "informing, consultation, and placation" (Timney 1998, 97) rather than enabling people to govern themselves.

Market-driven new managerialism can run counter to self-governance, as it is structured around the idea of happy consumers rather than involved citizens. This is a problem because government is not a business from which customers can voluntarily decide whether to purchase a product. It is, rather, a collective effort that includes every person within a defined geographic area (city, county, district, state, nation), and membership is involuntary unless a resident moves out of the jurisdiction. Mandatory membership carries with it a sense of the right to be involved if one so wishes in the process of deliberating and deciding on creation and implementation of public policy. As Barrett and Greene (1998) wrote, "Governments that buy too heavily into the idea that customers are a higher form of life than citizens risk losing the participation of taxpayers as partners" (62).

Customers, on the other hand, are people to be persuaded and sold an image, a product, or a service rather than people who deliberate and decide. Schachter (1997, 57–58) pointed out that only some public agencies can have customers in the manner of private–sector organizations. Many public agencies cannot easily identify their customers because the public they deal with is divided into a variety of individuals and groups with conflicting goals. Many others are regulatory or stewardship agencies for which the immediate client may not be the true beneficiary of the service. An example of the former would be a school district, for which the customers could be students, parents, or all adults in the community. Examples of the latter could include a restaurant regulated by the local health department (Is the department's customer the restaurant owner or the people who eat at the restaurant?), or the forest service (Is the customer the wood products industry or current and future generations who would use the forests?). These examples illustrate the fundamental difference between the market and the citizenship models of governance. The model of management formed around the market metaphor may lead to channeling resources into creating an image through public relations, surveying citizen opinion, and responding to perceived individual service preferences, rather than bringing citizens together to make their own decisions. This requires keeping the public at arm's length while operating in an entrepreneurial manner behind a facade that gives the appearance of involving citizens and making decisions in the general public interest.

## CONCLUSIONS: PRACTICING PUBLIC SERVICE IN THE MARKET ENVIRONMENT

Few would argue that government should be inefficient on purpose or inattentive to the needs and desires of its clients. In this sense, reinvention, privatization, entrepreneurism, customer service, and other such techniques are good things, bringing a breath of fresh air, challenge, and constructive change to the public sector. But market-like techniques may become problematic when they overwhelm values traditionally associated with the public sector and with public service. The economic assumptions of individual self-maximization, the public interest as the aggregate of private interests, and the public sector as just another form of market are powerful, focused, elegantly simple tools of analysis. Like other powerful and narrow theoretical constructs, they draw appropriate attention to matters of importance, but they also insist on their way of knowing the world while excluding other valuable theoretical orientations.

James March (1992) argued that in the past few decades economic theorists have softened the pure application of their ideas to the public sector, moving from methodological individualism to recognition of the fabric of institutional and structural relationships that make up a community. They now take into account, along with their original assumptions, "a rich, behavioral interpretation attentive to limited rationality, conflict, ambiguity, history, institutions, and multiple equilibria" (228). He also noted that this softer, more subtle application of economic concepts has not yet penetrated into the world of applied theory because "the news of the transformation of rational theory spreads rather slowly from the inner temples of microeconomics to the rationalizing missionaries in [the rest of an economizing society and social science" (229). As this news spreads, the current reforms will fade and reformers will move on to new ideas, but like earlier reforms, they are likely to leave a legacy. The legacy of economistic theory in the public sector may include greater attention to "performance-motivated administration" and the integration of economic concepts into the traditional intellectual matrix of public service (Lynn 1998, 232).

Turning toward application of these ideas to the size and scope of government, the historical American attitude toward the public sector has been that it should not compete with the private sector but should provide services that the private sector will not. However, with time, the clarity of the public-private distinction has faded as people ask government to do more and citizens grow accustomed to things as they are. In the past two decades, this combination of a preference for limited governmental scope with incremental accumulation of services in violation of that preference has been complicated by expansion of economics-based theory and practice. The public-choice side of the running government like a business metaphor suggests shrinking government by contracting out services or returning them to the private sector on the premise that the private sector is more efficient (in the case of contracting) or the assertion that the public sector should simply be smaller (in the case of true privatization). Meanwhile, the entrepreneurial side of the metaphor suggests that government may retain its traditional services and operate them like a business, plus operating services ordinarily thought of as private in order to make money. One way government officials are able to accommodate these diverse demands without making government smaller is to make it appear to be morphing into a publicly owned business by charging user fees, contracting parts of its services, or adopting the language and practice of the private sector by, for example, calling certain services companies or businesses and using a variety of private-sector internal management techniques.

In the area of application of market concepts to the conduct of public service, the potential impacts are significant. The prevailing American attitude about the nature of public service has been to expect market-like efficiency and businesslike operation but in combination with public service values such as accountability, fair and equal treatment, democratic self-governance, social justice, protection of the physical environment, and others. Schachter (1997) pointed out that progressive-era reformers, although striving for a more efficient government, also advocated informing citizens so they could be more active in governing. Today's expansion of economic thinking and the potential separation of expert service provider (public-service professional) from customer (citizen) may be one of the most serious threats to public-service values Americans have experienced.

This leaves contemporary academicians and practitioners with the task of defining preferred normative balances of public and private and of market-like management and public service. We can identify four broad areas in which economic thinking prompts reexamination of substantive assumptions about the public-private relationship and public service. As we do so, we recognize that these assumptions about public institutions are unique understandings that incorporate our history, institutional development, interpersonal interactions, and the surrounding political, social, and economic environment. They are unique because they vary and change by place, time, and human action; their "structural properties" exist as "practices and memory traces orienting the conduct of knowledgeable human agents" (Giddens 1984, 17) rather than as fixed and fully understood phenomena. Taken together, the narrative of these four areas outlines a framework for discourse about the nature of the American public sector and public service. This framework cannot provide clear normative answers to the challenge of the economistic environment, but it can point toward ways of preserving a public-regarding essence of citizenship and public service while responding constructively to the contemporary economic-political environment.

*Services the public sector should provide.* In every community, region, state, and at the national level, there are at a given time services that a majority of citizens believe should be provided by the public. This belief may not be based on extensive knowledge and could change if people were to have more information (this is the problem of improving the quality of public judgment; see Yankelovich 1991), but it is possible to identify attitudes about what services should be public. There are likely a range of reasons that people would give for wanting to have certain services provided by the public sector rather than by the private sector or not at all, but there may be a primary characteristic of public services that most Americans would agree forms a sound decision rule for determining what is public and what is private.

We may hypothesize that, asked to consider a particular service that they think should be publicly provided, people would generally agree that they want certainty that the public has the ability to maintain or change the service in keeping with what the majority thinks to be in the public interest (however defined; in this case, it can be assumed to be the long-term interests of the greatest number, when the public is provided adequate information to make a determination). The standard example of national defense is one on which strong majority agreement can be found and other examples would draw varying responses according to place, time, and the sampled population.

The decision rule of ability to maintain or change a service in accord with a majority view of the public interest is different from the market-driven service rule that uses individual preferences as the basis for governmental response. It focuses not only on efficiency or businesslike operation but also on citizen beliefs about the public interest, the good community, whatever it is that citizens think is best for themselves and others, acting collectively. How to inform and involve citizens in making such decisions may be unclear, but it is clear that this is a decision milieu driven by different values than those of the market. It is also a process that includes collective public deliberation and assistance from public-service professionals.

*Services the public sector should produce.* Within the category of services that people believe should be provided by the public, there are services involving discretion and accountability such that the public is uncomfortable with an arm's length contractual relationship and the possibility of the profit motive rather than public interest determining outcomes. Examples could include police patrol and crime investigation, protective services for children, land-use regulation, and some human resources functions.

These services can be contrasted with a range of things that do not involve the same level of discretion and accountability and are good candidates to be contracted out or fully privatized for reasons of cost efficiency, purchase of specialized expertise that would cost too much to maintain on staff, or greater flexibility in staffing levels. Examples could include operating a police impoundment facility for seized vehicles, conducting psychological evaluations of defendants awaiting trial, and constructing valid test instruments for jobs that draw large numbers of applicants.

*Democratic governance.* There are processes of governance that most Americans expect will be maintained as purely public, rather than being contracted, privatized, or operated by public employees in a closed, unilateral, market-like manner. Though people like good customer service when they need to pay their water bill or have a street repaired, they want to know that they have the option, whether exercised or not, to take part in determining policy and assessing implementation. This goes beyond Osborne and Gaebler's idea of steering rather than rowing, as the issue is not just what government does (steering, or making decisions, versus rowing, or carrying them out) but who has the right and ability to make policy and implementation decisions. This is at the heart of citizenship and self-governance.

Thus, although most would agree that government should use efficient business methods in technical, operational areas, this does not mean that business principles of efficiency, scientific management, or closed and centralized decision making should dominate the creation or evaluation of public policy, or exclude citizens from self-governance. To accept a broader view of governance is to assert that government is not ultimately guided by a market model of competition and efficiency but by a citizenship model of governance. This broader view places businesslike management techniques in an instrumental position subordinate to the larger sphere of governance. It draws citizens, elected officials, and public service professionals together in the joint project of creating and implementing public policy.

*The role of the practitioner.* It might be assumed, in the manner of the old politics-administration dichotomy or the current policy-management split of the market metaphor, that public-service practitioners should not play an active part in shaping issues, debates, and decisions on the questions of what is public or private or whether public policy and services should be approached using the market or the citizenship models. However, there is little doubt today that practitioners are an important part of policy formulation and implementation, providing information needed by citizens and elected officials to frame policy decisions and generating proposals that often form the basis for public action.

Thus, practitioners fill multiple roles in addition to the traditional bureaucratic role, serving as expert advisers and as facilitators of citizen discourse. The open question is how this is to be done in a society that expects nonelected public servants to maintain a position clearly subordinate to elected officials and citizens. This question involves issues of legitimacy and leadership. Is it possible to be an important actor in the creation and implementation of public policy without straying outside the legislative mandate or becoming dominating, self-serving, and causing restriction of public access and freedom to act?

Public practitioners are, because of proximity and knowledge, deeply involved in the broad issue of the extent to which the market metaphor should guide public governance. They exercise influence in discussions about what services should be public, how they should be operated, and whether the public practitioner serves customers or citizens. Though there will always be concern about the legitimacy of this role, many practitioners are in a position to shape the public sector by offering their knowledge to peers, citizens, and elected representatives trying to meet the challenge of governing in an economics–driven political culture. In doing so, they can serve the interests of public service and democratic will formation by keeping in mind the shifting and dynamic nature of the relationship of the market to the public sector and the importance of their actions in shaping the future.

## REFERENCES

Adams, G.B., P.V. Bowerman, K.M. Dolbeare, and C. Stivers. 1990. "Joining Purpose to Practice: A Democratic Identity for the Public Service." In *Images and Identities in Public Administration*, ed. H.D. Kass and B.L. Catron, 219–240. Newbury Park, CA: Sage.

Anderson, T.T., and R.G. Crawford. 1998. "Unsettling the Metaphysics of Neo-classical Micro-economic and Management Thinking." *International Journal of Public Administration* 21: 645–690.

Arnold, P.E. 1995. "Reform's Changing Role." *Public Administration Review* 55: 407–417.

Barber, B.R. 1984. *Strong Democracy: Participatory Politics for a New Age.* Berkeley: University of California Press.

———. 1998. *A Place for Us: How to Make Society Civil and Democracy Strong.* New York: Farrar, Straus, and Giroux.

Barrett, K., and R. Greene. 1998. "Customer Disorientation." *Governing* 11 (March): 62.

Bellah, R.N., R. Madsen, W.M. Sullivan, A. Swidler, and S.M. Tipton. 1985. *Habits of the Heart: Individualism and Commitment in American Life.* New York: Harper and Row.

Bellone, C.J., and G.F. Goerl. 1993. "In Defense of Civic-Regarding Entrepreneurship or Helping Wolves to Promote Good Citizenship." *Public Administration Review* 53: 396–398.

Bowles, S., and H. Gintis. 1986. *Democracy and Capitalism: Property, Community, and the Contradictions of Modern Social Thought.* New York: Basic Books.

Box, R.C. 1998. *Citizen Governance: Leading American Communities into the 21st Century.* Thousand Oaks, CA: Sage.

Campbell, I. 1996. *Understanding John Dewey: Nature and Cooperative Intelligence.* Chicago: Open Court Publishing.

Cheung, A.B.L. 1997. "The Rise of Privatization Policies: Similar Faces, Diverse Motives." *International Journal of Public Administration* 20: 2213–2245.

Cohn, D. 1997. "Creating Crises and Avoiding Blame: The Politics of Public Service Reform and the New Public Management in Great Britain and the United States." *Administration and Society* 29: 584–616.

Cope, G.H. 1997. "Bureaucratic Reform and Issues of Political Responsiveness." *Journal of Public Administration Research and Theory* 7: 461–471.

Downs, A. 1957. *An Economic Theory of Democracy.* New York: Harper and Row.

Dryzek, J.S. 1996. *Democracy in Capitalist Times: Ideals, Limits, and Struggles.* Oxford, UK: Oxford University Press.

Eberly, D.E. 1994. *Building a Community of Citizens: Civil Society in the 21st Century.* New York: University Press of America.

Eisenhardt, K.M. 1989. "Agency Theory: An Assessment and Review." *Academy of Management Review* 14: 57–74.

Fox, C., and H.T. Miller. 1995. *Postmodern Public Administration: Toward Discourse.* Thousand Oaks, CA: Sage.

Gerth, H.H., and C.W. Mills. 1958. *From Max Weber: Essays in Sociology.* New York: Oxford University Press.

Giddens, A. 1984. *The Constitution of Society: Outline of the Theory of Structuration.* Berkeley: University of California Press.

Golembiewski, R.T. 1977. "A Critique of 'Democratic Administration' and Its Supporting Ideation." *American Political Science Review* 71: 1488–1507.

Gurwitt, Rob. 1994. "Enterpreneurial Government: The Mornig After." *Governing* (May): 34–40.

Hood, C. 1996. "Beyond 'Progressivism': A New 'Global Paradigm' in Public Management?" *International Journal of Public Administration* 19: 151–177.

Hummel, R., and C. Stivers. 1998. "Government Isn't Us: The Possibility of Democratic Knowledge in Representative Government." In *Government Is Us: Public Administration in an Anti-Government Era,* ed., C.K. King and C. Stivers, 28–48. Thousand Oaks, CA: Sage.

Jennings, E.T., Jr. 1991. "Public Choice and the Privatization of Government: Implications for Public Administration." In *Public Management: The Essential Readings,* ed., J.S. Ott, A.C. Hyde, and J.M. Shafritz, 113–129. Chicago: Nelson-Hall.

Johnson, D.B. 1991. *Public Choice: An Introduction to the New Political Economy.* Mountain View, CA: Mayfield.

Kamensky, J.M. 1996. "Role of the 'Reinventing Government' Movement in Federal Management Reform." *Public Administration Review* 56: 247–255.

Kaplan, R.D. 1997. "Was Democracy Just a Moment?" *Atlantic Monthly* 280 (December): 55–80.

Kass, H.D., and B.L. Catron. 1990. *Images and Identities in Public Administration.* Newbury Park, CA: Sage.

Kelly, R.M. 1998. "An Inclusive Democratic Polity, Representative Bureaucracies, and the New Public Management." *Public Administration Review* 58: 201–208.

Kettl, D.F. 1997. "The Global Revolution in Public Management: Driving Themes, Missing Links." *Journal of Policy Analysis and Management* 16: 446–462.

King, C.S., and C. Stivers, C. 1998a. *Government Is Us: Public Administration in an Anti-Government Era.* Thousand Oaks, CA: Sage.

———. 1998b. "Introduction: The Anti-Government Era." In *Government Is Us: Public Administration in an Anti-Government Era,* ed. C.S. King and C. Stivers, 3–18. Thousand Oaks, CA: Sage.

Lan, Z., and D.H. Rosenbloom. 1992. "Public Administration in Transition?" *Public Administration Review* 52: 535–537.

Lasch, C. 1996. *The Revolt of the Elites and the Betrayal of Democracy.* New York: Norton.

Lynn, L.E., Jr. 1998. "The New Public Management: How to Transform a Theme into a Legacy." *Public Administration Review* 58: 231–237.

Lyons, W.E., and D. Lowery. 1989. "Governmental Fragmentation Versus Consolidation: Five Public-Choice Myths About How to Create Informed, Involved, and Happy Citizens." *Public Administration Review* 49: 533–543.

March, J.G. 1992. "The War Is Over, the Victors Have Lost." *Journal of Public Administration Research and Theory* 2: 225–231.

McSwite, O.C. 1997. *Legitimacy in Public Administration: A Discourse Analysis.* Thousand Oaks, CA: Sage.

Mikesell, J.A. 1995. *Fiscal Administration: Analysis and Applications for the Public Sector.* 4th ed. Belmont, CA: Wadsworth.

Moe, T.M. 1984. "The New Economics of Organization." *American Journal of Political Science* 28: 739–777.

Niskanen, W.A. 1971. *Bureaucracy and Representative Government.* Chicago: Aldine Atherton.

———. 1991. "A Reflection on Bureaucracy and Representative Government." In *The Budget-maximizing Bureaucrat: Appraisals and Evidence,* ed. A. Blais and S. Dion, 13–31. Pittsburgh: University of Pittsburgh Press.

Olson, M. 1965. *The Logic of Collective Action.* Cambridge, MA: Harvard University Press.

Osborne, D., and T. Gaebler. 1993. *Reinventing Government: How the Entrepreneurial Spirit Is Transforming the Public Sector.* New York: Penguin.

Ostrom, V. 1977. "Some Problems in Doing Political Theory: A Response to Golembiewski's 'Critique.'" *American Political Science Review* 71:1508–1525.

———. [1973] 1991. *The Intellectual Crisis in American Public Administration.* Tuscaloosa: University of Alabama Press.

———. [1991] 1994. *The Meaning of American Federalism: Constituting a Self-Governing Society.* San Francisco: ICS.

Peters, B.G., and J. Pierre. 1998. "Governance Without Government? Rethinking Public Administration." *Journal of Public Administration Research and Theory* 8: 223–243.

Phares, D. 1989. "Bigger Is Better, or Is It Smaller?" *Urban Affairs Quarterly* 25: 5–17.

*Public Administration Review.* 1993. "Forum on Public Administration and the Constitution." *Public Administration Review* 53: 237–267.

Ramos, A.G. [1981] 1984. *The New Science of Organizations: A Reconceptualization of the Wealth of Nations.* Toronto, Canada: University of Toronto Press.

Ross, B.H., and M.A. Levine. 1996. *Urban Politics: Power in Metropolitan America.* 5th ed. Itasca, FL: F.E. Peacock.

Savas, E.S. 1987. *Privatization: The Key to Better Government.* Chatham, NJ: Chatham House.

Schachter, H.L. 1997. *Reinventing Government or Reinventing Ourselves: The Role of Citizen*

*Owners in Making a Better Government.* Albany: State University of New York Press.

Schmidtz, D. 1991. *The Limits of Government: An Essay on the Public Goods Argument.* Boulder, CO: Westview.

Scott, W.G., and D.K. Hart. 1979. *Organizational America.* Boston: Houghton Mifflin.

Simon, H.A. [1945] 1997. *Administrative Behavior: A Study of Decision-Making Processes in Administrative Organizations.* 4th ed. New York: Free Press.

Skowronek, S. 1982. *Building a New American State: The Expansion of National Administrative Capacities, 1877–1920.* Cambridge, UK: Cambridge University Press.

Stillman, R.J. II 1991. *Preface to Public Administration: A Search for Themes and Direction.* New York: St. Martin's Press.

Terry, L.D. 1993. "Why We Should Abandon the Misconceived Quest to Reconcile Public Entrepreneurship with Democracy." *Public Administration Review* 53: 393–395.

———. 1998. "Administrative Leadership, Neo-Managerialism, and the Public Management Movement." *Public Administration Review* 58: 194–200.

Timney, M.M. 1998. "Overcoming Administrative Barriers to Citizen Participation: Citizens as Partners, Not Adversaries." In *Government Is Us: Public Administration in an Anti-Government Era,* ed. C.S. King and C. Stivers, 88–101. Thousand Oaks, CA: Sage.

Waldo, D. 1981. *The Enterprise of Public Administration: A Summary View.* Novato, CA: Chandler and Sharp.

Walters, J. 1998. "Did Somebody Say Downsizing?" *Governing* 11 (February): 17–20.

Waterman, R.W., and K.J. Meier. 1998. "Principal-Agent Models: An Expansion?" *Journal of Public Administration Research and Theory* 8: 173–202.

Weintraub, J. 1997. "Public/Private: The Limitations of a Grand Dichotomy." *Responsive Community* 7: 13–24.

White, L.D. 1954. *The Jacksonians: A Study in Administrative History, 1829–1861.* New York: Macmillan.

Wood, E.M. 1996. *Democracy Against Capitalism: Renewing Historical Materialism.* Cambridge, UK: Cambridge University Press.

Yankelovich, D. 1991. *Coming to Public Judgment: Making Democracy Work in a Complex World.* Syracuse, NY: Syracuse University Press.

# "THE BIG QUESTIONS OF PUBLIC ADMINISTRATION IN A DEMOCRACY"

## JOHN J. KIRLIN

*What are the big questions which should concern practitioners and students of public administration? Behn recently offered three big questions of public management, involving micromanagement, motivation, and measurement. Kirlin argues that the big questions of public administration in a democracy are different from those of public management and develops four criteria by which to judge big questions. Seven big questions of public administration in a democracy are offered, concerning: tools of collective action supporting a democratic polity; appropriate roles of nongovernmental collective action; tradeoffs between designs based on function versus geography; national versus loca l political arenas; when decisions are isolated from politics; balance among neutral competence, representativeness, and leadership; and societal learning.*

Behn's (1995) recent delineation of the "big questions of public management" makes an important and compelling argument that any field of inquiry should focus on major questions and should be driven by those questions, not diverted to more tractable questions nor limited by methodological orthodoxy. This is a strong critique of much of the contemporary public administration and public management literature, both in terms of the questions addressed and efforts to establish an orthodoxy of methods somehow judged to be most appropriate. Behn is careful to limit his suggestions to public management and to invite others to offer alternative definitions of big questions.

In this [reading], I respond to this invitation, arguing that the big questions of public administration in a democracy are quite different from the big questions of public management, a position also recently suggested by Newland (1994). To begin, I identify Behn's big questions, give an initial preview of the critique more fully developed later, and offer a listing of the seven big questions of public administration in a democracy.

From *Public Administration* Review 56, no. 5 (September/October 1996): 416–423. Copyright © 1996 American Society for Public Administration. Reprinted with permission.

## BIG QUESTIONS

Behn's three big questions for public management (1995, 315) are:

1. Micro management: *How* can public managers break the micromanagement cycle—an excess of procedural rules, which prevents public agencies from producing results, which leads to more procedural rules, which leads to . . . ?
' 2. Motivation: *How* can public managers motivate people (public employees as well as those outside the formal authority of government) to work energetically and intelligently toward achieving public purposes?
3. Measurement: *How* can public managers measure the achievements of their agencies in ways that help to increase those achievements?

These questions, asking "how" public managers can address each of the three big questions, place the public manager (implicitly operating from a public bureaucracy) at the center of the enterprise of governmental action. This approach, in common with others focused on public management, and much traditional public administration focused on public agencies, fails to confront adequately the issues of public administration in a democracy.

It gives management of organizations primacy over the democratic polity, a position effectively critiqued by Appleby (1949) nearly half a century ago. It similarly fails to address the argument of Rosenbloom (1983) that public administration theory includes three distinctive approaches—managerial, political, and legal—all of which must be incorporated if public administration theory is to be legitimate in this nation.

Primary attention here is focused on the important questions for public administration in a democracy, particularly the United States. Four criteria the big questions of public administration in a democracy must satisfy are: (a) achieving a democratic policy; (b) rising to the societal level, even in terms of values also important at the level of individual public organizations; (c) confronting the complexity of instruments of collective action; and (d) encouraging more effective societal learning.

Seven big questions emerge from the analysis:

1. What are the instruments of collective action that remain responsible both to democratically elected officials and to core societal values?
2. What are the roles of nongovernmental forms of collective action in society, and how can desired roles be protected and nurtured?
3. What are the appropriate tradeoffs between governmental structures based on function (which commonly eases organizational tasks) and geography (which eases citizenship, political leadership, and societal learning)?
4. How shall tensions between national and local political arenas be resolved?
5. What decisions shall be "isolated" from the normal processes of politics so that some other rationale can be applied?
6. What balance shall be struck among neutral competence, representativeness, and leadership?
7. How can processes of societal learning be improved, including knowledge of choices available, of consequences of alternatives, and of how to achieve desired goals, most importantly, the nurturing and development of a democratic polity?

## CRITIQUES OF MAKING PUBLIC BUREAUCRACY THE STARTING POINT OF PUBLIC ADMINISTRATION

Four critiques of making public bureaucracy the starting point of public administration in a democracy are offered here. These critiques are based on fundamental criteria to be met by any list of big questions of public administration in a democracy. Development of criteria by which any listing of big questions can be evaluated provides a foundation for this effort and a framework within which dialogue about questions central to the field can unfold. Indeed, development of the four criteria receives more attention here than do the seven big questions.

### The Big Questions of Public Administration in a Democracy Must Be Rooted in Achieving a Democratic Polity

One schism in the study and practice of public administration concerns the starting point: Is it public bureaucracy or a democratic policy? Public bureaucracy and democratic polity should be seen as complementary; both are needed in our society. But analysis and advocacy often start with and emphasize one perspective over the other. Those who make public bureaucracy the starting point focus largely on their operations. POSDCORB is an early iteration of this orientation, concerns with (organizational) economy, efficiency, and effectiveness are constants, and contemporary studies of public management are rooted in this tradition. Much of the contemporary reinvention effort seeks to improve performance of public bureaucracies (Osborne and Gaebler 1992; Gore 1993; Carroll 1995). Simon (1947) issued a challenge to the conventional proverbs of public administration on the basis of method, but he did not challenge the focus on organization as the core of the field. This remains the dominant subject focus of the field, as measured by articles appearing in *Public Administration Review* (Bingham and Bowen 1994).

The primary alternative starting point of a democratic polity may be more diffuse and less coherent, but it is also a major current in our history. It can be seen in the attention paid to citizenship in early education for public service. For example, the two first university-based programs with their own deans both included "citizenship" in their names: The Maxwell School of Citizenship and Public Affairs of Syracuse University (founded in 1924) and the School of Citizenship and Public Administration of the University of Southern California (founded in 1929) (Stone and Stone 1975). Waldo (1948) concluded that over-attention to perfecting administrative processes was harmful to democracy. Advocacy of the council-manager form of government, sought by early reformers, included both hopes for increased efficiency and effectiveness and enhancing democratic norms of citizen participation and political accountability (Stillman 1974, 9).

Appleby (1949, 43) argued strongly that politics and policy making interpenetrate public administration. He characterized common processes of public administration as an "eighth" political process:

> Arguments about the application of policy are essentially arguments about policy. Actual operations are conducted in a field across which mighty forces contend; the forces constitute policy situations. Administration is constantly engaged in a reconciliation of these forces, while leadership exerts itself in that process of reconciliation and through the interstices of the interlacing power lines that cut across the field.

Some of the lament of those troubled by the dominance of narrow methods of inquiry occasioned by embrace of behavioral social science approaches to the generic study of administration includes loss of the nuanced appreciation of public institutions operating in democratic polities (Fesler 1975). Vincent Ostrom (1974) critiqued the embrace of Woodrow Wilson as the founder of the field, arguing that the defining feature of the American political system is its constitutional design.

Moe and Gilmour (1995) pose the issue in terms of legitimate foundations for actions of public agencies, finding them in public law rather than in management theories. Business management practices developed within the constraints of judge-made common law intended to protect the rights and establish the obligations of private parties pursuing private interests. Practices in the governmental sector are" . . . founded on the body of the Constitution and the Bill of Rights and articulated by a truly enormous body of statutory, regulatory and case law to ensure continuance of a republican form of government and to protect the rights and freedoms of citizens at the hands of an all-powerful state" (Moe and Gilmour 1995, 135).

This position is similar to that of Rosenbloom (1983) who analyzed the legal perspective as distinct from the managerial and political perspectives. He found that the origin and values, suggestions for organizational structure, and views of individuals differ among the three perspectives. The legal perspective, derived from constitutional law, administrative law, and the "judicialization" of public administration, embodies three central values: procedural due process, individual substantive rights, and equity.

One of the fundamental flaws in making public bureaucracy the starting point of public administration is that it easily supports substitution of organizational concerns and measures of performance for those of a democratic polity, including the rule of law. Organizations may focus on effectiveness, efficiency, or economy. They may also focus on the impacts of organizations on their members or consumer satisfaction. But the ultimate value underpinning organization theory is organizational survival; any other values or constraints must be imposed from an external framework, intellectual, political, or legal.

Democratic polities must focus on: the sustained capacity of the political system itself to make and act on collective choices, opportunities for effective citizenship and political leadership, ensuring a limited government, nurturing the civic infrastructure necessary for collective action without public authority, providing the institutional structures necessary for operations of the economy, and protecting individual freedoms and rights. These are very different issues than those seen at the organizational level. How public administration can contribute to sustaining democratic polities is an issue long central to public administration, contributing several traditional big questions.

**The Big Questions of Public Administration in a Democracy**
**Must Rise to the Societal Level, Even in Terms of Values Also**
**Important at the Level of Individual Public Organizations**

There is reason to doubt that improving performance of individual public bureaucracies, or the operations of all agencies of the national government, for example, will aggregate to economy, efficiency, or effectiveness judged from a societal point of view. Even if each individual public organization approaches perfection, the totality of their effects may be found wanting. This is a consequence of the necessity of organizations to develop specialized competencies, to limit the range of their actions, and thus to have limited, partial effects. This phenomenon is well recognized in organization theory as occurring in the process of goal formation (March and Simon 1958). For James D. Thompson (1967), the critical issue confronting organization theory is how to reconcile organizations' drive for internal certainty, accomplished by limiting information, technology, structures, and processes, with the uncertainties and changes encountered externally.

A central theme of the policy implementation literature is the difficulty of achieving coherent actions in complex systems consisting of organizations and political entities each with independent capacity for action (Pressman and Wildavsky 1973; Ingram 1990). A constant complaint of states and local governments about the national government, and of local governments about states, is that fragmentation of policies, programs, and funding flows from above makes effective action at the point of impact extraordinarily difficult. Analysts of federalism wrestle with these issues frequently (Walker 1995), as do those who analyze complex systems of collective action (Dahl and Lindblom 1954; Kaufmann 1991). Advocates of deregulating government address some of these issues but commonly focus on individual public bureaucracies without addressing the total, societal impact of many such agencies (Wilson 1989, 365–378).

A simple question challenges the emphasis upon single organizations dominant in much public management and public administration literature, revealing that such a focus is ultimately inadequate. That question is: "If each and every single public organization performed ideally as seen in your theory, would the results be societally desirable?" The response must be, No, as public organizations cannot be assumed to be subject to a blind hand of external forces. Public organizations require external direction and constraint to achieve societally desired results. How this can be achieved is the focus of several traditional big questions.

## The Big Questions of Public Administration in a Democracy Must Confront the Complexity of Instruments of Collective Action

While traditional public administration and public management focus largely on government, the instruments available for collective action to a modern society characterized by a limited, democratic government and a market economy are not limited to government. They include also the market itself (Lindblom 1977), the rich tapestry of institutions that comprise the civic infrastructure (Putnam 1993), and regulation, grants-in-aid, government corporations, and other approaches (Salamon 1989). Without government, society lacks the property rights, monetary system, legally enforceable contracts, forms of business organization and legally sanctioned practices, or processes to resolve conflicts, required of a modern economy. Government can also encourage civic infrastructure, through legalizing collective action without an explicit grant of governmental authority, through supportive provisions of tax codes, or through contractual and coprovision arrangements for action.

At the most fundamental level, public action creates the institutional frameworks through which individuals are born; live; marry and divorce; parent; worship; enter into work relationships; create and run businesses; buy and sell property; exercise political voice and choice; join with others to pursue recreation, art, or a vision of the good society; are held accountable for their actions; and resolve conflicts. In essentially any dimension of human activity, ranging from housing and education through transportation and personal safety or recreation to environmental quality, examination reveals some areas of direct governmental production of a service but much greater impacts through governmental shaping of the legal forms of collective action, establishment of rights and responsibilities, boundaries of acceptable behaviors and practices, and tax codes. Within the framework constructed by these governmental policies and the relevant constraints of critical "private" institutions such as financial institutions, themselves working within frameworks of public policy, private industries emerge, usually making direct private expenditures larger than those made by government in the area. Of course, private interests, firms, and associations influence the public policy frameworks within which they operate, but they do so within a political process in which other parties also participate.

Moreover, social institutions are active, being altered purposively in response to changed constraints, opportunities, and preferences. For example, the system of state and local public finances in California has evolved through ten identifiable iterations since passage of proposition 13 in 1978. One major adaptation involved local governments, investment banks, and developers, creating new processes and instruments to pay for infrastructure required to accommodate growth. Another involved a voter-approved initiative pushed by school interests requiring allocation of increasing percentages of state revenues to K–14 education, which stimulated a voter-approved initiative supported by cities, counties, and other interests to moderate those effects (Kirlin, Chapman, and Asmus 1994).

An area where public administration needs a better conceptualization of its roles is precisely in the contributions government and public administration make to the creation, nurturing, and restructuring of complex functional systems. But even within the constraints of traditional public administration, the variety of available instruments of collective action received considerable recognition (Fesler 1975), and this variety is recognized in traditional big questions.

### The Big Questions of Public Administration Must Address
### Processes by Which Societal Learning Is Made More Effective

The discussion above suggests an important challenge for advocates of increasing the importance of research in public administration (White and Adams 1994). The challenge emerges directly from the necessity to improve the capacity to achieve desired results in complex systems, where governments are creators and shapers but have severe limits on their direct actions, and the central values are those of democracy in which citizen values and choices are ultimately controlling. In this situation, science—as organized, structured inquiry—is useful but limited. Adding subjectivist, interpretist, or critical styles of research to public administration may well be desirable, but these proposals remain focused on improving the enterprise of science rather than improving the broader processes by which society learns.

It is well known that policy makers and public administrators often do not make effective use of available knowledge derived from scientific research (Lindblom and Cohen 1979). However, the challenge posed here is more fundamental, concerning limits upon the insights available from science. In some instances, available science requires fragmenting problems in ways that do not provide ready insight into operations of the relevant real systems. Separation of Environmental Protection Agency (EPA) science (and programs) into air, water, pesticides, and toxic and solid waste categories provides an example (National Academy of Public Administration 1995, 16–18). Partial insight and some progress toward policy goals occur under these divisions, but uncertainty and conflict among sciences and programs are common in application to specific geographical places. In other instances, continuing controversy among scientists suggests that knowledge is not perfect. Examples can be found in the disputes over global cooling or global warming (Stone 1993) or about how best to teach children.

When the question moves from what should be done to how, the issues of complexity of action will frequently confound science with the literature on public policy implementation bearing witness to the difficulties encountered (Mazmanian and Sabatier 1983). The ultimate challenge arises when increasing citizen understanding is confronted. Citizens do gain information from science, with general acceptance of the risks of smoking providing an example. But the linkages among information acquisition, attitude, personal action, and political voice and choice are complicated. Some issues are "wicked," characterized by conflicting values among citizens and imperfect or conflicting understanding among analysts. Yankelovich (1991) distinguishes between public opinion and public judgment, with the former being more fleeting. Where the public reaches judgment citizens believe themselves to understand the issue as it affects them, they have a preferred outcome (and, often, associated governmental actions), they are willing to accept the consequences of these choices, and these judgments endure.

Citizen understandings are important not only because they influence or control eventual policies in a democratic polity. When human behaviors and the actions of organizations or institutions are involved, Lindblom (1990) argues that social sciences are intimately rooted in categories and values of ordinary lives and language. From this perspective, social sciences, and much of whatever science is available to public administration, cannot escape close relationship with their nominal subjects, humans.

In this situation, the interests encompassed in public administration must expand beyond traditional science and also beyond processes to increase the use of science in policy making and implementation, to encompass how society at large learns. Only as society, broadly defined, learns what it wishes to pursue and how to achieve those desired outcomes more reliably can citizens participate effectively in policy choices and in collective action by their informed, as opposed to coerced, bought, or manipulated, actions (Dewey 1927).

Appleby (1949, 155–156) again provides a strong rationale for the importance of expecting public administration in a democracy to positively contribute to societal learning.

> In every case, the principal roles of the especially responsible citizens who are also public officials are: to bring into focus—to resolve and integrate—these popularly felt needs; to give specific form to responses of the government . . . to inject foresight and concern for factors nor readily visible to citizens at large. . . . The process produces a kind of political logic unlike any other logic, the validity of which is tested or attested by popular consent and governmental survival. . . . But it is constantly adjusted by repetitive phases. . . .
>
> There would be a grave danger, for example, in straining too hard for "rationality" and minimizing the political, for it is the political that makes room for the whole of human potential, including the rational potential.

A framework to analyze the range of societal institutions within which societal learning occurs can be developed from an approach suggested by Weschler (no date). He identifies faith, tradition, mass media, science, politics, and ideology as arenas in which societal learning can occur, with different mixes operative in any society at any time. This framework can be expanded by addition of the categories "markets" and "professional practices," resulting in these categories: faith, tradition, mass media, professional practices, science, markets, politics, and ideology.

These arenas are shown in Table [4.3.1] with the type of proof offered for each for the veracity of an insight. To illustrate the relevance of the various arenas in the practice and study of contemporary public administration, an easily understood example of learning from each arena is offered. Of course, not all accept the truthfulness of all the illustrative examples offered, but each is a recognizable, powerful factor in the practice (and study) of public administration.

The important conclusion of this discussion and examination of Table [4.3.1] is that science is *one* of several arenas in which societal learning occurs. Increasing the influence of science is a strategy to increase rationality in society but likely to be constrained by limits on scientific understanding and by resistance to turning contentious choices over to elites. For most issues confronting society, any political decision-making body, or any public administrator, science is unlikely to provide definitive guidance. A perspective on societal learning that includes more than science is needed just to understand the factors shaping decisions and actions in the public sector.

Dewey (1927) provides a classical philosophical foundation for the position that societal learning is critical. The issue is addressed in seminal works on civic culture (Almond and Verba 1965) and in contemporary examinations of how cities encourage or discourage citizen participation, with Berry, Portney, and Thomson (1993) finding that cities which institutionalize neighborhood participation in policy processes generate more informed, efficacious, and participative citizenry whose inputs are heard by both administrators and elected officials. The Kettering Foundation has supported work on effectively involving citizens in public issues: examples include the writing of its president (Mathews 1994) and the work of Yankelovich (1991), which it supported. The National Civic League (1994) has long been committed to development of strong citizenship, developing a ten-item "Civic Index," which is the template used for its annual "All-American Cities" competition. Putnam (1993) offers a theoretically sophisticated and empirically supported rationale for the critical contribution of social capital developed by citizens' joint activities outside of government to successful functioning of government and administrative processes.

Beyond mere passive understanding, the argument advanced here is that an important challenge for public administration in a democracy is to improve the whole of societal learning. This provides even stronger reason that discussion regarding inquiry in the field must not be limited to any narrow focus on method. It is possible to "improve" the learning that can occur in society, that is, make it more likely to improve effective understanding of what is going on, appreciation of choices, and of strategies for action that improve odds of achieving desired goals. The big questions public administration in a democracy must engage this role.

## BIG QUESTIONS OF PUBLIC ADMINISTRATION IN A DEMOCRACY

The big questions of public administration in a democracy must satisfy the four criteria developed above. In general, what is required is moving up in levels of abstraction beyond a favored instrument of collective action (the public bureaucracy) and a favored approach to inquiry (science) to broader processes. Public bureaucracies are *one* instrument of collective action; our commitment should be to develop, manage, nurture, change, and improve the range of instruments of collective action which achieve societal goals. Similarly, science is *one* approach to social inquiry and our commitment should be to develop, manage, nurture, change, and improve the range of processes through which societal learning occurs. To the extent public administration limits its scope of action and learning to public bureaucracies and science, it limits its relevance and impact. To limit scope of learning and action is to elevate instrument above purpose, a foolish choice. By analogy, a society focused on instrument rather than goal would have remained limited to walking as its mode of transportation and to consulting oracles as its approach to learning.

Table 4.3.1

**Arenas of Societal Learning Relevant to Public Administration**

| Arena | Proof offered | Illustrative example |
| --- | --- | --- |
| Religion | Faith (no effort to verify) | High value of individual lives |
| Tradition | Societal custom | Hierarchy is preferred organizational structure |
| Literary and mass media | "Good" values | "No more Vietnams" |
| Professional practice | Expert status; license current services base | Building budgets from |
| Science | Verifiability | Rule making in EPA or FDA |
| Market | Price, demand, and supply | Preference for single family housing (in USA) |
| Politics | Political decision | Intergenerational redistribution is good (Social Security, Medicare) |
| Ideology | Fits ideology | Business practices are better than governmental practices |

To move beyond a general exhortation to satisfy the four criteria, students and practitioners of public administration have developed seven big questions identified at the beginning of the article, which are easily recognized as long central to American public administration:

1. What are the instruments of collective action that remain responsible both to democratic political processes and to core societal values? This question is posed at the organizational level, for example, in the positions argued by Friedrich (1940) and Finer (1941) on internal norms of professionalism versus external controls in achieving accountability in public agencies. At the policy level, a recurring choice is between instruments which rely more on elite judgment and authoritatively directed compliance behaviors versus those that emphasize citizen judgment and empowering strategies (Dahl and Lindblom 1954; Shonfield, 1965).

2. What are the roles of nongovernmental forms of collective action in society and how can desired roles be protected and nurtured? This is a central question of any limited form of government. Some theorists emphasize the market as the primary nongovernmental alternative for collective action, with Lindblom (1977) providing a reasoned examination of this alternative. Other theorists emphasize nonmarket forms of nongovernmental collective action, as seen in the contemporary work of Berger and Neuhaus (1977), Elinor Ostrom (1990), Putnam (1993), Etzioni (1993), and Mathews (1994), among others.

3. What are the appropriate tradeoffs between governmental structures based on function (which commonly eases organizational tasks) and geography (which eases citizenship, political leadership, and societal learning)? The decades-long debates about

structures for collective action in metropolitan areas offer an example of these issues (Erie, Kirlin, and Rabinovitz 1972; Kirlin 1996a) as do the similar long-lived debates about block grants (Posner 1995). Recent efforts to add civic (John 1994) or community-based (Hansen 1995) perspectives to environmental policy making are another example of this tension.

4. How shall tensions between national and local political arenas be resolved? This is a classic question central to the adoption of our Constitution. Elazar (1974) and Vincent Ostrom (1987) are among those who regularly address this question. The journal *Publius* takes this issue as its central focus as has the U.S. Advisory Commission on Intergovernmental Relations. Administration of virtually all domestic policies and programs includes features, dynamics, and conflicts rooted in this question (Kincaid 1993).

5. What decisions shall be "isolated" from the normal processes of politics so that some other rationale can be applied? This is a recurring issue regarding expertise and how to break through paralysis on contentious issues. Nathan (1995) reports that the National Commission on State and Local Public Service (1993) made the explicit decision that some decisions needed to be isolated from the turmoil of pluralist politics. The base-closure process is an example of this approach. Appleby (1949, 162) critiques the move to take issues out of politics as often being intended to take choice out of public control, to transfer power to special interests.

6. What balance shall be struck among neutral competence, representativeness, and leadership? This is the classic formulation of Kaufman (1956), central to the design of institutions and policies. The *PAR* Symposium on Public Administration in Europe (Kickert et al. 1996) illustrates the continuing relevance of Kaufman's three contending forces in seven European democracies, without use of his terminology.

7. How can processes of societal learning be improved, including knowledge of choices available, of consequences of alternatives, and of how to achieve desired goals, most importantly, the nurturing and development of a democratic polity? Schachter (1995) has demonstrated how the Bureau of Municipal Research advocated citizens as owners of government and efficient citizenship in 1908–1913. More recently, Gawthrop (1984) and Barber (1984) address the importance of encouraging effective citizenship.

These seven big questions of public administration in a democracy are both researchable and actionable. Each has a long legacy in the field of public administration, enduring through changes in political regimes or academic fashions. Collectively, they satisfy the four criteria; indeed, each question touches on at least some facet of all four criteria.

In contrast, elevating public management to primacy in our field invites design and management of government to satisfy internal organizational needs of public agencies, sometimes constrained by directives of legislators, budget allocations and courts, and sometimes seeking to respond to customers. In this nation, public administration is not only subordinated to the values and constraints of a democratic polity but has a responsibility to protect, nurture, and to develop that policy. Appleby concludes *Policy and Administration* with the sentence: "Public administration is one of a number of basic political processes by which this people achieves and controls governance" (1949, 170).

From this perspective, questions derived from a public management perspective can only become "big" as they are cast within the values of a democratic polity. The Behn (1995) questions can be reposed as interesting, and even reasonably big:

1. "Micro management" becomes "Function Bias": How can institutions be developed that overcome the function bias cycle—excessive use of single-function-focused policies, programs, organizations, regulations, funding flows—that enfeeble geographically based political systems and civic infrastructure, while significantly increasing uncertainty and transaction costs for those who must live lives and conduct business and community affairs across functional boundaries?

2. Motivation: How can institutions and policies be developed that empower citizens, individually and in civic organizations, businesses and nonprofits, and also governments and their employees, to work energetically, intelligently, and collaboratively, toward politically legitimated and socially valued purposes?

3. Measurement: How can society measure its overall progress and the contribution or hindrance contributed by major institutions, including business, civic infrastructure and government, and various policies, toward desired goals and use that knowledge to learn regarding future choices of goals and strategies of action?

These reformulations of Behn's three questions cut across the seven big questions identified above, with number 1 linking respectively to questions one, three, and four; number 2 with one and two; and number 3 with question seven. Questions five and six are missed.

## BIG QUESTIONS ENDURE

Public administration is characterized by periodic changes in dominant conceptualizations of what government does, of the roles of public administration, and of appropriate styles of inquiry. Big questions endure, and recent reconceptualizations leave the seven big questions as critical and central to the field of public administration in a democracy. Lan and Rosenbloom (1992, 537) conclude that even "marketized" public administration would retain features of democracy, including, for example, legislatures, responsiveness to citizens, courts to adjudicate conflicts, constitutional integrity, robust substantive rights, and equal protection. John et al. (1994) analyze experiences with reinvention to develop prescriptions, which include an emphasis on strengthening state-local capacity and engaging and empowering citizens. Kirlin's (1996b) more integrated, abstract perspective explicitly makes democratic political attributes of places a central focus of the ways in which government and public administration create value for society.

As long as democracy is valued, the big questions of public administration must go beyond the big questions of public management. Even the contemporary antigovernment rhetoric does not abandon democracy. However, public administration cripples its role in society if understood primarily in terms of managing public agencies.

These suggestions concerning the four criteria by which big questions should be judged and the seven big questions of public administration in democracy are offered with knowledge that other perspectives exist. Other formulations of essentially similar arguments are possible. These four criteria and seven questions do serve to clearly demarcate the big questions of public administration in a democracy as distinct from the big questions of public management.

## NOTE

The author is indebted to Ross Clayton, Alexis Halley, Mary Kirlin, Chester Newland, Lou Weschler, and anonymous referees for comments

## REFERENCES

Almond, Gabriel A., and Sidney Verba. 1965. *The Civic Culture: Political Attitudes and Democracy in Five Nations, an Analytical Study.* Boston: Little, Brown.

Appleby, Paul H. 1949. *Policy and Administration.* University: University of Alabama Press.

Barber, Benjamin. 1984. *Strong Democracy: Participatory Politics for a New Age.* Berkeley: University of California Press.

Behn, Robert D. 1995. "The Big Questions of Public Management." *Public Administration Review* 55 (July/August): 313–324.

Berger, Peter L., and Richard John Neuhaus. 1977. *To Empower People: The Role of Mediating Structures in Public Policy.* Washington, DC: American Enterprise Institute.

Berry, Jeffrey M., Kent E. Portney, and Ken Thomson. 1993. *The Rebirth of Urban Democracy.* Washington DC: The Brookings Institution.

Bingham, Richard D., and William M. Bowen. 1994. "Mainstream Public Administration Over Time." *Public Administration Review* 54 (March/April): 204–208.

Carroll, James D., 1995. "The Rhetoric of Reform and Political Reality in the National Performance Review." *Public Administration Review* 55 (May/June): 302–312.

Dahl, Robert A., and Charles E. Lindblom. 1954. *Politics, Economics and Welfare.* Chicago: University of Chicago Press.

Dewey, John. 1927. *The Public and Its Problems.* Chicago: Swallow Press.

Elazar, Daniel J., ed. 1974. *The Federal Polity.* New Brunswick, NJ: Transaction Books.

Erie, Steven P., John J. Kirlin, and Francine F. Rabinovitz. 1972. "Can Something Be Done? Propositions on the Performance of Metropolitan Institutions." In *The Governance of Metropolitan Regions, no. 1,* series ed. Lowdon Wingo. Baltimore: Johns Hopkins University Press.

Etzioni, Amitai. 1993. *The Spirit of Community.* New York: Crown.

Fesler, James W. 1975. "Public Administration and the Social Sciences: 1946 to 1960." In *American Public Administration: Past, Present, Future,* ed. Frederick C. Mosher, 97–141. University: University of Alabama Press.

Finer, Herman. 1941. "Administrative Responsibility in Democratic Government." *Public Administration Review* 1 (Summer): 336–350.

Friedrich, Carl J. 1940. "Public Policy and the Nature of Administrative Responsibility." In *Public Policy: A Yearbook of the Graduate School in Public Administration, Harvard University, 1940,* ed. C.J. Friedrich and Edward S. Mason. Cambridge, MA: Harvard University Press.

Gawthrop, Louis. 1984. "Civis, Civitas and Civilitas: A New Focus for the Year 2000." *Public Administration Review* 44, special issue (March): 101–107.

Gore, Albert. 1993. *Creating A Government that Works Better and Costs Less: Report of the National Performance Review.* Washington, DC: U.S. Government Printing Office.

Hansen, Fred. 1995. "Community-Based Environmental Protection—Key Attributes and Next Steps." Memorandum from the deputy administraror to assistant administrators, general counsel, regional administrators, and associate administrators, August 24. Washington, DC: Environmental Protection Agency.

Ingram, Helen. 1990. "Implementation: A Review and Suggested Framework." In *Public Administration: The State of the Discipline,* ed. Naomi B. Lynn and Aaron Wildavsky, 462–480. [Chatham, NJ: Chatham House.]

John, DeWitt. 1994. *Civic Environmentalism: Alternatives to Regulation in States and Communities.* Washington, DC: Congressional Quarterly Press.

John, DeWitt et al. 1994. "What Will the New Governance Mean for the Federal Government?" *Public Administration* Review 54 (March/April): 170–175.

Kaufman, Herbert. 1956. "Emerging Conflicts in the Doctrines of Public Administration." *American Political Science Review* 50 (December): 1057–1073.

Kaufmann, Franz-Xaver, ed. 1991. *The Public Sector-Challenge of Coordination and Learning.* New York: Walter de Gruyter.

Kickert, Walter J.M. et al. 1996. "Changing European States; Changing Public Administration." *Public Administration* Review 56 (January/February): 65–103.

Kincaid, John. 1993. "From Cooperation to Coercion in American Federalism. Housing, Fragmentation, and Preemption, 1780–1992." *Journal of Law and Politics* 9 (Winter): 333–433.

Kirlin, John J. 1996a. "Emerging Regional Organizational and Institutional Forms: Strategies and Prospects for Transcending Localism." In *Globalization and Decentralization: Institutional Contexts, Policy Issues and Intergovernmental Relationships in Japan and the United States,* ed. Jon Jun and Deil Wright, 107–133. Washington, DC: Georgetown University Press.

———. 1996b. "What Government Must Do Well: Creating Value for Society." *Journal of Public Administration Research and Theory* 6 (January): 161–185.

Kirlin, John J., Jeffrey I. Chapman, and Peter Asmus. 1994. "California Policy Choices: The Context." In *California Policy Choices, vol. 9,* ed. John J. Kirlin and Jeffrey I. Chapman, 1–23. Sacramento: University of Southern California Press.

Lan, Zhiyong, and David H. Rosenbloom. 1992. "Public Administration in Transition?" *Public Administration Review* 52 (November/December): 535–537.

Lindblom, Charles E. 1977. *Politics and Markets. The World's Political-Economic Systems.* New York: Basic Books.

———. 1990. *Inquiry and Change: The Troubled Attempt to Understand and Shape Society.* New Haven, CT: Yale University Press.

Lindblom, Charles E., and David K. Cohen. 1979. *Usable Knowledge: Social Science and Social Problem Solving.* New Haven, CT: Yale University Press.

March, James G., and Herbert A. Simon. 1958. *Organizations.* New York: Wiley.

Mathews, David. 1994. *Politics for People: Finding a Responsible Public Voice.* Chicago: University of Illinois Press.

Mazmanian, Daniel A., and Paul Sabatier. 1983. *Implementation and Public Policy.* Glenview, IL: Scott, Foresman.

Moe, Ronald C., and Robert S. Gilmour. 1995. "Rediscovering Principles of Public Administration: The Neglected Foundation of Public Law." *Public Administration Review* 55 (March/April): 135–146.

Nathan, Richard P. 1995. "Reinventing Government: What Does It Mean." *Public Administration Review* 55 (March/April): 213–215.

National Academy of Public Administration. 1995. *Setting Priorities, Getting Results. A New Direction for EPA.* Washington, DC: NAPA.

National Civic League. 1994. *All-America City Yearbook.* Denver: National Civic League.

National Commission on State and Local Public Service. 1993. *Hard Truths/Tough Choices: An Agenda for State and Local Reform.* Albany, NY: Nelson A. Rockefeller Institute of Government.

Newland, Chester A. 1994. "A Field of Strangers in Search of a Discipline." Public *Administration Review* 54 (September/October): 486–488.

Osborne, David, and Ted Gaebler. 1992. *Reinventing Government.* New York: Addison-Wesley.

Ostrom, Elinor. 1990. *Governing the Commons: The Evolution of Institutions for Collective Action.* New York: Cambridge University Press.

Ostrom, Vincent. 1974. *The Intellectual Crisis in American Public Administration.* University: University of Alabama Press.

———. 1987. *The Political Theory of a Compound Republic: Designing the American Experiment.* 2nd ed. Lincoln: University of Nebraska Press.

Posner, Paul L. 1995. *Block Grants: Issues in Designing Accountability Provisions.* Washington, DC: General Accounting Office (GAO/AMID-95226, September).

Pressman, Jeffrey L., and Aaron Wildavsky. 1973. *Implementation.* Berkeley: University of California Press.

Putnam, Robert D. 1993. *Making Democracy Work.* Princeton, NJ: Princeton University Press.

Rosenbloom, David H. 1983. "Public Administration Theory and the Separation of Powers." *Public Administration Review* 43 (May/June): 219–227.

Salamon, Lester M., ed. 1989. *Beyond Privatization: The Tools of Government Action.* Washington, DC: Urban Institute Press.

Schachter, Hindy Lauer. 1995. "Reinventing Government or Reinventing Ourselves: Two Models for Improving Government Performance." *Public Administration Review* 55 (November/December): 530–537.

Shonfield, Andrew. 1965. *Modern Capitalism: The Changing Balance of Public and Private Power.* London: Oxford University Press.

Simon, Herbert A. 1947. *Administrative Behavior.* New York: Macmillan.

Stillman, Richard J. II. 1974. *The Rise of the City Manager: A Public Professional in Local Government.* Albuquerque: University of New Mexico Press.

Stone, Alice B., and Donald C. Stone. 1975. "Appendix: Case Histories of Early Professional Education Programs." In *American Public Administration: Past, Present, Future,* ed. Frederick C. Mosher, 268–290. University: University of Alabama Press.

Stone, Christopher D. 1993. *The Gnat Is Older than Man: Global Environment and Human Agenda.* Princeton, NJ: Princeton University Press.

Thompson, James D. 1967. *Organizations in Action.* New York: McGraw Hill.

Waldo, Dwight 1948. *The Administrative State.* New York: Ronald Press.

Walker, David B. 1995. *The Rebirth of Federalism: Slouching Toward Washington.* Chatham, NJ: Chatham House.

Weschler, Lou. No date. "Forms of Social Inquiry." Tempe: Arizona State University, class teaching handout.

White, Jay D., and Guy Adams, eds. 1994. *Research in Public Administration: Reflections on Theory and Practice.* Thousand Oaks, CA: Sage.

Wilson, James Q. 1989. *Bureaucracy.* New York: Basic Books.

Yankelovich, Daniel. 1991. *Coming to Public Judgment.* Syracuse, NY: Syracuse University Press.

# PART V

# THE PUBLIC SERVICE PRACTITIONER IN A DEMOCRATIC SOCIETY

In a sense, Part V contains the "punch line" of the book. From the beginning, we have been searching for answers to the question, *As public service practitioners, whom do we serve, and for what purposes?* The argument has been offered that knowledge of the history and nature of the surrounding society is useful for understanding the nature of contemporary public administration and deciding what actions to take in the future.

The readings here join the social, economic, and political environment of public administration with administrative theory and practice, expressing the result both as description of, and normative prescriptions for, the role of administrators in society. This material may in part be thought of as building on the book *Images and Identities in Public Administration* (1990), edited by Henry Kass and Bayard Catron. In their concluding "Epilogue," Kass and Catron summarize perspectives on the field in seven "images" of public administrators: functionary, opportunist/ pragmatist, interest broker/market manager, professional/expert technician, agent/trustee, communitarian facilitator, and transformational social critic. Like Kass and Catron's book, much writing in public administration is about relationships between career public professionals and the society they serve. The point here is to integrate knowledge of the development of American society with awareness of these relationships, helping practitioners improve their work and society as well.

In a field of applied practice, the value of ideas is in part determined by their usefulness in assisting people as they carry out their professional tasks. In the ways they engage citizens, and organize and deliver services on a daily basis, many public-service practitioners are in a position to directly affect the quality of democracy, citizenship, and public policy. In doing so, they become part of an historical chain of people doing public work within the social framework of a nation in which, as outlined by the authors in this book, aversion to governmental control is mixed with desire for a sense of community. In the first reading, Richard Box outlines the development of the public service role in the United States, with particular attention to the local government practitioner. Contemporary full-time, career public service grew out of volunteer community work, accumulation of knowledge of governmental practices, and the increasing size and complexity of urban areas.

Questions asked today about the role of public practitioners in society, such as whether they should be neutral functionaries, keepers of expert knowledge, equal partners in governance with citizens and elected officials, or facilitators of democratic discussion and public decision making, are symptomatic of the difficulty of the task of balancing values inherent in society with the need for competent and effective delivery of public services. There is no "right" answer to these questions. Each person develops role preferences over time, what is appropriate in one situation may not fit another, and it is common for practitioners to use elements of more than one role type, for example serving as an expert while also facilitating citizen discussion and self-determination.

The second reading, by Richard Box, Gary Marshall, B.J. Reed, and Christine Reed, continues a theme from Part IV, the effects of marketlike economic thought on the public sector. It links this theme to concepts of democracy and the possibility that public practitioners might participate with citizens in creation and implementation of public policy with these ideas in mind. Finally, Cheryl Simrell King and Camilla Stivers, in a chapter from their book, *Government Is Us: Public Administration in an Anti-Government Era*, portray public practitioners as finding ways to preserve a sense of the value of public service in the midst of citizen mistrust of government and the emphasis on running government like a business. This constructive and optimistic thought is a fitting way to end this book.

## REFERENCE

Kass, Henry D., and Bayard L. Catron. 1990. "Epilogue." In *Images and Identities in Public Administration*, ed. Henry D. Kass and Bayard L. Catron, 241–251. Newbury Park, CA: Sage.

# Reading 5.1

# "PRACTITIONERS"

## Richard C. Box

## THE CONTEXT OF PUBLIC SERVICE

Today's community political environment is a challenging one for the public service practitioner. As we have seen, practitioners must navigate between entrenched elites and demands for decision-making access by citizens. They must find their way between the representational failure of governing-body dominated systems and the lack of accountability and rationality in decentralized or citizen-dominated systems. They are called on, simultaneously, to offer expert advice to elected officials, exhibit deference to those same officials, provide efficient daily administration of public services with decreasing resources, and present a pleasing customer-service face to citizens. This is taking place in a nationwide macro-environment of dislike of government, tax revolt, downsizing, privatization, and bureaucrat bashing fueled by political leaders at all levels of government.

These phenomena are not surprising, given the return to values of the past . . . . With movement back toward localism, small and responsive government, and the professional as adviser rather than controller of public agencies, public bureaucracies at all levels are experiencing pressure to change. The size and complexity of American government makes it difficult to generalize about the impact of these trends in a particular agency or on a specific public service practitioner.

In each local governmental unit, practitioners work within a particular community orientation toward public policy and governance . . . an accessible and open or excluding and closed governance system; an emphasis on marketplace or living space values; a desire for a large or restricted role for government; and acceptance of, or resistance to, professionalism. These orientations affect how practitioners approach their work and the success they have in serving their communities. Many people who have served in several local governments have observed elements of each of these orientations and, within a particular community, they have watched orientations change over time.

The complexity of the overall environment of public service, the variation in community policy orientation, and the rate of change in both the overall environment and in communities make it difficult to generalize about the nature of public service. Despite this, meaningful generalizations can be made in the contemporary setting that are helpful in sorting out how practitioners can best serve the public. These generalizations are based on the history of the practice of public service and its meaning to both practitioners and citizens. As we consider the Citizen Governance concept of the practitioner role toward the end of the chapter, it becomes apparent that the principles of community governance are well served by professional practice solidly grounded in the practices developed in the past, combined with a willingness to adapt to the future.

## EVOLUTION OF THE PUBLIC SERVICE PRACTITIONER

Early in the nation's history there were no administrative agencies to perform community services. Individuals took care of themselves and, in situations that required collective action, citizens or groups of citizens volunteered their time or were paid small stipends for their efforts. In the typical New England community, citizens were selected by their peers at the annual town meeting to perform tasks such as keeping the peace, returning wayward livestock, maintaining the roads, collecting taxes, and looking after other community needs (Cook 1976, 23–62).

Through the eighteenth century and into the nineteenth, it became common for people who took on such tasks to be paid enough to seek the work rather than to perform it only from a sense of duty. Gradually, as cities grew and volunteer or part-time workers could not handle the quantity and complexity of the work to be done, formal organizational structures emerged to provide fire, police, public health, infrastructure, schools, and other services. This history includes solving problems of technical methods, transitioning from volunteers to specialists through training and acquired experience, and dealing with occasional corruption and lack of resources. Today, we expect technical capability and efficiency from local public administrative bodies. During the twentieth century they have developed into the familiar form of well-organized bureaucratic agencies with a body of professional knowledge, expertise, and the funds to carry out their missions.

The contemporary positions held by career public service practitioners evolved from the community-minded work of early citizen volunteers. Full-time, trained dedication to public service is a model of community activity that grew slowly and with difficulty through trial and error over a period covering more than a century. The local practitioner's role evolved from that of citizen volunteer, but because of the complexity of contemporary governance and the full-time nature of the role, practitioners are now and forever different from citizens. Cooper (1984, 307) argued that "the ethical identity" of the public administrator is "that of the citizen who is employed as one of us to work for us; a kind of professional citizen ordained to do that work which we in a complex large-scale political community are unable to undertake ourselves."

This view recognizes the evolutionary connection between the citizen and the public professional. It also reflects the trend in the twentieth century . . . toward separating public service work from citizens, making it the province of public professionals. It relates in large part to the national level of governance, where the size of agencies and remoteness from citizen accountability often make a more direct citizen-practitioner connection impractical. At the local level, organizational size is less limiting to the role of the citizen, so that the practitioner, rather than acting instead of citizens as a surrogate, can become a facilitator of direct citizen action.

## THE PUBLIC SERVICE ROLE

The institution of community governance is changing, from domination by the ideal of professionalized, scientific administration to the ideal of citizen self-determination. This is part of the long-term change nationwide in citizen expectations of government we have identified as returning to values of the past. The impact of this long-term change on the practitioner can be significant, though it is somewhat limited by the day-to-day reality of delivering services such as repairing the streets, investigating crimes, building parks, and so on. Except for advances in technology (such as paving machines, sophisticated crime detection equipment, and computers), the routine performance of much public work is as it always has been.

Changes in the surrounding macro-level political and economic environment mean that pressures for cost efficiency and responsiveness to citizen preferences are increasing, pushing the practitioner into new ways of approaching old tasks. Though the work is not much different, the process of deciding what to do and how to fund it is very different. It includes dialogue, deliberation, and negotiation requiring interpersonal and administrative skills seemingly unrelated to the technical tasks of service delivery.

Our contemporary understanding of the role of the public service practitioner in this setting does not spring fully formed from the events of today or the recent past, but comes from the history of community governance. This understanding includes boundaries on the *role frames* (Schon 1983, 309–14) of public service practitioners, who must carry out the basic functions of the community and receive direction, monitoring, and evaluation. Those who direct, monitor, and evaluate practitioners include interested citizens or citizen groups, economic or political leaders (*elites*), and professional peers and coworkers. This is the practitioner's *role set*, those who, over time, create "role expectations" of performance for the practitioner and give feedback when they think performance improvement or change in the role is necessary.

Our understanding of the roles of public service practitioners in communities nationwide must be broad and flexible to allow for local variation, individual choice, and vocational specialization. And it must change over time with the institution of community governance, because a view of the practitioner's role that made sense in relation to the challenges of the 1920s or 1930s can be inappropriate entering the twenty-first century.

In addition to long-term, broad change in the role of the public service practitioner, local political conditions can change rapidly, leading to variability in the community policy orientations (accessible or closed governance system, emphasis on marketplace or living space values, a large or restricted role for government, attitude toward professionalism). As policy orientations change, so do expectations of public professionals. Though there are boundaries on the range of choices practitioners can make about their roles, each practitioner can choose from among many ways to respond to the expectations of those she or he serves.

An example of the impact of local role expectations on the practitioner role can be found in the selection of city managers by city councils. Saltzstein (1974) and Flentje and Counihan (1984) found that city councils sometimes hire a relatively weak and compliant city manager after a strong one leaves for another job or is fired. In replacing a strong manager with a weak one, a council may hire a manager with a weaker commitment to professionalism or one who is not a professional city manager but an insider from the organization who is a known quantity and can be controlled more easily. Like politics at the national level, where we see fluctuations over long periods of time between conservative and liberal or Republican and Democrat administrations, local government can go through cycles of attitudes toward professionalism that have quite an effect on public administrators.

Another way to think of the role of a particular public service practitioner is to imagine a "sphere of discretion," a defined zone of possible action. Each position in a public agency occupied by a practitioner carries with it such a role sphere of discretion. Most people in the role set would find the actions within this sphere to be a logical and desirable part of the practitioner's role. The sphere of discretion rests in a larger area of possible actions that are not regarded by the role set as appropriate to this particular practitioner's role. Although the practitioner's daily activities stay within the sphere of discretion, citizens, representatives, and peers, though they may not always agree with things the practitioner does, will believe that his or her actions are those they expect of someone filling this specific role.

If the practitioner takes actions that fall outside the sphere or does less than expected within the sphere, members of the role set will become concerned and begin to question the practitioner's performance. Actions outside the sphere could include making decisions about matters usually dealt with by superiors or elected representatives, working directly with citizens in a way that makes organizational superiors or representatives feel threatened, or strongly advocating an idea or program before citizens or representatives have had an opportunity to discuss or consider it. Over time, the role sphere of discretion can change, growing larger or smaller. This can be caused by changes in the role set as members come and go, the issues and challenges facing the community change, role set responses to practitioner actions, or a practitioner request to renegotiate the contents of the role.

In an interview I conducted as part of a research project, a community planning director described his mental image of his sphere of discretion. He saw it as a "ball and chain," with the ball as the "center of political gravity," and the end of the chain attached to his ankle. For him, this was not a negative image; he was a thoughtful person who spent time mentally calculating the sphere of discretion scribed by the length of the chain. He felt he had plenty of "room to roam," or freedom to act, within his sphere and was careful about attempting to enlarge it. When he thought it necessary to take an action that might be seen as falling outside the sphere, one creating a significant change from current practice, he would discuss his ideas with role set members, discovering how receptive they might be. He preferred to have the role set incrementally incorporate such ideas into their own thinking, thus claiming ownership for the ideas and moving the focus of attention away from the planner.

## THE GOALS OF PRACTITIONERS

Relationships between practitioners and people in role sets can be influenced by a variety of personal motivations and substantive goals. Public service practitioners are motivated by a range of preferences about public policy and governance and individual motivations to act are as complex and varied as the people who work in public administration. There are substantive goals specific to areas of practice within local public administration, such as law and order in police work, improving the lives of economically disadvantaged people in social work, and so on. Despite this complexity, we can identify perspectives on the relationship of the role set and the practitioner that cut across vocational boundaries and are of broad interest to the field of public administration. Before we turn to a discussion of these perspectives, we should take into account individual practitioner calculation of the personal risks of professional action.

**Personal Motivations.**

Practitioners are motivated to some extent by personal concerns, such as career advancement or financial security. Though these may seem like purely private, individual matters, decisions in the professional workplace are often affected by them. For example, imagine two police officers confronted with a risky and difficult choice. They have, separately, observed other officers using excessive force during on-the-street arrest situations. Both officers believe strongly that such behavior must be reported so that it does not continue, and both know that officers who report it can be subject to various kinds of retaliation from some of their peers, people who think that officers should stick together and protect each other. It could easily happen that one officer chooses to report the incident and the other chooses to be silent. In such a situation, the first officer decided that potential difficulty in the workplace could not offset the belief that reporting excessive force is the right thing to do, and the second officer decided that the consequences of displaying apparent disloyalty to peers outweighed the ethical commitment to reporting serious misbehavior.

Another relatively common example is that of the community public practitioner who believes that her or his livelihood could be threatened by speaking out on an issue. This often involves a situation in which the practitioner believes that a decision should be based on rational, professional knowledge of an issue, or it should be the result of a process that involves the community, whereas influential people would rather decide it on the basis of potential impact on their interests and without public input. The practitioner faces the choice of acting on norms of professional conduct or acting out of a desire to preserve personal career status and economic security. It may seem clear that the right choice is the professionally correct one, but practitioners can only be expected to assume a certain level of risk and uncertainty in their personal lives as a result of their commitment to public service.

Many times, the personal motivations of practitioners agree with the demands of organizations. When they do not, practitioners must calculate the balance between the risk and the value of the goals to be achieved. Practitioners may not be aware of this calculation in some situations. But in others, they are very much aware of the stress and difficulty of deciding whether to be safe or take the risky actions they believe to be right.

**Value-Free Neutrality**

Turning from personal motivations for action to substantive professional concerns, a good way to begin is with the substantive goal that is the least assertive or outwardly apparent to the role set. Many public professionals view themselves as neutral, highly proficient implementers of policy determined by others. Neutral professionals are presumably "value-free," because they do not decide what should be done, only how to do it. The idea of neutrality comes from the reform impulse of the late nineteenth and early twentieth centuries and the challenges of administering the emerging urban-industrial society of that time.

Despite the declared lack of values in the idea of neutrality, supposedly neutral practitioners often espouse the paired values of economy and efficiency, which are long-accepted measures of successful community governance (Stillman 1974, 20–2). Especially in this time of scarce resources and public scrutiny of government services, economy and efficiency are important practitioner goals. Privatization, reinventing, reengineering, and similar trends in the field are evidence of the pervasiveness of these goals and, although they can help produce better public services at reduced cost, they can also replace other goals and have impacts on the democracy and accountability principles if practitioners strive to fulfill them at the cost of including citizens in decision making.

In public administration we often use an essay written in 1887 by Woodrow Wilson, then a college professor, as an indicator of the concerns of the field at the beginning of what would become modern public administration. Wilson urged adoption of "scientific" methods of public management and greater separation between the process of determining policy and the process of carrying it out. As Dwight Waldo (1981, 65) has noted, the "politics-administration" relationship is the focus of much thinking and writing in public administration, as well as being a major part of the daily work life of many public practitioners, and any study of the role of the public administrator must take it into account.

We know today that the neutral professional model cannot fully explain the reality of public service. This is because many, if not most, practitioners are involved in helping to shape public policy as well as implement it. Even so, the idea of neutrality fits well with an image of democratic governance by the people rather than by career professionals, an image that is deeply embedded in American political culture. For this reason, it remains a powerful way to describe the idea that public professionals are different from citizens or representatives, even when professionals are as much a part of the process of creating policy as are citizens and representatives.

In the last half of the twentieth century, the "public choice" perspective on the role of the professional became well known. It advocates policy neutrality for the practitioner, but with greater emphasis on control of the bureaucracy. In the public choice view, public professionals are "agents" to the representatives and citizens who determine policy and should not themselves take part in the policy process. The term "public choice" does not mean choices made by citizens deliberating together over public policy issues. Instead, following basic economic assumptions that focus on individual preferences and self-interest, it means that individual citizen choices about policy or political candidates are aggregated (through voting) to determine the majority will.

Public choice theorists view the public sector as simply another kind of market; the medium of exchange in the private sector is money and in the public sector it is votes (Downs 1957). Though many of us think of the motivations of people in public service as being different from those of people in the private sector, the public choice theorist sees little difference. Thus, "if the individual is motivated by personal benefits and costs when making decisions as a consumer, worker, or investor, that individual is going to be motivated by personal benefits and costs when making decisions in the voting booth, in the halls of Congress, and in the conference rooms of the bureaucracy" (Johnson 1991, 13).

The public choice view of career public service practitioners is not flattering. Public choice theorists tend to see them as "budget maximizers" out to increase the size of their agencies so they can have larger salaries, bigger offices, more benefits, and other things of value to them personally (Niskanen 1971; 1991). To avoid this problem, the ideal public employee should function as a delegate of the elected representatives of the people. This view of the practitioner role fits with the antigovernment attitudes of the 1980s and 1990s. It can be seen in the movement toward smaller and more efficient government, for example through privatization and applying private-sector management techniques to government, such as Total Quality Management, the customer service orientation, pay-for-performance, "reinventing" and "reengineering."

Management fads are not new in public administration; the public sector often imports ideas from the private sector for a time, then either discards them or incorporates parts of them in modified form to suit the unique circumstances of public organizations. This can be said of program-planning-budgeting (PPB), management-by-objectives (MBO), zero-base budgeting (ZBB), and quality circles. However, economic and private-sector concepts of organizational purpose and management have had a significant impact on public administration in the last two decades of the twentieth century and, given the general climate of public attitude toward government, they may be expected to influence public practice into the twenty-first century.

Paradoxically, the use of economic concepts in the public sector has also produced a demand for public practitioners to break out of their bureaucratic confines, managing the public's business like private sector businesses, behaving like "entrepreneurs" rather than tightly controlled agents. These public entrepreneurs are expected to act like business executives, finding ways for their organizations to get "a piece of the action" (Osborne and Gaebler 1993, 200) by becoming involved in public–private partnerships and speculation in shopping centers, golf courses, hotels, convention centers, and similar typically private-sector ventures. The public agency's ability to restrict and direct the use of land, create special taxing districts for improvements, and sell bonds to generate development capital allows it to intervene in the market, creating surplus resources that can be invested in community amenities or new enterprises.

For public professionals to mimic the behavior of their private-sector counterparts, they must adopt some of their characteristics, such as "autonomy, a personal vision of the future, secrecy, and risk-taking" (Terry 1993, 393). This is not what Americans traditionally expect from public professionals or from public agencies. The impulse to blend the public and private sectors, to "run government like a business," ignores the gulf of values and purposes that separates the two sectors. There is no question that good public managers look for ways to improve the delivery of public services, and always have. But carried too far, running government like a business can be a threat to the democracy and accountability principles as agencies and practitioners take action on their own to maximize efficiency and financial return rather than involving citizens in shaping communities that serve the citizens' values and purposes. Ultimately, the question is whether community residents are consumers of an efficiently delivered product or citizens collectively deciding the fate of their communities.

There is a danger associated with the traditional reform-era model of neutrality and a separation of policy making and administration. This danger is the tendency of practitioners to resist the value shifts described in this book as returning to values of the past by keeping elected representatives and citizens as far away from their work as possible. Highly skilled practitioners may come to regard their work as beyond the understanding of citizens who, if they ask questions or request a meaningful share in the decision-making process about public programs, are viewed as a nuisance and waste of time. Practitioners who resist citizen claims to govern programs in their communities often think of citizen self-governance as a violation of the principle of separation of politics and administration.

Resistance to perceived intrusion or micromanagement by nonprofessionals into administrative affairs is not uncommon in government, though it becomes harder to sustain in the contemporary environment of a return to earlier, citizen-driven values of governance. Though we need to recognize that citizen involvement in administration must be moderated through use of professional methods and respect for due process of law, pushing citizens and their representatives away from the conduct of their government violates the democracy and accountability principles.

## Legitimacy

One way to counter the "downslope of the wave of reform" trend of skepticism about government and resulting efforts toward downsizing and control is to elevate the status of public administration. In this view, public professionals, instead of the agents of elected representatives and citizens, become a "legitimate" part of government with approximately equal standing in the formation and implementation of public policy alongside legislators and citizens. Legitimacy would mean giving greater autonomy in public governance to public service practitioners, and it has become an important model in the literature of public administration.

Those desiring increased legitimacy as a goal for public administration often focus on the national level of government. The question of legitimacy centers on the relationship of public administration to the Constitution and the nature of the founding period (Rohr 1986, 1993; Spicer and Terry 1993; Stivers 1993; Wamsley et al. 1987). Though there is confusion about what is meant by legitimacy (Warren 1993, 250–2), a legitimized public administration would seem to be one that is respected by the public, has more control, authority, and discretion to act independently than at present, and is given the status of a somewhat equal partner in relation to elected leaders and other parts of government.

Legitimacy may seem to be a good idea because the complexity of modern society causes political leadership to be ineffectual and because citizens have lost the capacity to carry out their citizenship responsibilities as a result of ignorance or indifference. If this is true, the expertise of public professionals can be used to steer government out of its morass. Carried to its logical extreme, the search for legitimacy might culminate in a guardian class of administrators who make decisions that they believe citizens choose not to make or cannot make competently (Fox and Cochran 1990).

Because of fundamental American political values, the impact of the legitimacy model of the practitioner's role seems to have been minimal. Asking for greater practitioner status in the policy-making process may soothe people irritated with what they perceive as the unfortunate effects of bureaucrat bashing or conservative ideology. The problem with the idea of greater "legitimacy" for public professionals is that most Americans simply will not accept it. It is not necessary to characterize the time in which we live as one of great skepticism about government to reach this conclusion. At any time, Americans, faced with the idea of a powerful and independent administrative class, would ask questions such as, "What would change if we had it?" "Why should we expect these people to do any 'better' than our politicians, or for that matter the bureaucrats of totalitarian states?" and "Why should I give away whatever ability I have to decide the fate of my community to a clique of non-accountable technicians?"

The idea of legitimacy runs into the same difficulties as practitioner resistance to citizen self-governance. It is in opposition to the three values of the past and could violate the democracy and accountability principles. If the purpose of local public administration is to help people create the communities they want, demanding equal status with those who legitimately "own" communities may accomplish little more than generating citizen opposition. Public service practitioners need to be experts ready to assist citizens in fulfilling their vision of the future but, as Charles Goodsell noted,

> Our craft knowledge must not become a source of technological arrogance, according to which only we know. Our official responsibilities and authority must never develop into an

egotism that refuses to accept opposition or criticism. . . . We must always regard part of our job to consist of engaging in open and authentic dialogue with as many people as possible who are affected by our work. (1996, 49)

**Development Versus Sustainability**

There is increasing pressure on many public professionals to be skilled in "economic development," an area of practice that involves techniques for making the community attractive to new or expanding business. The goal of economic development is linked conceptually to the growth machine phenomenon . . . in which elites who benefit from land speculation and development push their communities to compete with one another for economic advantage. It can be good for professional careers for practitioners to orient their relationship to the role set around the substantive goal of economic development. By doing so, they help the economic elite and can become highly "marketable," eligible for promotion or advancement to better jobs in other places. In the past decade or so, a trend toward emphasis on economic development has been evident in the advertisements for certain public management specialties, such as planning and community development and city management.

The substantive goal of environmental protection or community sustainability can be directly opposed to the goal of economic development. The practitioner who chooses to act based on the idea of "fostering quality of person, community, and environment" (Nelson and Weschler 1996, 13) simultaneously takes on a task that runs counter to much of the political essence of American communities. The linkage between community policy making and the powerful people who benefit from physical and economic growth is a strong one. There are many community practitioners who contribute to this goal through their work in areas such as parks, social services, and economic improvement that involves strengthening local businesses that employ local people. However, it is not often that practitioners have the opportunity to assess the community as an integrated system that, if managed properly, can serve the needs of people while using land and resources in a way that preserves their productive capacity and aesthetic beauty for future generations.

**Social Equity**

Attention given by public administration academicians and practitioners to the goal of social equity grew out of a belief in the 1960s that America had become dangerously divided by class and economic status. The counter-cultural revolution, racial uprisings in American cities, and resistance to the Vietnam War served as evidence for some that social institutions and leaders were preserving gross inequities in wealth and opportunities for self-determination that had to be changed by radical action.

Some public administration academicians responded to these conditions with proposals to change the relationship of public professionals to their political superiors and society at large. According to Dwight Waldo, these proposals included *client-centered* or *street-level* services to reduce bureaucratic obstacles for citizens, participation of client populations in governing public programs, and creating a public bureaucracy that was *representative,* in the sense that public professionals would be similar to the citizenry in relation to geography, class, race, and other factors (1981, 95–6).

These ideas came to be known as the *New Public Administration.* To the extent that one believes that existing political structures and leaders favor the wealthy and powerful at the expense of the weak and powerless, the New Public Administration shift of decision-making focus from elected representatives to street-level bureaucrats and their clients is logical. It also can be threatening to those elected representatives and to those who view them as the rightful and legal locus of the power of the people to govern themselves. In his book *Without Sympathy or Enthusiasm,* Victor Thompson found the premises of New Public Administration to be "a most amazing effort to establish a new claimant [to public power] in place of the owner (that is, in place of the public). It is a brazen attempt to 'steal' the popular sovereignty" (1975, 66).

### Facilitating Citizen Discourse

For practitioners who do not wish to advocate a particular policy outcome but believe in the democracy principle, an appropriate goal is to encourage open public discourse and decision making (this goal is the basis for the "helper" role, discussed in the section that follows). Pursuing this goal is not passive or risk free, because it carries with it the dual risks of retaliatory action by the elite or citizen-chosen policy outcomes with which the practitioner may strongly disagree. Action toward fulfilling the goal of citizen self-determination may rest on the conceptual foundation of critical theory . . . . This foundation includes enlightening citizens, giving them knowledge and the opportunity to use it. This leads to "emancipation," as citizens have the tools needed to make a difference in the future of their communities.

The reader may think of other practitioner goals common across the field of public administration. For the purposes of this book, the goals we have discussed provide a good foundation for thinking about types of roles practitioners adopt in their daily work.

### THREE PRACTITIONER ROLE TYPES

Earlier in this chapter we discussed concepts related to the role of the public service practitioner, including the role set, role expectations, and the sphere of discretion. Now, we can create a typology of roles that incorporates these concepts and makes it easier for us to apply them to specific situations. The literature of public administration contains many role typologies for public practitioners, organized around a variety of issues, such as practitioner motivations, how they interact with citizens, the practitioner's legal or constitutional obligations, how they organize agencies and supervise their employees, and so on. In this book, we are especially concerned with the relationship of practitioners to citizens and elected representatives and the way they become involved in the creation and implementation of public policy. (For an excellent overview of public administration roles, see Kass and Catron 1990.)

Practitioner involvement in public policy is the subject of the typology offered here, which describes three practitioner roles. One of these is the *implementer* role: Implementers are "neutral" practitioners who avoid significant involvement in shaping policy. Another, the *controller* role, includes practitioners who seek to influence the outcomes of the policy process. Implementers and controllers are located at polar ends of a continuum we could label *practitioner intent to influence public policy.* Between the polar ends of this continuum may be found an infinite number of intermediate positions, but for our purposes we will identify just one, the *helper* role. Helpers take an active part in policy creation and implementation by serving to interpret public wishes for representatives, presenting professional knowledge of organizational and technical practices to citizens and representatives, and monitoring decision making and implementation to ensure that citizens have opportunities to participate.

The apparent goals of implementers are to dutifully carry out lawful policy directions in a professional and competent manner and to avoid the potential trouble and risk associated with direct involvement in determining public policy. They may have strong feelings about the substantive issues dealt with in the previous section, but they keep such thoughts to themselves. Where professional service values of efficiency or effectiveness are at stake, implementers will take care that appropriate people hear of their concerns, but they avoid going outside accepted formal channels of communication. Many people who serve in public agencies are implementers or have observed implementers. Often, implementers are valued colleagues, providing competent service without causing disturbance or pushing for change. Sometimes, when they serve in positions that require leadership or innovation, their lack of initiative and assertiveness can hamper agency effectiveness.

Controllers seek to guide the policy process and outcomes by influencing the attitudes of their superiors or of elected officials or citizens. Though we know it is unrealistic to expect a clear separation between policy making and administration, controllers sometimes push the boundaries of accepted professional practice by going outside the normal hierarchy, interacting directly with high-level organizational officials, elected representatives, or citizens to mobilize opinion in favor of a particular policy or to influence the outcome of a policy creation process already under way. They are "true believers" crusading for their vision of how the community should be organized and what the end result of public action should be. It is with the controller role that the substantive goals discussed previously become especially apparent and important.

The behavior of controllers can be risky for themselves and for the community. It is important for those of us in public administration to understand this role behavior because it occurs at the boundaries of what is commonly accepted as appropriate for public practitioners in a democratic society. Controller behavior can be an exciting and worthwhile source of new ideas and policy innovation, but it can also be a threat to control of public governance by citizens and their representatives, placing the professional in a situation much like that of the elected official, leading the way in shaping public opinion to conform with the controller's vision of the community future.

Practitioners who act as "helpers" often are interested in certain public policy outcomes in the community, but their focus is on the process of dialogue and deliberation that leads to policy decisions and program implementation. Like controllers, helpers may go outside the formal hierarchy, but they do so to reach the goal of citizen self-governance, allowing citizens to select the substantive results they prefer. The section that follows devotes more attention to the helper role, as it is the one advocated in the Citizen Governance model.

Of course, the implementer, helper, and controller are "pure" types, drawn to illustrate a range of intent to shape policy; a given individual may exhibit mixtures of the role types and may change from one role type to another in different situations. There is another important dimension of practitioner behavior in relation to citizens that cuts across the imp lementer-helper-controller policy to influence typology. This is the practitioner's desire to limit access to the everyday machinery of administration by citizens or elected representatives. We are just as likely to see an implementer try to limit citizen access to the tasks and processes of administration as we are a controller, and a helper may do this as well. The rationale for keeping "outsiders" away from the details of administration (aspects of personnel actions, budgeting, and specific decisions about programs) is that they are not professionals, cannot understand the technical realities of program administration, and would only cause confusion, delay, and irrationality if they "intrude" into expert administration.

There are two very important problems with excluding citizens and representatives from administration. . . . The first is that, without knowledge of administration, citizens and representatives cannot make informed decisions about the creation and implementation of policy; this violates the rationality principle.

Second, if the public or its representatives are not informed about how the affairs of the members of the community are being administered, there is potential for abuse of administrative discretion. However well-intentioned public professionals may be, distance from the watchful eye of the "owners" of the community can lead to actions that strike the ordinary citizen as odd, inappropriate, biased in favor of particular groups or people, or simply incompetent. The actions that come to light, possibly on the pages of the local newspaper, often seem trivial in themselves. Recent examples from the author's community include paying for managers to have four-wheel-drive vehicles for personal use, paying a former mayor a large sum for the privilege of building a water line across his property, and giving unusually large salary bonuses to senior staff at a public hospital. Even if relatively minor on their face, situations like these can convey an impression of abuse, arrogance, and insensitivity to the values and expectations of the tax-paying public.

In addition to these minor actions that can be irritating to public opinion, on occasion actions may be taken away from public view that have serious, possibly far-reaching implications. These could include decisions about how to plan for growth, such as whether the public or developers should pay the costs of new infrastructure, or decisions about whether to reallocate resources (street improvements, policing, education and daycare assistance for single parents, and so on) into troubled neighborhoods, or crucial safety-related decisions such as where to locate fire stations and how to staff them. Excluding the public from important community-interest policy issues breaks the chain of accountability, violating the accountability principle.

Of course, it is possible for citizens or representatives to become so involved in the details of administration that they really do impede effective action; the rationality principle can be violated if administration becomes deprofessionalized, just as it can if citizen or representative decision making is based on insufficient information. This is a matter of balance, of finding an understanding about the practitioner role that fulfills the rationality and accountability principles in a given community at a specific point in time.

## THE HELPER AND THE PARADOX OF "GIVING AWAY"

With others in public administration, I have argued that it no longer makes sense for the goals of neutrality or legitimacy to be primary guides to public service roles (Adams et al., 1990; Box 1995a, 1995b; Fox and Miller 1995). This is because the social and political context of the three values of the past puts practitioners in a situation in which

> One, professionals can survive by opting out of the policy dialogue and playing safe. Two, professionals can resist the current cycle of change, clinging to the traditional control-oriented model as long as possible as events pass them by. Three, professionals can enter into the stream of change and dialogue, by their participation tempering outcomes with professional rationality. (Box 1995b, 89)

When we discuss role options for practitioners, we need to be careful how we use the concepts of time and change. Preferred practitioner roles will change in the long term. They are changing now, as we move from an era of reform, institution building, and technical expertise, into an era of citizen self-governance. This movement is accompanied by pressures for change in the role of the public service practitioner. Change is always with us and roles will change with the times.

In the short term, individual practitioners have always been able to choose role positions based on circumstances. The implementer, helper, or controller roles each can be appropriate at different times during interactions with the public, elected officials, or peers, and in varying circumstances. As a practitioner works on a project or handles a series of events, he or she thinks about desired outcomes for the current situation and considers the ways members of the role set will react to various action choices. These thoughts shape the practitioner's role choices.

Because of long-term role change and the reality of short-term flexibility, we need to accept the importance of all possible role choices to public service. However, the theme of the Citizen Governance model is change, understanding and adapting to societal trends and needs within the framework of the history and characteristics of the institution of community governance. I believe that as we move fully into the twenty-first century and the era of Citizen Governance, the implementer and controller roles will continue as useful alternatives for public service practitioners. I also believe that the helper role will become the most important in our thinking about how practitioners serve their communities.

If we deemphasize the quest for professional neutrality or greater legitimacy, the task is to build better relationships between public service practitioners and citizens and representatives. The issue is how to combine the practitioner's unavoidably subservient position in the hierarchy of public organizations with a proactive stance in relation to public policy formulation and implementation. The position of the public service practitioner is unavoidably subservient because practitioners are employees, agents of the public through their elected representatives, and these elected representatives have the power to remove and replace practitioners. Rather than protest this position or try to wall ourselves off from the world with claims of a policy–administration separation, we need to seek constructive ways to be a part of the exciting institutional changes underway.

The logic of my argument for the helper role as a central feature of Citizen Governance is, on a superficial level, that of process of elimination: The implementer and controller roles are inadequate in the contemporary situation, so the helper role is what remains, in the middle of the continuum of intent to affect policy creation and implementation. But the helper role of the public service practitioner is not only what remains after other possibilities are eliminated. Rather, it is a role model built on a foundation of critical theory and the history of the institution of community governance, a history of citizens striving for a balance of administrative rationality and citizen self-determination.

The Citizen Governance model explicitly recognizes the political constraints inherent in the ecology of public administration practice. We recognize that the community practitioner is usually employed directly or indirectly by people who wield political power because of their control of the private market. Though the partially democratic nature of our political system allows some public access to the policy process, the reality of elite control based on power and wealth cannot be avoided. Because members of the public are often not aware of this control of the policy process, they cannot effectively participate in public governance nor step outside their situation to evaluate the need for change.

In contrast to the controller role in which practitioners become the focus of meaningful change, advocating for substantive goals such as social equity, the helper aims to create conditions in which a fully conscious public enacts change. This involves pursuing the critical theory goals of enlightening citizens and giving them access to the policy dialogue, thereby empowering, or "emancipating," them to act. In this way, helpers function as "facilitators, educators, and coparticipants, rather than deference-demanding experts or independently responsible decision makers" (Adams et al. 1990, 235–6). They "learn humility and respect for the developmental potential of others and to enjoy apparently self-limiting transfers of authority and responsibility to citizens" (Adams et al. 1990, 236).

Helpers do not strive for greater power, autonomy, and recognition. Instead, they give away knowledge and thus the power to make decisions to the people who are affected by those decisions. Paradoxically, this giving away of control makes practitioners more rather than less effective, as community residents, informed by practitioners, understand the issues and insist on meaningful change. The knowledge the helper gives away is knowledge of the practice of community governance that has been gathered over decades, centuries, by all the people who have worked to build a better future for their communities. It is a rich legacy to pass on to those who join today in the stream of people experimenting with new ideas in the process of changing the institution of community governance.

Giving away control and knowledge can be a hard thing to do. Many practitioners have been trained to maintain a separation between themselves and the public, because they are the experts and citizens are the "customers," people who know little about public services except the end products they receive and the fees or taxes they pay. Such practitioners use "mystery and mastery" of knowledge (Schon 1983, 229) to maintain separation between themselves and their clients and it can be frightening to think of letting go of this control.

There also can be significant risks to the helper role. These risks can be of two types. One is that helpers may be disciplined or fired for stirring up political currents disliked by representatives or by powerful citizens. The other is that the public may not do what practitioners think it ought to do. Once citizens are given the information needed to take part in policy dialogue and the access to the process needed to take meaningful action, they may make choices that directly contradict the substantive goals held by the public professionals who enabled them to make those choices. For example, citizens may take a stand against construction of public housing when involved practitioners may be concerned about social equity. Another example would be a citizen-driven policy process that results in development decisions that could harm the environment, whereas the practitioners who helped citizens study the options and make choices are committed to the substantive goal of environmental protection and sustainability. This latter risk is one the helper must be prepared to take, because emancipation means empowerment to act, to allow people to make their own choices when fully informed of the alternatives available to them. In so doing, as Cooper put it, the public practitioner "is responsible for upholding the sovereignty of the people while also making available to them certain technical skills and knowledge" (1991, 167).

The helper's role posture of assisting in the development of a meaningful citizen dialogue (for those citizens who wish to become involved) can be especially risky in situations in which the motives and interests of economic and political elites are substantially different from those of citizens generally. Citizens may not realize that such a difference exists, thus there is relative peace in the community; but if public practitioners supply them with the knowledge needed to understand their situation and enter the policy dialogue, elites may come to regard both citizens and practitioners as a threat to the established order (Box 1995a). In so doing, public service practitioners become an active part of the substance as well as the process of governance, communicating to citizens the essence of the institution including its practices and available policy options, helping them arrive through discourse at outcomes that are acceptable to the majority and respectful of the minority.

The evolving Citizen Governance practitioner role described in this book carries with it the burden of special knowledge and the responsibility to serve not only a person, a group, or the practitioner's personal preferences or opinion based on experience, but rather the goal of offering the opportunity for open community discourse on matters of collective importance. Recognizing that every discourse opportunity is bounded by the specific organizational-political-legal setting in which it rests and that coercion by the powerful is always a possibility, this evolving role allows the best chance at free and open discourse in search of fulfilling the democracy and rationality principles.

## REFERENCES

Adams, Guy B., Priscilla V. Bowerman, Kenneth M. Dolbeare, and Camilla Stivers. 1990. "Joining Purpose to Practice: A Democratic Identity for the Public Service." In *Images and Identities in Public Administration,* ed. Henry D. Kass and Bayard L. Catron, 219–40. Newbury Park, CA: Sage.

Box, Richard C. 1995a. "Critical Theory and the Paradox of Discourse." *American Review of Public Administration* 25 (March): 1–19.

———. 1995b. "Optimistic View of the Future of Community Governance." *Administrative Theory & Praxis* 17(1): 87–91.

Cook, Edward M. 1976. *The Fathers of the Towns: Leadership and Community Structure in Eighteenth-Century New England.* Baltimore: Johns Hopkins University Press.

Cooper, Terry L. 1984. "Public Administration in an Age of Scarcity: A Citizenship Role for Public Administrators." In *Politics and Administration: Woodrow Wilson and American Public Administration,* ed. Jack Rabin and James S. Bowman, 297–314. New York: Marcel Dekker.

———. 1991. *An Ethic of Citizenship for Public Administration.* Englewood Cliffs, NJ: Prentice-Hall.

Downs, Anthony. 1957. *An Economic Theory of Democracy.* New York: Harper and Row.

Flentje, H. Edward, and Wendla Counihan. 1984. "Running a 'Reformed' City: The Hiring and Firing of City Managers." *Urban Resources* 2 (Fall): 9–14.

Fox, Charles J., and Clarke E. Cochran. 1990. "Discretion Advocacy in Public Administration: Toward a Platonic Guardian Class?" *Administration & Society* 22 (August): 249–71.

Fox, Charles J., and Hugh T. Miller. 1995. *Postmodern Public Administration: Toward Discourse.* Thousand Oaks, CA: Sage.

Goodsell, Charles. 1996. "A Memo to the Public Employees of America." *Administrative Theory & Praxis* 18(1): 48–49.

Johnson, David B. 1991. *Public Choice: An Introduction to the New Political Economy.* Mountain View, CA: Mountain View.

Kass, Henry D., and Bayard L. Catron, eds. 1990. *Images and Identities in Public Administration.* Newbury Park, CA: Sage.

Nelson, Lisa S., and Louis F. Weschler. 1996. "Community Sustainability as a Dimension of Administrative Ethics." *Administrative Theory & Praxis* 78(1): 13–26.

Niskanen, William A. 1971. *Bureaucracy and Representative Government.* Chicago: Aldine Atherton.

———. 1991. "A Reflection on Bureaucracy and Representative Government." In *The Budget-Maximizing Bureaucrat: Appraisals and Evidence,* ed. Andre Blais and Stephane Dion, 13–31. Pittsburgh: University of Pittsburgh Press.

Osborne, David, and Ted Gaebler. 1993. *Reinventing Government: How the Entrepreneurial Spirit Is Transforming the Public Sector.* New York: Penguin Books.

Rohr, John A. 1986. *Ethics for Bureaucrats: An Essay on Law and Values.* New York: Marcel Dekker.

———. 1993. "Toward a More Perfect Union." *Public Administration Review* 53 (May/June): 246–49.

Saltzstein, Alan L. 1974. "City Managers and City Councils: Perceptions of the Division of Authority." *The Western Political Quarterly* 27 (March): 275–88.

Schon, Donald A. 1983. *The Reflective Practitioner: How Professionals Think in Action.* New York: Basic Books.

Spicer, Michael W., and Larry D. Terry. 1993. "Legitimacy, History, and Logic: Public Administration and the Constitution." *Public Administration Review* 53 (May/June): 239–46.

Stillman, Richard J. 1974. *The Rise of the City Manager: A Public Professional in Local Government.* Albuquerque: University of New Mexico Press.

Stivers, Camilla. 1993. "Rationality and Romanticism in Constitutional Argument." *Public Administration Review* 53 (May/June): 254–57.

Terry, Larry D. 1993. "Why We Should Abandon the Misconceived Quest to Reconcile Public Entrepreneurship with Democracy: A Response to Bellone and Goerl's 'Reconciling Public Entrepreneurship and Democracy.'" *Public Administration Review* 53 (July/August): 393–95.

Thompson, Victor A. 1975. *Without Sympathy or Enthusiasm: The Problem of Administrative Compassion.* University: University of Alabama Press.

Waldo, Dwight. 1981. *The Enterprise of Public Administration: A Summary View.* Novato, CA: Chandler and Sharp.

Wamsley, Gary L., Charles T. Goodsell, John A. Rohr, Camilla M. Stivers, Orion F. White, and James F. Wolf. 1987. "The Public Administration and the Governance Process: Refocusing the American Dialogue." In *A Centennial History of the American Administrative State,* ed. Ralph C. Chandler, 291–317. New York: Free Press.

Warren, Kenneth F. 1993. "We Have Debated Ad Nauseum the Legitimacy of the Administrative State—But Why?" *Public Administration Review* 53 (May/June): 249–54.

Reading 5.2

# "NEW PUBLIC MANAGEMENT AND SUBSTANTIVE DEMOCRACY"

Richard C. Box, Gary S. Marshall, B.J. Reed, and
Christine M. Reed

*The authors are concerned that a remaining refuge of substantive democracy in America, the public sector, is in danger of abandoning it in favor of the market model of management. They argue that contemporary American democracy is confined to a shrunken procedural remnant of its earlier substantive form. The classical republican model of citizen involvement faded with the rise of liberal capitalist society in the late nineteenth and early twentieth centuries. Capitalism and democracy coexist in a society emphasizing procedural protection of individual liberties rather than substantive questions of individual development. Today's market model of government in the form of New Public Management goes beyond earlier "reforms," threatening to eliminate democracy as a guiding principle in public-sector management. The authors discuss the usefulness of a collaborative model of administrative practice in preserving the value of democracy in public administration.*

## INTRODUCTION: THE CHALLENGE

This is a [reading] about the relationship between American democracy and public administration in a time when the public sector is under considerable pressure to adopt the values and operational techniques of the private market sector. We are concerned about the nature of contemporary American democracy and the effect it has on people and the physical environment. Today, despite the success of American democracy in securing individual liberties and the material success of a wealthy society that provides more goods and services to a broader range of Americans than ever before, vexing problems remain: poverty, poor-quality education, inequalities of race, gender, and wealth, crime and violence, destruction of forest, farmland, wildlife habitat and other natural resources, and pollution of air and water.

These are not trivial, new, or surprising problems, and the technology and resources are, for the most part, available to make significant improvements. One reason the problems persist is that the public lacks the knowledge and political influence to give public administrative agencies a

From *Public Administration Review* 61, no. 5 (September/October 2001): 608–619. Copyright © 2001 American Society for Public Administration. Reprinted with permission.

mandate to solve them. Various barriers stand in the way, such as control of information and the policy-making process by interest groups and economic elites, inertia in bureaucratic organizations, and resistance by "experts" to democratic governance (McSwite 1997; Yankelovich 1991). The resulting disconnect between the potential wishes of an informed populace and the condition of society is as old as the idea of democracy. In 1927, John Dewey called it *the* problem of the public" (1985, 208, emphasis in original).

One could argue that "the people" have chosen an equilibrium in the balance between democracy and efficiency (Okun 1975) that includes an instrumental, efficient public administration, one that does not challenge the status quo or unilaterally set out to solve problems. But this argument is based on the questionable assumption that citizens possess relatively complete knowledge of the condition of society, along with the ability to effectively wield the available political and institutional tools to effect change. Instead, the contemporary situation appears to us to be the result of constraints imposed upon public action by what may be termed "liberal democracy in a capitalist setting." By "liberal" we mean the classical, Lockean view of the relationship of the individual to society from the Enlightenment of the seventeenth and eighteenth centuries, one that emphasizes protecting the individual from society. It is the polar opposite of the older, classical, republican tradition emphasizing the social nature of the individual in constructing society jointly with others. By "capitalist" we mean a society based on the idea that each individual economic actor should be relatively free to accumulate wealth independently from social control or collective determination of the "public good."

Liberal democracy "is capable of fiercely resisting every assault on the individual—his privacy, his property, his interests, and his rights—but is far less effective in resisting assaults on community or justice or citizenship or participation" (Barber 1984, 4). This negative and procedural, rather than positive and substantive, conception of democracy has a solid foundation in American political thought, but so does a more substantive view of democracy as "a quality pervading the whole life and operation of a national or smaller community, or if you like a kind of *society*, a whole set of reciprocal relations between the people who make up the nation or other unit" (Macpherson 1977, 5–6, emphasis in original). However, the problems remain, the public sector moves to fashion itself after private business, and we ask what it is about American democracy that allows this to occur. We further ask what the position of a "self-aware" (Waldo 1981, 10–11) public administration should be in such circumstances. As the problems are not new, our questions are time-worn companions to the study of public administration, though the answers have changed over the decades. Public administrators play an important role in the formulation and implementation of public policy; if they do not value and promote a substantive model of democracy, the likelihood of constructively dealing with pressing public problems decreases significantly.

In this [reading], we argue that democracy as we know it is a shadow of the ideal, and modeling the public sector after the private may aggravate this problem. After examining the history and condition of democracy, we explore the nature of the current wave of governmental "reform." Placing it in historical context, we show that, while earlier reform efforts included democracy as a central value (even if as a facade), today's reform efforts have largely sidestepped the question of democracy altogether, weakening the connection between citizens and the broader community. Next, we discuss the contemporary meanings of "community" and "democracy" using managed health care as an example, discovering that imposing the market model on citizens and administrators does not support democratic self-determination.

Recently, Jane Mansbridge (1999, 706) argued that we are in a "holding pattern" in relation to democracy: Today, "not many Americans care about making this country more democratic," but "at some point a larger fraction of the populace will come to care deeply about democracy again. When they do, several scholarly traditions have ideas that will help." Though Mansbridge may not have meant to include public administration in the list of traditions that might help, we believe our field has something to offer in the recovery of substantive democracy. Thus, we conclude our analysis of the condition of democracy with thoughts about implications for public administration and the possibility of moving toward a collaborative, as opposed to a market-oriented, model of public practice.

## REDISCOVERING SUBSTANTIVE DEMOCRACY

In the twentieth century, Americans have largely come to accept the procedural view of democracy associated with classical liberalism, which "as a philosophy is rooted in the twin ideas of individualism—negative liberty—and a distrust of government. . . . In this context, anything and everything, including democracy, take second place" (Hollinger 1996, 7). In liberal democracy, the role of citizen consists of voting for representatives who act on behalf of their constituents. Substantive issues of social justice, economic inequality, and the relationship of capitalism and the physical environment are addressed in the "public space" when problems become so severe they threaten stability or safety (for example, social conditions during the Depression, or destruction of the ozone layer).

Separating the procedural and substantive spheres in democracy leaves unanswered questions about "outcomes, conditions of people's lives, and realization of all people's political potential that made democracy a politically explosive concept in the past" (Adams et al. 1990, 220). Issues that affect the whole citizenry are dealt with in the context of liberal democracy, tightly controlling the extent to which the public, through institutions of government, can take action. The contemporary definition of democracy is characterized by the split between procedure and substance, with the public sphere being limited largely to questions of process. As Ellen Wood puts it, "The very condition that makes it possible to define democracy as we do in modern liberal capitalist societies is the separation and enclosure of the economic sphere and its invulnerability to democratic power" (1996, 235).

Bowles and Gintis (1987, 41–62) suggest that American democracy has gone through several "accommodations" that provided temporary equilibria between property and personal rights. The first, the Lockean accommodation, limited political rights to the propertied classes, who would not be a threat to the economic order. This was followed in the nineteenth century by the Jeffersonian accommodation, which was based on abundant land and the idea that every free-born male would have a chance to be a landowner and share in the economic advance of the nation.

The Madisonian accommodation of the late nineteenth century and into the mid-twentieth century protected the "few from the many" by allowing pluralist competition to cancel out demands by the masses that might threaten the elite. After World War II, the Keynesian accommodation placated citizens with economic success and egalitarian economic policies. Today, it is difficult to predict how the globalization of economics and the expansion of market concepts into the public sector will affect democracy and public service, or to foresee the nature of the current and future accommodation.

During the latter half of the nineteenth century and most of the twentieth century, citizens ceded control of public-sector policy making and implementation to bureaucratic professionalism. This made sense as part of building an administrative state to meet the demands of a growing nation, wars, depression, and so on. But now people mistrust the public sphere, regarding politicians as corrupt, bureaucrats as self-serving and inefficient, and governing as "a matter of invisible negotiations conducted in government offices by public officials and private interests" (King and Stivers 1998b, 15).

This gloomy view is countered somewhat by the long history of substantive democracy in American thought, indicated by the views of the anti-Federalists and other founding-era figures such as Thomas Jefferson. Jefferson's view of democracy included both classical liberal protection of individual liberty and a classical republican element, drawn from the ancient Greeks and from eighteenth-century Scottish moral philosophers, in which democracy begins with people actively shaping a society grounded in social relationships (Sheldon 1991; Wills 1979). For Jefferson government is not top-down, but begins with the individual in a "pyramid structure . . . in which each higher level is held directly and immediately accountable to its next lower level" (Matthews 1986, 126).

The republican philosophy included "the idea that liberty is participation in government and therefore is self-government" (Dagger 1997, 17). Jefferson's "radical democracy" required "an egalitarian redistribution—and redefinition—of the social good(s) on an ongoing basis . . . and governments must either be structured or dissolved and restructured" to achieve that goal (Matthews 1986, 122). The ideal of citizen self-governance can be found in the twentieth century as well. In the early part of the century, John Dewey envisioned a future democracy in which the political and economic spheres would be joined. Democracy would be an ongoing process of citizens working toward cooperative, shared governance of social institutions, including those of the market (Campbell 1996, 177–84). In *Dilemmas of a Pluralist Democracy* (1982), Robert Dahl argued that it might be possible to mitigate some of the problems of liberal democracy by ensuring a fair distribution of wealth and making some economic decisions subject to democratic control. In his book, *Strong Democracy* (1984), Benjamin Barber advocated a shift from the "weak," liberal version of democracy to a form he described in this way: "Strong democracy is a distinctively modern form of participatory democracy. It rests on the idea of a self-governing community of citizens who are united less by homogeneous interests than by civic education and who are made capable of common purpose and mutual action by virtue of their civic attitudes and participatory institutions rather than their altruism or good nature" (117).

Scholars have cautioned against assuming that a more active, substantive, "communitarian" democracy will result in a "better" community (Conway 1996). They also note that a pure, classical, republican society may have serious consequences for individual liberty. Societies of the past that exhibited a greater commitment to shared governance often did so at the expense of groups excluded from the definition of citizenship, such as women, outsiders, and slaves in ancient Athens (Phillips 1993). However, it may not be necessary to abandon hard-earned progress on pluralism and individual rights and liberties to gain ground on substantive democracy (Dagger 1997, 3–7). Nor need substantive democracy represent an extreme departure from what we know and feel comfortable with today.

There is not space here, even if we felt equipped for the task, to construct a fully developed description of what our society would be like if it were more oriented toward substantive democracy. With many other thinkers, the authors cited in the preceding paragraphs have examined aspects of the economy, social life, the voluntary sector, and government, offering ideas that emphasize substance and the normative character of governance as well as process and protection of rights. For the limited purposes of this essay, our conception of the ideal of substantive democracy may be summarized as a setting in which people may, if they choose, take part in governing themselves with a minimum of interference or resistance (for example, from economic or other elites, or administrative "experts"), and without being required to assume in advance a uniform or universal set of constraints (such as representative systems of decision making, or the normative, classical liberal view of the proper sphere of citizen action). This is a setting that allows people to create a society and future through informed dialogue and exchange of ideas (the classical republican model), in addition to the traditional American concern with defining rights and protecting and distinguishing the individual from the collectivity (the classical liberal model). It allows people to freely discuss their values and preferences absent the limitations of a predetermined split between the public (political) sphere and the private (economic) sphere. Thus, substantive democracy involves rekindling a public discourse about the purposes of collective action, accepting a role for citizens and public administrators in shaping the future.

Public administration must be a key actor in any effort to rediscover substantive democracy because of the complexity of providing public services in contemporary society. Creating new forms of public discourse and implementing the policy outcomes requires attention to the administrative apparatus of government and to the interplay of policy formulation and implementation. The task of rediscovering substantive democracy is made more difficult by the spread, over the past three decades, of economic theory throughout the social sciences, "a phenomenon commonly referred to as 'economic imperialism'" (Udehn 1996, 1). Over the past decade or two in the field of public administration, economic theory has become an important normative influence on the management of public organizations and their relationship to the broader society. As a result, elements of New Public Management (NPM) are the expected mode of operation for many public agencies in the United States and in a number of other nations (Kettl 1997).

This market-based model includes the familiar elements of shrinking government and making administration more efficient through use of private-sector performance-management and motivation techniques. It advocates treating citizens like customers, separating public administrators from the public policy process, and convincing both that government is nothing more than a business within the public sector. The assumption is that people are rational self-maximizers who compete with others and respond primarily to economic incentives. When such behavior occurs, it may be efficient in some sense, but it may also pose a threat to democratic governance (Terry 1998).

This, then, is the problem of democracy and public service in a postindustrial, liberal capitalist society. It is a society in which democracy is equated with equal procedural and personal rights, but not democratic determination of economic property rights. To the extent this situation is at variance with the American ideal of democracy, today we have something of a false democracy. Liberal capitalism and procedural democracy displaced the earlier republican vision (Sandel 1996) as Americans built the professional, bureaucratic, administrative state. Today the trend continues as the public sphere of life is increasingly occupied by the behaviors and values of the individualistic economic market. The effect on public administration is that the ideal public sector is thought to be small in size, efficient, and subservient, while simultaneously providing a broad range of effective, expertly run services. This is paradoxical and frustrating, but not surprising given the political culture associated with liberal capitalist democracy.

## THE DIFFERENCE OF NEW PUBLIC MANAGEMENT

In many ways, New Public Management has characteristics of previous management reform efforts, particularly in the twentieth century. The progressive movement included the rise of the city manager form of government, the Hoover and Brownlow Commission recommendations, management by objectives, and program, planning, and budgeting systems, all of which spoke to values of management efficiency, effectiveness, and performance improvement that are so much a part of NPM. Rosenbloom refers to this and the "public administration orthodoxy" reflected in the politics-administration dichotomy (Rosenbloom 1993, 503). However, we argue that these were all couched or justified within the framework of broader democratic values. Moe (1994) makes a similar point about the National Performance Review (NPR), noting that NPR was fundamentally different from previous reforms, which "all emphasized the need for democratic accountability of departmental and agency officers to the President and his central management agencies and through these institutions to the Congress" (112). Though the NPR is different in several ways from NPM, they share a focus on economic, market-based thinking in government.

The progressive movement began as a reaction against political machines and perceived subversion of democratic values through corruption and patronage systems that controlled who was elected to political office and who was rewarded with government employment. The rise of management reforms during this period focused on broadening participation and increasing access to elected office and the political process. Judd argues that there was a clear class bias in this effort, in that "municipal reformers shared a conviction that it was their responsibility to educate and instruct the lower classes about good government" (Judd 1988, 89). Judd also links this movement to the rise of "municipal experts." Similar arguments were made in the South in advocating for wresting control of the political process from the segregationist elements of the Democratic Party and "reforming" the political process.

Richard Childs, a founder of the city manager form of government and an excellent example of the progressive management-reform spirit, said the purpose of the council-manager plan is "not good government . . . but democratic government" (Childs 1952, 141). Childs's intent was for the city manager form to achieve a "practical working of the democratic process" that would include "sensitive responsiveness" that will "diligently cater to the sovereign people" (141). This is not to say that progressive reformers embraced these values or were even sympathetic to them in operation. Stivers (1995) has been critical of the motives of "bureau men" whose concerns about economy and efficiency, in her view, far outweighed social welfare interests. While one can debate whether the primary focus of progressive reformers was service or administrative efficiency (Schachter 1997), there is no doubt that preserving democratic values was a key argument used to justify these efforts. Efficiency was always offered as a way to help achieve democratic accountability. Many of the management reforms proposed by the Taft, Brownlow, and Hoover commissions were also couched in terms of preserving democratic values. Luther Gulick, the driving force behind the Brownlow Commission, was focused on how to link democratic leadership and accountability (Wamsley and Dudley 1998, 329). The Brownlow Commission, describing government efficiency, stated the following: "The efficiency of government rests upon two factors: the consent of the governed and good management. In a democracy consent may be achieved readily, though not without some effort as it is the cornerstone of the Constitution. Efficient management in a democracy is a factor of peculiar significance" (President's Committee 1937, 2–3).

The Hoover Commission framed its recommendations primarily in terms of the executive branch's accountability to Congress and the need to fix responsibility to the people, noting that "responsibility and accountability are impossible without authority" (Commission on Organization 1949, 154). Mosher and Appleby both note the concerns, however, that existed over the rise of professional management during this period. Mosher wrote that threats to public service and the "morality" of the service during this time included the potential move toward "the corporate, the professional perspective and away from that of the general interest" (1982, 210). Appleby (1952) expressed concern about protecting democratic values and argued that two factors were most critical: exposing administrators and their decisions to the electoral process, and a bureaucratic hierarchy that forces managerial decisions to be reviewed by broader and more politically aware upper level administrators.

As Arnold (1998) notes, the NPR reflects a very different orientation than previous reforms, even those that occurred under Carter and Reagan. As with NPM, this difference is that NPR makes little or no distinction between the role of government and the role of the marketplace. In fact, NPR moves beyond the concept of managing government organizations like a business to the idea that business itself should perform governmental functions. Arnold (1998) and Rosenbloom (1993) both note that NPR has a distinctively populist cast combined with a heavy focus on public choice economics. Rosenbloom refers to NPR's use of "neo-populist" prescriptions that advocate "decentralization, competition, deregulation, load-shedding, privatization, user fees, and 'enterprise' culture" (506).

Managerialism and New Public Management have been worldwide phenomena. Democratic regimes in New Zealand, Australia, and the United Kingdom have all implemented some range of reforms consistent with NPM (Eggers 1997; Pollitt 1993; Stewart and Kimber 1996). Malta and Austria have also implemented NPM elements (Maor 1999). Each of these initiatives has had some combination of elements including cost cutting, creating of separate agencies or "business enterprises" to eliminate traditional bureaucracies, separating the purchaser of goods from the provider of those goods, introducing market mechanisms, decentralizing management authority, introducing performance-management systems, moving away from tenure-like civil service systems to contractual and pay-for-performance personnel systems, and increasing use of customer-focused quality improvement systems (Armstrong 1998, 13).

Credit for the impetus of these reforms is given to American ideas, "particularly the ideas of American public choice economists" (Orchard 1998, 19–20). Pollitt (1993) links managerialism to Frederick Taylor and to Luther Gulick. While the ideas may have come from intellectual traditions in the United States, their implementation has primarily occurred in other countries. They are being implemented in very different ways, largely as a result of legal, social, political, and historical traditions that exist in each country. New Zealand is most often cited as "leading the way" in implementing NPM beginning in 1984. However, Pollitt notes that the United Kingdom had actually begun implementing such reforms in the mid- to late 1970s (52).

It is clear that New Zealand's reforms have been the most substantial and ongoing, for several reasons. New Zealand's initiative started with a Labour government and not with the more conservative National Party, but the NPM initiatives were supported by both. Second, New Zealand has no written constitution, a unitary rather than federated political system, a unicameral legislature, and a nonpartisan civil service. All these factors made implementation much easier to accomplish (Eggers 1997, 35–7). Countries with federal systems like Australia have had mixed experience with implementing NPM. This is the primary reason managerialism reforms have had less impact in the United States, where the federal structure is the most decentralized in the world. Also, NPM initiatives in the United States started locally, whereas in other nations they started at the national level (Osborne and Gaebler 1993).

If success is defined by the elements of NPM, some success has been achieved. Privatization of traditional government functions has been dramatic in New Zealand. In addition, there has been a clear demarcation between civil service managers and policy decisions made by political executives (Eggers 1997; Maor 1999). Australian public service has become less bureaucratic in terms of layers of hierarchy, rigidity of duties, and centralization of functions (Stewart and Kimber 1996, 47–9).

However, if one defines success as substantive involvement of citizens in shaping the direction of policy that affects their lives, there is little indication of such involvement beyond what existed before NPM implementation began. As Pollitt (1993) notes, citizenship is an awkward concept for those promoting managerialism, where the term "customer" is more common. He argues that the collectivist view of citizenship is "alien to an individualist model where the market is the chief focus of transactions and values" (125–6). Armstrong (1998) notes in his assessment of Australian implementation of NPM that the concept of meeting customer needs "ignores the ability of customers to articulate their needs or make choices, either because they are uninformed or do not have the resources to do so" (23–5). Rhodes further argues that in Australia, "there is no evidence to show that (NPM) has provided customers with any means whatever of holding the government to account" (1996, 106–10). Those claiming success for NPM have focused on short-term effects and on issues of efficiency. While it may be too early to assess the long-term impact of NPM in countries such as New Zealand and Australia, the evidence supporting democratic accountability and citizen engagement is not encouraging.

This concept of management has little to do with democracy and democratic values, shedding the reality or the facade of democracy found in earlier public-sector reforms. What is left is a core of market orientation to economic efficiency in the public sector.

## WHAT DO PEOPLE WANT FROM THEIR GOVERNMENT?

Thus far, we have argued that the market model of administration evident in NPM hinders any return to substantive democracy and limits the degree to which citizens can meaningfully affect policy and administration. New Public Management claims to make government customer-centered and therefore more responsive in its delivery of services. We suggest, however, that recent reforms fail to understand the basic foundation of public administration in democratic practice. As Borgmann (1992) argues, when citizens are recast as consumers, they operate within an attenuated form of democracy: "But to extol the consumer is to deny the citizen. When consumers begin to act, the fundamental decisions have already been made. Consumers are in a politically and morally weak position. They are politically weak because the signals that they can send to the authorities about the common order are for the most part ambiguous. Does the purchase of an article signal approval, thoughtlessness, or lack of a better alternative?" (115).

The issue of treating citizens solely as customers has also been addressed by others (deLeon and Denhardt 2000: Kettl 1997; Terry 1998). However, a deeper issue is the underlying debate about what people want from government. In our view, the market model of public administration reflects a disenchantment with the modern welfare state. The market model symbolically saves society from the bureaucratic leviathan to which the public service is wed and provides a clean, seemingly apolitical, solution to the messiness of social life. Though much of the critique of the welfare state is on target, there is value in holding on to seemingly anachronistic ideas such as citizenship, the public interest, social responsibility, and dialogue. In other words, we want to continue to claim there is a connection between public administration, governance, and social life (White and McSwain 1993).

It is quite evident that the highly individualized, technologically dynamic society in which we live is congruent with the market model of administration. For example, in the United States we are more and more likely to see ourselves as individualized users of discrete public goods and services. To a large degree, this has led to an evacuation of public life. Public managers, in turn, are focused more on the management of performance-based contracts than with face-to-face contact with the citizen. Clarke and Newman argue,

> [This] means that the capacity of organisations and management processes to respond to critical issues facing public services is very limited. Such issues . . . include crime, poverty, community safety, the care of the elderly and of people with disabilities, economic regeneration, environmental issues, transport, child protection and a host of others. . . . The pursuit of unconnected initiatives as organisations or government departments pursue an ever narrowing agenda and set of programmes defined around their core businesses serves to exacerbate, rather than address, such complex social issues and problems. The combined managerial and policy deficits in a dispersed field of power militates against the development of a capacity to address issues which resist being neatly defined as managerial problems. (1997, 148)

They go on to warn that the market model is very weak in its notions of citizenship and community: "The increasing adoption of consumerist discourse involves the dismantling of notions of collective power in favor of individualized users of services. It is the very power of this symbolism that leads to attempts to incorporate other formulations alongside it, as in organizational mission statements which talk of 'serving communities' as well as 'serving customers,' and the deployment of the language of 'citizens' to fill the spaces in the impoverished individualism of the discourse" (128).

Public administration is in an interesting position, as people no longer look to models of democratic practice to solve public problems. Ironically, by using the language of management, we are relegated to using technique to represent the democratic pole of the tension between bureaucracy and democracy. In this situation the market model is primarily about transforming the bureaucratic state, appealing to the public at large and to those in government, whose charge is to make the government "run better" and "cost less." To us, this appears to be too easy a solution because the market model assumes a return to homogenous society with very stable social institutions—a realm of social experience that no longer exists (what Clarke and Newman term "regressive modernization"). For example, welfare-to-work programs assume that all families are capable of getting off the welfare rolls if they just have a bit of "moral fiber." This model implicitly claims that welfare recipients have a responsibility to become self-sufficient because of the financial obligation that welfare payments place on other members of the community.

What this view does is erase an entire generation of social and political research that has identified structured differences and inequalities in society—class, gender, race and ethnicity, "ableness," and so forth. The result is a tendency to reduce complex social and economic issues to the management of diversity at the level of the individual organization. What we are suggesting is that the managerial state, that is, the market model of administration, avoids addressing the underlying social issues that affect society. Substantive issues at the core of contemporary life, such as racism, poverty, and disability, become individual problems rather than matters to be addressed through substantive democracy.

According to Clarke and Newman, the managerial state is an inherently unstable solution to the problems of the welfare state. Their argument, while written in the context of Great Britain's public management reform, applies equally to the United States:

> The imagery of the nation [as mono-cultural and free of conflict] is constantly interrupted by questions of the care of black elders, by the question of pension benefits for non-married couples and gay or lesbian partners, by employment tribunals confronted by evidence of the racist, sexist or homophobic organizational cultures of public services, by disability activists demanding citizenship rights, and by the long-running—and multi-faceted—"crisis of the family." In these and many more ways the unresolved crisis of the social settlement ensures that the formation of a new relationship between the state and the public will remain embattled—and unstable. (Clarke and Newman 1997, 155)

As citizens increasingly identify themselves as individual consumers and discrete users of government services, social issues are also cast in the same language and framework. For those in public administration to "buy in" to such a model seals their fate as managers of technique. In this way, the politics–administration dichotomy reasserts itself quite clearly. However, as we see it, society is being reconstructed with the political dimension being recast as issues of individual choice. People as individuals, and not society as a collective entity, are now responsible for solving complex problems. The result is a superficial gloss in the name of efficiency, while substantive issues—often those in which public administrators are most engaged—remain hidden but not solved. Thus, we maintain that public administration, as a crucial and unavoidable part of the public policy arena, is inherently about the social construction of society.

## DEMOCRACY AND PUBLIC ADMINISTRATION TODAY

We are concerned about the condition of democracy in American society and the resulting impact on public administration. The current environment of public institutions has deteriorated beyond procedural democracy to a market model in which citizens' primary action outside the household is earning money, to make product and service choices in the market economy to maximize the satisfaction of their desires. The result is a distancing of the citizen from her or his public-service institutions and a tacit assumption that interactions in the public sphere (determining what issues will be on the public agenda and how they will be addressed, for example) should also be left to the invisible hand of the market.

We argue, with Curtis Ventriss (1998), that public administrators and academics should play a part in the recovery of a substantive democratic ideal. Moving toward a substantive democratic ideal seems difficult because such a powerful die has been cast. James March suggests that we are locked into a social order that is based on "rationality and exchange rather than history, obligation, reason and learning" (1992, 230). Much of the conceptual development in public administration seems to be locked into this exchange mind-set as well, particularly in the public management movement.

The underlying approach offered by proponents of New Public Management is even more restrictive than the current trend in economic thinking. March notes, it "reflect[s] not so much an application of contemporary economic theory to government as a naive adaptation of an obsolescent version of that theory to modern political ideology" (230). More recent iterations of the economic approach argue that the traditional economic exchange model is complicated by history, "socially constructed institutions," and trust. Two examples of this perspective are North's *Institutions, Institutional Change and Economic Performance* (1990) and March and Olson's *Democratic Governance* (1995).

This newest version of neoinstitutionalism, while still clearly embedded in the functionalist paradigm (Burrell and Morgan 1979), incorporates the language of its critics. In this regard, writers like March and Olson have already responded to the next step in the development of this language game. This post-exchange perspective could eventually surface as an important aspect of mainstream public administration. Some examples include Lynn (1996) and Kettl (1997), both of whom express concern about the narrow foundation and the seeming hyperbole of New Public Management. Their responses, however, call for more rigorous methods of research, consistent with the post-exchange view of economics. This is a new and improved functionalism, a rationalized model that emphasizes predicting and controlling the behavior of institutions by determining the institutional rules. The emphasis is still on prediction and control, which results in an abstracted empiricism (Mills 1967) that is not particularly useful in an applied professional field like public administration (Box 1992).

The reality of our social experience is a hyper-rationalized world in which democracy is equated with consumer choice. The problem we face, then, is this: In what ways might we reassert a meaningful democratic context for the practice of public administration in light of such a social experience? We are aware that we cannot find solace in absolute, foundational principles that hold across time, space, and culture, nor can we return to the "certainties" of an earlier era. As Fox and Miller (1997, 88) put it, "the toothpaste cannot go back into the tube."

We are not pining for some overarching set of democratic values that will put us back on an imagined high moral plane of Democracy. We are mindful however, that American democracy made possible the idea of collective self-governance as a political end. As Gardbaum (1992, 760) notes, "For the first time, public life—previously closed to all but the political class—became an arena in which ordinary individuals could through participation and dialogue with others, define and realize themselves."

Let us review what is at stake. We want to assert that public administration does play a role in the social order. At every level of government—federal, state, and local—people in public agencies not only deliver services but also serve as facilitators, interpreters, and mediators of public action (Barth 1996; King and Stivers 1998a; Marshall and Choudhury 1997; Wamsley et al. 1990). This role has clear linkages to the founding period of the nation (McSwain 1985; Rohr 1986) and the substantive form of democracy outlined earlier in this essay. It should not be shelved in favor of a limited role for public administration involving satisfaction of consumer demands through focus groups and customer surveys, in which largely uninformed public opinion is equated with the public interest. Such an attenuated model of democracy suggests that humans are truly economic beings in search of narrow personal satisfaction; we do not regard this as an effective model of governance.

While it is easy enough to make this normative argument, the prevailing mind-set of human action is undoubtedly the market perspective. Many argue that the discourse of the Enlightenment and particularly the themes essential to democratic public administration, such as the public interest, justice, and progress, have been discarded because they have no empirically demonstrable foundational justification (Fox and Miller 1995; Marshall 1996; McSwite 1999). The market mind-set has surfaced in its place as the legacy of modernism. Much of our argument in this [reading] laments the passing of the normative basis on which many public administration writers have legitimated the role of public administration in the governance process. This normative basis is also, we have argued, the tradition of substantive democracy in America.

It is important to consider that the market model has its own modernist limitations (McSwite 1999, 8). March and others have already identified earlier economic approaches, such as public-choice economics, as naive and outdated. These writers seek to adjust economic approaches in light of current social experience. Indeed, the term "socially constructed reality," once marginalized by mainstream social science, is now commonly used across most disciplines (Barber 1984; March and Olson 1995). Many see the rise of New Public Management as the discipline's chance to regain influence and legitimacy, and there is significant political pressure on public administrators to adopt its approach.

However, it is encouraging to note the beginnings of resistance to what Stivers (2000) calls "the ascendancy of public management." In their criticism of the National Performance Review, which has market-based elements similar to those of the New Public Management movement in the United States, Moe and Gilmour (1995) argue that public administration must not ignore its normative grounding in public law. Kettl (1997) notes that customer satisfaction surveys would not serve as an effective proxy for the public interest. And Terry (1998) makes the point that both the entrepreneurial and the market-driven models of management displace the democratic foundation that is essential to public-sector leadership.

## AN ALTERNATIVE

The reader may reasonably expect the authors to introduce some better alternative, one that reasserts the normative democratic context of public administration. Why not merely reassert the democratic context we hold dear, redressing the imbalance that exists? In the current environment, we are readily able to talk about public management, but we have difficulty discussing the democratic context of public administration. According to Kirlin (1996, 417). this context includes achieving a democratic polity; addressing the nexus between larger societal issues and decision making in public organizations; confronting the complexity of instruments of collective action; and encouraging more effective societal learning.

In the search for an alternative or addition to the market-based model of public administration, we wish to affirm an emerging view that a central element should be a collaborative relationship between citizens and public administrators. This relationship is based on shared knowledge and decision making rather than control or pleasing and placating. It assumes that citizens have the ability to self-govern, even in these complex and confusing times. Further, this relationship between public employees and the public assumes that, while many people choose not to take part in public decision making, all citizens want to believe they could participate and could make a difference if they chose to do so.

This collaborative model of administration has been discussed in various forms by several authors over the past 10 years or so (Adams et al. 1990; Box 1998; King and Stivers 1998a; McSwite 1997; Stivers 1994; White and McSwain 1990). The emphasis in the collaborative model is giving citizens the knowledge and techniques they need to deal with public policy issues and providing an open and nonthreatening forum for deliberation and decision making (Box and Sagen 1998). This model is only one way to enhance substantive democracy, but we focus on it here because it presents a well-developed alternative that could be especially useful and powerful.

We are not arguing for greater legitimacy for public administration or a different view of public administration in the democratic order (Spicer and Terry 1993; Wamsley et al. 1990). There are elements of our call for substantive democracy that echo the New Public Administration of the 1970s (Frederickson 1980), with its emphasis on social equity, but our vision lacks that movement's sense of large-scale changes in the purpose and practice of public administration. We do, however, recognize that the old politics–administration dichotomy, born of a vision of public administrators as value-neutral implementors of public policy that is determined elsewhere, has long since been found to be a false description of the world of creation and implementation of public policy (Svara 1999). This is a world in which career public professionals interact with elected officials and citizens as they sense the "public interest," however that may be perceived, and work to solve problems and deliver services.

A collaborative model of the administrative role is not universal, in relation to individual public employees or to administrative tasks and situations. Not every career public employee interacts with elected officials or citizens or helps others who do. Many carry out technical and professional tasks within public organizations, tasks that are important to the public welfare but do not offer opportunities for the sort of collaboration discussed here. However, public service practitioners who interact with citizens (whether those citizens are leaders or everyday people concerned about the quality of life in their neighborhoods), can take incremental steps toward improving the quality of democracy by actively helping people govern themselves.

Adopting one particular model of administrative practice will not automatically result in substantive democracy. We need to move beyond describing our situation to taking what steps we can take now, through the practice of public administration, to recapture the values of substantive democracy. This requires the courage to share rather than control knowledge and administrative processes, to create opportunities for meaningful dialogue and decision making, and to listen and facilitate growth of individual understanding of public issues and the people involved in them (Box 1998; King and Stivers 1998c, 203).

Over time, this approach may enable public professionals to shift the balance, bit by bit, from the metaphor of the market to the values of substantive democracy. Action in this direction might involve, for example, choosing to create an ongoing structure for citizen administration of a particular program rather than using focus groups to sense relatively uninformed opinion. Or it might involve staffing an office that assists neighborhood organizations, rather than funding a public relations office that seeks to sell an image. It could mean allocating resources for infrastructure or school improvements on the basis of the seriousness of local problems, rather than formulas intended to spread funds evenly over political districts. And it could mean taking the initiative to fully inform elected officials about available action alternatives and their consequences for real people, instead of waiting for policy direction.

In the twenty-first century, there is much for public administration to do beyond the mandate of perfecting efficient mechanisms for service delivery. Our intent is to contribute to recrafting a public administration that supports the values of substantive democracy in a time of significant change in the public sector. We are aware of the normative weight we place upon public administration, which is often thought by citizens and people working in the field to have little normative meaning or purpose (this is highlighted by the subtext of the message of New Public Management, which is to strip administration of disabling idealism and "cut to the chase"— economically efficient results). Despite the odds, we believe the stakes are well worth the effort.

## REFERENCES

Adams, Guy B., Priscilla V. Bowerman, Kenneth M. Dolbeare, and Camilla Stivers. 1990. "Joining Purpose to Practice: A Democratic Identity for the Public Service." In *Images and Identities in Public Administration,* ed. Henry D. Kass and Bayard L. Catron, 219–40. Newbury Park, CA: Sage.

Appleby, Paul H. 1952. *Morality and Administration in Democratic Government.* Baton Rouge: Louisiana State University Press.

Armstrong, Anona. 1998. "A Comparative Analysis: New Public Management—The Way Ahead?" *Australian Journal of Public Administration* 57(2): 12–26.

Arnold. Peri E. 1998. *Making The Managerial Presidency.* 2nd ed. Lawrence: University Press of Kansas.

Barber, Benjamin. 1984. *Strong Democracy: Participatory Politics for a New Age.* Berkeley: University of California Press.

Barth, Thomas J. 1996. "Administering the Public Interest: The Facilitative Role for Public Administrators." In *Refounding Democratic Public Administration: Modern Paradoxes, Postmodern Challenges,* ed. Gary L Wamsley and James F. Wolf, 168–97. Thousand Oaks, CA: Sage.

Borgmann, Albert. 1992. *Crossing the Postmodern Divide.* Chicago: University of Chicago Press.

Bowles, Samuel, and Herbert Gintis. 1987. *Democracy and Capitalism: Property, Community, and the Contradictions of Modern Social Thought.* New York: Basic Books.

Box, Richard C. 1992. "An Examination of the Debate over Research in Public Administration." *Public Administration Review* 52(l): 62–9.

———. 1998. *Citizen Governance: Leading American Communities Into the 21st Century.* Thousand Oaks, CA: Sage.

Box, Richard C., and Deborah A. Sagen. 1998. "Working With Citizens: Breaking Down Barriers to Citizen Self-Governance." In *Government Is Us: Public Administration in an Anti-Government Era,* ed. Cheryl S. King and Camilla Stivers, 158–72. Thousand Oaks, CA: Sage.

Burrell, Gibson, and Gareth Morgan. 1979. *Sociological Paradigms and Organizational Analysis.* London: Heinemann.

Campbell, James. 1996. *Understanding John Dewey: Nature and Cooperative Intelligence.* Chicago: Open Court Publishing.

Childs, Richard. 1952. *Civic Victories.* New York: Harper and Brothers.

Clarke, John, and Janet Newman. 1997. *The Managerial State: Power, Politics and Ideology in the Remaking of Social Welfare.* Thousand Oaks, CA: Sage.

Commission on Organization of the Executive Branch of Government. 1949. *Report of the Commission.* Washington, DC: Government Printing Office.

Conway, David. 1996. "Capitalism and Community." *Social Philosophy and* Policy 13(1): 137–63.

Dagger, Richard. 1997. *Civic Virtues: Rights, Citizenship, and Republican Liberalism.* Oxford, UK: Oxford University Press.

Dahl, Robert A. 1982. *Dilemmas of Pluralist Democracy: Autonomy vs. Control.* New Haven, CT: Yale University Press.

deLeon, Linda, and Robert Denhardt. 2000. "The Political Theory of Reinvention." *Public Administration Review* 60(2): 89–97.

Dewey, John. [1927] 1985. *The Public and its Problems.* Athens, OH: Swallow Press Books.

Eggers, William D. 1997. "The Incredible Shrinking State." *Reason* 29(1): 35–42.

Fox, Charles J., and Hugh T. Miller. 1995. *Postmodern Public Administration: Toward Discourse.* Thousand Oaks, CA: Sage.

———. 1997. "Can the Toothpaste Be Pushed Back into the Tube?: The Return of Foundationalism to Public Administration." *Administrative Theory & Praxis* 19(1): 88–91.

Frederickson, George H. 1980. *New Public Administration.* University: University of Alabama Press.

Gardbaum, Stephen A. 1992. "Law, Politics and the Claims of Community." *Michigan Law Review* 90(4): 685–760.

Hollinger, Robert. 1996. *The Dark Side of Liberalism: Elitism vs. Democracy.* Westport, CT: Praeger.

Judd, Dennis 1988. *The Politics of American Cities.* New York: HarperCollins.

Kettl, Donald F. 1997. "The Global Revolution in Public Management: Driving Themes, Missing Links." *Journal of Policy Analysis and Management* 16(3): 446–62.

King, Cheryl Simrell, and Camilla Stivers. 1998a. *Government Is Us: Public Administration in an Anti-Government Era.* Thousand Oaks, CA: Sage.

———. 1998b. "Introduction: The Anti-Government Era." *In Government Is Us: Public Administration in an Anti-Government Era,* ed. Cheryl Simrell King and Camilla Stivers, 3–18. Thousand Oaks, CA: Sage.

———. 1998c. "Conclusion: Strategies for an Anti-Government Era." In *Government Is Us: Public Administration in an Anti-Government Era,* ed. Cheryl Simrell King and Camilla Stivers, 193–204: Thousand Oaks, CA: Sage.

Kirlin, John J. 1996. "The Big Questions of Public Administration in a Democracy." *Public Administration Review* 56(5): 416–23.

Lynn, Laurence E., Jr. 1996. *Public Management as Art, Science and Profession.* Chatham, NJ: Chatham House.

Macpherson, C.B. 1977. *The Life and Times of Liberal Democracy.* Oxford, UK: Oxford University Press.

Mansbridge, Jane. 1999. "The Holding Pattern." *Political Theory* 27(5): 706–16.

Maor, Moshe. 1999. "The Paradox of Managerialism." *Public Administration Review* 59(2): 5–18.

March, James G. 1992. "The War Is Over, The Victors Have Lost." *Journal of Public Administration Research and Theory* 2(3): 225–331.

March, James G., and Johan P. Olson. 1995. *Democratic Governance.* New York: Free Press.

Marshall, Gary S. 1996. "Deconstructing Administrative Behavior: The 'Real' as Representation." *Administrative Theory & Praxis* 18(1): 117–27.

Marshall, Gary S., and Enamul Choudhury. 1997. "Public Administration and the Public Interest: Re-presenting a Lost Concept." *American Behavioral Scientist* 41(l): 119–31.

Matthews, Richard K. 1986. *The Radical Politics of Thomas Jefferson: A Revisionist View.* Lawrence: University Press of Kansas.

McSwain, Cynthia J. 1985. "Administrators and Citizenship: The Liberalist Legacy of the Constitution." *Administration & Society* 17(2): 131–48.

McSwite, O.C. 1997. *Legitimacy in Public Administration: A Discourse Analysis.* Thousand Oaks, CA: Sage.

———. 1999. "On the Proper Relationship of the Theory Community to the Mainstream Public Administration Community." *Administrative Theory & Praxis* 21(1): 4–9.

Mills, C.W. 1967. *The Sociological Imagination.* London: Oxford University Press.

Moe, Ronald C. 1994. "The 'Reinventing Government' Exercise: Misinterpreting the Problem, Misjudging the Consequences." *Public Administration Review* 54(2): 111–22.

Moe, Ronald C., and Robert S. Gilmour. 1995. "Rediscovering Principles of Public Administration: The Neglected Foundation of Public Law." *Public Administration Review* 55(2): 135–46.

Mosher, Frederick C. 1982. *Democracy and the Public Service.* 2nd ed. Oxford, UK: Oxford University Press.

North, Douglass C. 1990. *Institutions, Institutional Change and Economic Performance.* New York: Cambridge University Press.

Okun, Arthur M. 1975. *Equality and Efficiency: The Big Tradeoff.* Washington, DC: The Brookings Institution.

Orchard, Lionel. 1998. "Managerialism, Economic Rationalism and Public Sector Reform in Australia: Connections, Divergences, Alternatives." *Australian Journal of Public Administration* 57(1): 19–33.

Osborne, David, and Ted Gaebler. 1993. *Reinventing Government: How the Entrepreneurial Spirit Is Transforming the Public Sector.* New York: Penguin Books.

Phillips, Derek L. 1993. *Looking Backward: A Critical Appraisal of Commmunitarian Thought.* Princeton, NJ: Princeton University Press.

Pollitt, Christopher. 1993. *Managerialism and the Public Services.* Cambridge, MA: Blackwell.

President's Committee on Administrative Management. 1937. *Report of the Committee.* Washington, DC: Government Printing Office.

Rhodes, Rod. 1996. "Looking Beyond Managerialism." *Australian Journal of Public Administration* 55(2): 106–10.

Rohr, John A. 1986. *To Run a Constitution: The Legitimacy of the Administrative State.* Lawrence: University Press of Kansas.

Rosenbloom, David H. 1993. "Editorial: Have an Administrative Rx? Don't Forget the Politics!" *Public Administration Review* 53(6): 503–6.

Sandel, Michael J. 1996. *Democracy's Discontent: America in Search of a Public Philosophy.* Cambridge, MA: Harvard University Press.

Schachter, Hindy Lauer. 1997. "Settlement Women and Bureau Men: Did They Share a Usable Past?" *Public Administration Review* 57(1): 93–4.

Sheldon, Garrett W. 1991. *The Political Philosophy of Thomas Jefferson.* Baltimore: Johns Hopkins University Press.

Spicer, Michael W., and Larry D. Terry. 1993. "Legitimacy, History, and Logic: Public Administration and the Constitution." *Public Administration Review* 53(3): 239–46.

Stewart, Jenny, and Megan Kimber. 1996. "The Transformation of Bureaucracy? Structural Change in the Commonwealth Public Service 1983–93." *Australian Journal of Public Administration* 55(3): 37–49.

Stivers, Camilla. 1994. "The Listening Bureaucrat: Responsiveness in Public Administration." *Public Administration Review* 54(4): 364–9.

———. 1995. "Settlement Women and Bureau Men: Constructing a Usable Past for Public Administration." *Public Administration Review* 55(6): 522–9.

———. 2000. "Resisting the Ascendancy of Public Management: Normative Theory and Public Administration." *Administrative Theory & Praxis* 22(1): 10–23.

Svara, James H. 1999. "Complementarity of Politics and Administration as a Legitimate Alternative to the Dichotomy Model." *Administration & Society* 30(6): 676–705.

Terry, Larry D. 1998. "Administrative Leadership, NeoManagerialism, and the Public Management Movement." *Public Administration Review* 58(3): 194–200.

Udehn, Lars. 1996. *The Limits of Public Choice: A Sociological Critique of the Economic Theory of Politics.* London: Routledge.

Ventriss, Curtis. 1998. "Radical Democratic Thought and Contemporary American Public Administration: A Substantive Perspective." *American Review of Public Administration* 28(3): 227–45.

Waldo, Dwight 1981. *The Enterprise of Public Administration: A Summary View.* Novato, CA: Chandler and Sharp.

Wamsley, Gary L., and Larkin S. Dudley. 1998. "From Reorganization to Reinventing: Sixty Years and We Still Don't Get It." *International Journal of Public Administration* 21(2/4): 323–74.

Wamsley, Gary, Robert L. Bacher, Charles T. Goodsell, Phillip Kronenberg, John A. Rohr, Camilla Stivers, Orion F. White, Jr., and James F. Wolf. 1990. *Refounding Public Administration.* Newbury Park, CA: Sage.

White, Orion F., Jr., and Cynthia J. McSwain. 1990. "The Phoenix Project: Raising a New Image of Public Administration from the Ashes of the Past." In *Images and Identities in Public Administration,* ed. Henry D. Kass and Bayard L. Catron, 23–59. Newbury Park, CA: Sage.

———. 1993. "The Semiotic Way of Knowing and Public Administration." *Administrative Theory & Praxis* 15(1): 18–35.

Wills, Garry. 1979. *Inventing America: Jefferson's Declaration of Independence.* New York: Vintage Books.

Wood, Ellen M. 1996. *Democracy Against Capitalism: Renewing Historical Materialism.* Cambridge, UK: Cambridge University Press.

Yankelovich, Daniel. 1991. *Coming to Public Judgment: Making Democracy Work in a Complex World.* Syracuse, NY: Syracuse University Press.

READING 5.3

# "CITIZENS AND ADMINISTRATORS: ROLES AND RELATIONSHIPS"

CHERYL SIMRELL KING AND CAMILLA STIVERS

Thus far in our attempts to understand the American people's anti-government feelings, we have looked at the history of popular attitudes toward government, the current political economy and its impact on citizens' lives, and the knowledge gap that lies at the heart of representative government. To fill out the picture, one more element is important: the respective ways in which citizenship and administration have been understood and practiced in the United States. It is safe to say that, in general, neither citizens nor administrators are happy with their roles and relationships in the late 20th century, nor are they happy with each other. For every citizen cry against the bureaucracy, there is a matching administrative response that disparages a lazy, apathetic, and uncommitted citizenry. As a Harwood Group study for the Kettering Foundation (1991) states, although Americans have always been distrustful and cynical about government, never before have officials and citizens been so disconnected from each other.

As we will see, throughout our history, ideas about citizenship have given it a constricted and instrumental role, one borne out in the rather limited and distanced part citizens actually play in government today. On the other hand, ideas about the proper role of administration and its actual part in U.S. governance have expanded during the past 200 years. After reviewing the development of these respective roles and how they appear today, this [reading] concludes with a discussion of their implications for public administrators and their sense of their working lives, drawing on research conducted especially for this [reading].

## CHANGING IDEAS OF CITIZENSHIP AND ADMINISTRATION

The citizen role in U.S. governance has been defined narrowly since the founding. As the *Federalist Papers* made clear, the founders believed that the extended geographic scope and social complexity of the new American state made direct participation by citizens unworkable. James Madison argued that popular governance could only work in "a small spot" (Cooke 1961, 84). More important, however, in Madison's view, was the propensity of popular governments to the

---

"violence of faction" and their tendency to produce decisions based on "the superior force of an interested and overbearing majority" (57). Representation would not only make it possible to extend government over a large area but, by restricting citizen involvement to the selection of representatives, would "refine and enlarge the public views, by passing them through the medium of a chosen body of citizens, whose wisdom may best discern the true interest of their country" (62–63). Thus, the founders' faith in the will of the people was tempered by acute awareness of the potentially negative effects of citizen power, particularly citizens who were not of the "chosen body."

The Federalists held that, by and large, ordinary people were neither qualified for nor interested in participating directly in governance. A well-run government would win the continuing allegiance of citizens and make their involvement, other than as voters, unnecessary. Alexander Hamilton argued that as people grew accustomed to national authority "in the common occurrences of their political life," familiarity would "put in motion the most active springs of the human heart" and will for the national government "the respect and attachment of the community" (Cooke 1961, 173). This argument, grounded in a view of the people as the legitimate source of legitimacy but in need of protection against their own errors and delusions, sounds a theme that persists throughout subsequent thinking about the relation between citizenship and administration. Instead of a democracy "of the people, by the people, and for the people," the founders' democracy would be an "elite republic of elected representatives who would deliberate together and speak *for* the people" (Fishkin 1995, 21).

The founders' skepticism about the governing abilities of ordinary people made them confine the citizen role to that of voter. Voting is, in fact, the only guaranteed citizenship activity in the United States. If you are of a certain age, were born in the United States or have been naturalized, and have not lost your citizenship rights, you can participate in the voting process. Although, as we have seen, electoral turnouts have varied over the course of U.S. history, voting remains the defining feature of citizenship. In this context, being a citizen means exercising the right to choose political leaders, nothing more. Citizenship is much more a status than a practice.

Perhaps because voting is such a restricted role, less that half of eligible voters today cast ballots in presidential elections; turnouts in state and local races are generally even lower. A 1992 national election study showed that 78% of white, 67% of black, and 61% of Hispanic registered voters actually cast ballots (Flanigan & Zingale 1994). Registration rates are lower for blacks and Hispanics than for whites, making these voting rates even more problematic. College graduates are more likely to vote than are those with a high school diploma. Those who do not complete high school are least likely to vote. Thus, in practice, there is a class and race bias to the electorate. And those who do vote may be more likely than in the past to feel disaffected, as was seen in 1992 and 1994, "Years of the Angry Voter."

During the early years of the United States, while the public role of ordinary people was restricted to voting, governing was left safely in the hands of wise leaders. From the time of George Washington through the presidency of John Quincy Adams, what Frederick C. Mosher (1982) called "government by gentlemen" prevailed. Although government during this period was generally small and weak, the ambiguity of the Constitution about the proper role of administration planted the seeds of what would become a continuing struggle between the executive and Congress for control of administrative agencies, one that created a space in which agencies "set about developing power resources of their own" (Nelson 1982, 755). During this period, however, the federal government mostly delivered mail, fought wars, secured new territories, and collected customs and excise taxes. Administrative agencies were organized in a semi-aristocratic rather than a bureaucratic fashion, one based on the sense that administration was a "mystery . . . [therefore] a fitting vocation for men who were supposed to carry the knack for governance in their blood and breeding" (Matthew Crenson, quoted in Nelson 1982, 756). Kinship and class shaped membership in the early administrative echelons.

Not until the presidency of Andrew Jackson did thinking about the capacities and proper role of citizens and administrators change. Jacksonian thinking closed the gap between the two, in theory and practice, at least for a time. Jackson asserted, "The duties of all public officers are, or at least admit of being made, so plain and simple that men of intelligence may readily qualify themselves for their performance" (quoted in Nelson 1982, 759). This philosophy greatly expanded the pool of fit candidates for administrative office, from the well-born to any citizen who had demonstrated his loyalty to the political party in power (still confined, during this period, to anyone white and male). Government jobs became, in Senator William Marcy's famous words, "the spoils" of electoral victory (Nelson 1982, 759). The door to direct involvement in governmental processes was opened to a great many ordinary citizens. Throughout the balance of the 19th century, citizens were not only voters, they could also see themselves as, potentially, occupants of administrative jobs; in other words, as participants in governance.

. . . this new view of citizenship led to an upwelling of public activity: meetings, rallies, parades, and the like. For those who qualified, the citizen role was not just a legal status but a performance. In small towns, citizens and their officials worked closely together to govern their communities, although only a select group of citizens was actually involved (Skowronek 1982; Wiebe 1967). In rural areas, citizen involvement was paternalistic, populist, and moralistic: "Daily life among the conflicts and inequalities in these intimate communities was eased by community values emphasizing neighborly morality as well as a nonconfrontational norm for public demeanor of exhibiting friendliness and egalitarianism" (Tauxe 1995, 473). In urban areas, participation was engendered through political machines, where party loyalty for the ordinary man or money and influence for the rich could buy participation in politics and administration. Yet for every person in a town, city, or rural area who felt a part of their government, there were handfuls more who were on the outside: slaves or those one step from slavery, women, immigrants, and other marginal people.

The administrative role, still limited by the restricted scope of government responsibilities, was defined largely by ties to political parties. Jackson, however, sought to systematize agency processes somewhat in order to make good on his statement that anyone could do government work (Morone 1990), thus beginning the slow march toward bureaucratization. Government jobs functioned as a career ladder for immigrants and other men of modest means who had proven their worth to party leaders. When one party was swept out of office at election time, all the workers who had gotten their jobs through that party disappeared, to be replaced by loyalists of the rival party.

The extent to which this broadly democratic approach to administration produced governments riddled with corruption and incapable of efficient execution of the laws is a matter of debate. Today our view of the Jacksonian approach and the machine governments it produced is colored by the aspersions of turn-of-the-century Progressive reformers who sought to dislodge party loyalists and install themselves, or people like them, in the halls of administration. In any event, in the late 19th and early 20th centuries another sea change occurred in views of citizenship and administration, one that reversed the emphasis on direct involvement of ordinary citizens and the simplicity of government work, arguing instead the need for administrative expertise.

Woodrow Wilson's (1887) famous essay, "The Study of Administration," is emblematic of the shift. Arguing that administration was the prime governmental challenge, Wilson asserted the need for business-like and expert methods. Echoing Hamilton, he believed that administrators should be given considerable latitude in the execution of their duties, a freedom that was defensible because, he argued, administration was not political. Administrators carried out the laws, holding themselves accountable to the citizenry at large, whose role was to serve as ultimate source of legitimacy and not to become meddlesome. Citizens became the source of something called public opinion, a factor in political life that ever since has been the object of keen interest, if not outright manipulation, on the part of governmental leaders, as ubiquitous opinion polls today on every conceivable topic and candidate for office attest.

Progressive reformers, concerned with rescuing governments, especially city governments, from what they saw as the ill effects of machine politics, called for administrative practice based on scientific knowledge. In their view, the proper role of citizens in the reform process was to inform themselves about issues and rally around the quest for efficient, expert government methods. Progressives sought, then, to improve public opinion by making it judicious rather than meddlesome. Citizens were assured that the experts and professional administrators were more capable of handling public problems and situations and were better able to make decisions than common folk. The public service, growing rapidly due to the need for infrastructure and services, particularly in cities, was developing into a government "of the technocrats, by the technocrats, and for the technocrats" (Kearney & Sinha 1988, 571).

Whereas the reformers saw the dissemination of technical knowledge as a means of improving the relationship between citizens and their government by informing citizens, which would lead to citizen understanding and support for government, over time, as Dwight Waldo (1948) observed, "research and facts have come to be regarded less and less as devices of citizen cooperation and control and more and more as instruments of executive management" (43n). The advent of scientific management and efforts to professionalize the public service transmuted facts from ammunition for what had been called "efficient citizenship" to the basis of increasingly specialized modes of public administration.

Vast increases in the size and scope of governments at all levels during the first half of the 20th century seemed to justify the continuing call for expert, professional administration. In this environment, citizens relied on policy makers and administrators to make decisions that "enhance[d] the greater good of the community" (Parr & Gates 1989, 55). The administrator was charged with implementing programs that met the policy directives of elected officials. Citizens, busy with the demands of an increasingly complex world and recovering from wars and economic depressions tended to trust, or at least to tolerate, decisions made by administrators. Trust in a knowledgeable elite grew out of increasing reliance on science, and on those who could practice it or apply it, to address difficult issues of industrialization and technological progress. Demands for an active citizen role were muted during this period.

In the 1960s, however, the dialogue about citizenship and administration shifted once again. Growing public distrust of governmental institutions, engendered by events such as Watergate and the war in Vietnam, drove many citizens to challenge the legitimacy of administrative as well as political decisions (Parr & Gates 1989). This distrust, coupled with federal mandates requiring more public participation, opened the door for citizens to become more involved in administrative processes. The Economic Opportunity Act of 1964, which launched the War on Poverty, called for "maximum feasible participation" by the poor in governmental programs aimed at solving their problems. It authorized a determinative role for citizens in deciding about sizable expenditures and the design and execution of significant programs, a role that met with vigorous resistance by established city governments. After only three years, the Green Amendment put poverty programs back under municipal control. Thus, the federal government flirtation with "power to the people" was extremely short-lived. Meaningful citizen involvement lived on here and there, for example, in the community health center program . . . and in a plethora of administrative regulations aimed at getting citizen input. In practice, however, many of these regulations were interpreted in a fashion that turned citizen participation into an instrument for the achievement of administrative objectives rather than genuine collaboration or the sharing of authority with citizens.

Generally speaking, citizens came to be seen as clients or consumers, whose needs and demands, although legitimate, tended to compromise the rational allocation of resources and the impartiality of standardized procedures Citizens were viewed as passive recipients of governmental services rather than active agents who could work with administrators to deal meaningfully with their predicaments. Requirements for citizen participation were generally treated in administrative agencies as a cost of doing business instead of as an asset to effectiveness or a responsibility worth carrying out for its own sake (Jones 1981; Mladenka 1981; Thomas 1995). At best, citizens were viewed as a constituency, the source of important political support (MacNair, Caldwell, & Pollane 1983) or of important values to guide policy decisions (Stewart, Dennis, & Ely 1984). Only in the notion of *co-production* (Sharp 1980) could an active citizen role be glimpsed. Coproduction was based on the idea that citizens could play an active part in the production and delivery of services; yet the idea faced uphill sledding in the context of widespread skepticism among administrators and politicians about the skills and wisdom of ordinary people. As a result, by the late 1970s, public participation began to be perceived as detracting from administrative expertise. Increasingly, administrators came to see citizen participation as a cost of doing public business. This attitude combined with a decreased level of community activism led to a decline in the practice of public participation.

In 1984, about 50 leading scholars and practitioners of public administration took part in a conference on citizenship and public administration held in New York City. Most of the participants were strongly committed to active citizenship but skeptical about its feasibility. Again and again throughout the discussions, speakers urged public administrators to broaden their base by encouraging members of the public to become active in agency work. A central part of the administrative role in these arguments was the education of citizens and their integration into administrative decision making and implementation. Yet many at the conference found it difficult to specify the particular forms and mechanisms through which such relationships could be nourished. "Cudgling my mind as I may," said Dwight Waldo (1984), "I cannot imagine what . . . it would look like in practice" (108).

In fact, few documented models exist of successful citizen participation in governmental agency activities. Most discussions of it focus on theoretical pros or cons rather than showing how it works. Yet, as Dennis Thompson (1970) said, "Ideals have to be grounded in reality" (30). The importance of living examples of active citizenship has been recognized at least since Phillip Selznick's classic case study, *TVA and the Grass Roots* (1949). Selznick argued that a commitment to democracy requires wrestling with the question of the concrete circumstances under which it can occur. Although Selznick's study is often cited as proof that citizens are inevitably co-opted by administrative agencies, his research in fact also lends strong support to the idea that the succcss of collaborative efforts to work with citizens depends on the nature of organizational arrangements:

> The tendency of democratic participation to break down into administrative involvement requires constant attention. This must be seen as part of the organizational problem of democracy and not as a matter of the morals or good will of administrative agents. . . . For the things which are important in the analysis of democracy are those which bind the hands of good men. We then learn that something more than virtue is necessary in the realm of circumstances and power. (266)

The study by Berry, Portney, and Thomson (1993) of citizen involvement systems in five U.S. cities not only offers one of the few available detailed descriptions of such systems, but echoes Selznick in its emphasis on bringing to light organizational and other factors that contribute to success . . . .

In the 1990s, environmental activism, new class social movements, neighborhood action in response to crime and other urban problems, and political organization around ideological issues led to a resurgence in public participation activity (Thomas 1995; Timney, 1996) and to changes in the citizen-administrator relationship. Ironically, although participation in voting is at an all-time low (less than 25% of the eligible voting population elected President Clinton in 1996), and observers are decrying a general lack of civic involvement (e.g., Putnam 1995), some citizens are demanding a place at the table in administrative decision making. According to a Kettering Foundation study (1991), citizens are not apathetic, as many claim, but rather, feel "impotent" (4). Apathy implies a voluntary, intentional choice. Impotence is involuntary; citizens believe their lack of participation has been thrust on them, against their wishes.

In the current political economy, citizenship tends to be equated with paying taxes and consuming benefits. From this perspective, government uses up people's money and gives them back certain goods and services, a view that restricts possible relationships between citizens and officials. As long as citizens see themselves as taxpayers first and members of a civil collective second (if at all), consumption, or the purchase and use of goods and services, becomes the main connection between citizens and government. Citizens judge government by whether they feel satisfied with the results of their consumption.

Politicians and administrators play to these citizen perceptions. For example, in 1996, Camden County, New Jersey, opened a "county store" in a local mall. In addition to offering county services, the store also makes it possible for people to pay utility bills, register for school, attend small-business classes, register to vote, return library books, secure a photo ID, or get various health-related tests. The county has endeared itself to its residents and won awards for the concept. But, a recent *PA Times* article referring to county citizens as "consumers" or "taxpayers" points out that the store was created to meet "taxpayer" demand for "value." For the store, success is measured by whether citizens feel they are getting their "money's worth" (Bezich 1997, 1). Thus, the citizen role is couched in terms of purchasing decisions rather than in terms of a share in the authority and dignity of public life. In such cases, administrators seem to have swallowed whole the assumption that the entire point of public administration is keeping the customer happy.

---

**The Spiritual Condition of Citizenship**

"Today, of course, many people who are disengaged from prevailing allegiances have not acquired new ones, and so are inattentive to political concerns of any kind. They are neither radical nor reactionary. They are inactionary. If we accept the Greek's definition of the idiot as an altogether private man, then we must conclude that many citizens of many societies are indeed idiots. This—and I use the word with care—this spiritual condition seems to me the key to much modern malaise among political intellectuals, as well as the key to much political bewilderment in modern society." (C. Wright Mills 1959, 41)

---

Seeing citizens as consumers, taxpayers, and customers, and encouraging them to see themselves that way, leads people to evaluate government according to what each individual receives rather than what the community as a whole receives. It also obscures aspects of public life that extend beyond who gets what—such as who decides, who participates, and the quality of relationships among citizens and between citizens and government. When one's only yardstick for assessing government is how long one had to wait in line or the size of one's own stock of government goodies in comparison to those of others, it is easy to turn individual events and allocations into a sense of isolation and discontent. Political philosophers like John Stuart Mill and Alexis de Tocqueville argued that one of the chief benefits of direct citizen involvement in government was that ordinary people would come to see how their own lives were interwoven with the lives and fortunes of others and be able to raise their sights from what they themselves received from government to the overall good of the community. When people think of themselves as consumers or taxpayers and have no say in how things are decided, there is little prompting for them to take the high road and put the public interest ahead of their own private wants. At the same time, administrators see the ideal citizen as one who understands citizenship as being a follower, supporter, and ratifier of government action, conforming to the administratively defined mandate and climate.

In such a climate, it is not surprising that some citizens get fed up with following, accepting, and receiving what administrators think they should have. Many believe that their concerns will only be heard if they organize into protest groups and vocalize angrily about administrative policy decisions (Timney 1996). Citizens involved in NIMBY (Not In My Backyard) actions believe they must turn to confrontation because administrators operate on the basis of their own (or their agency's) interests and aren't really concerned with the impact of agency actions on citizens (Kettering Foundation 1991). In this context, citizens become dissenters, moving from what might have been collaboration with administrators to confrontation that pits them against administrators.

Many citizens perceive the information they receive from agencies as managed, controlled, and manipulated in order to limit their capacity to participate. They see the techniques of participation (public hearings, surveys, focus groups) as designed, at best, to generate input but to keep citizens on the outside of the governance process. They are particularly sensitive to vacuous or false participation efforts that ask for and then discount public input. Such inauthentic processes simply lead to greater tension between administrators and citizens. It is better not to work with

### The Progressive Legacy: Neutral Competence, Expertise, and Managerialism

*Lisa Zanetti*

For all its emphasis on technical expertise and political neutrality, the field of public administration was also born out of a desire for reform and some version of social justice. The Pendleton Act had instituted the beginnings of a competitive civil service to combat patronage and corruption in government (Van Riper 1958). This reform set the stage for the social programs of the Progressive Era (Adams 1992; Stever 1986). "Adventurous pragmatists" such as Walter Lippmann, Herbert Croly, Thorstein Veblen, John Dewey, and, to some extent, Mary Parker Follett saw in the philosophy of progressivism a means for setting into motion a reform agenda that would rehabilitate political liberalism, using a combination of organic idealism and pragmatism to justify placing the administrator as an important, and legitimate, agent of reform (Stever 1986, 1990).

Organic idealism was most influential as an intellectual force in America in the northeast, particularly among the educated upper class. As a political philosophy, idealism originated in early 19th-century Europe, first among German philosophers such as Johann Fichte, Friedrich Schelling, and G.W. Friedrich Hegel, and later among the British idealists T.H. Green and his students, Bernard Bosanquet and L.T. Hobhouse. Whereas both the German and the British schools viewed the state as the visible, tangible expression of the achievement of abstract reason in a given nation, the British believed that the state could be constructed and administered democratically. The Germans tended to relinquish the construction and administration of the state to a strong central ruler and a dominant civil service class (Hollinger & Capper 1989; Sabine & Thorson 1973; Stever 1986, 1990).

Green was particularly influential in revising the view of liberalism to one that recognized the mutual relationship between the individual and the social community. This independence was an ethical, not simply a legal, conception. Although the state could not compel individuals to be moral, it could create the conditions necessary to develop a responsible moral character—access to education, reduction of poverty, and regulation of the market, portrayed as a social institution rather than a natural condition. Liberalism, and liberal policies, ought to be an effort to provide n humane way of living to the largest number of persons (Sabine & Thorson 1973).

Among public administrationists, the influence of idealism was most pronounced in the work of Wilson, Goodnow, and Follett.[a] Wilson and Follett were both exposed to the traditions of German idealism in their studies, but all proponents of an organic approach, in various ways, stressed the need for an evolutionary development of American liberal democracy and argued that the administrator could contribute to this evolution without being a disruptive influence. The vision was that of a nonthreatening administrator who could exercise the necessary technical competence within the democratic political framework (Stever 1986, 1990).

The organic administrator stressed the need for social change to fit within the native tradition. Professional administrators could contribute to the stable evolution of the American social and political system, but the future would necessarily have to be constrained by the past. The concept of planning was incompatible with such a perception of the administrative role. When adminis-

citizens at all than to work with them under false, purely instrumental pretenses (King, Feltey, & Susel 1998).

## THE PUBLIC ADMINISTRATOR'S ROLE

As we saw previously, since early in the 20th century, the administrator's role has shifted from party loyalist, ensuring government's responsiveness to the needs of supporters, to neutral bu-

trators did formulate and offer policy recommendations, these would be the carefully cultivated products of consensus (Stever 1990).

Other Progressive thinkers were more influenced by the philosophy of pragmatism, however, and took very different approaches to the problem of administration. Pragmatism emerged from the late 19th-century preoccupation with the advances of science and the connection with social progress to become the most original and well-known contemporary American philosophy, particularly as it was expressed through the political theories of Dewey. The core of the philosophy was the belief that individuals must act continuously in an experimental fashion, testing beliefs and concepts against existence. Pragmatists rejected transcendent, absolute ideas that could not be confirmed by experience and action. Propositions could only be judged by the results produced when put into practice[b] (Adams 1992; Lustig 1982; Skowronek 1982; Weinstein 1968).

One of the foremost articulators of the pragmatic method was Charles Sanders Peirce. Peirce contended that any reliable approach to knowledge must meet two requirements: empirical, rule-bound methodology, and public, agreement-bound verification. He sought both continuity and communication. To avoid complete intellectual fragmentation on one side, and tyranny of opinion on the other, he proposed applying scientific methodology to the "community of inquirers" whose collective judgment would control and validate the belief, in line with a "critical common-sensism" (Buchler 1955; Diggins 1994).[c] Peirce had great faith in the redemptive power of science and the scientific method, believing that individuals would subsume their differences of opinion in the cause of truth, likening the power of science to both religious faith and romantic love.[d] He believed that the rigorous methods of science were intrinsically moral—that the cooperative ethos characteristic of scientific inquiry would provide an antidote to the ruthless individualism of capitalism (Buchler 1955; Diggins 1994).

Rather than viewing history and social change as a smooth, seamless, and evolutionary process, pragmatists partitioned history into periods of stability interspersed with rapid, dramatic, and sometimes revolutionary change. Sudden, sometimes disruptive, adjustments of social structure, law, custom, and interaction often became necessary during these periods. Progressive pragmatists contended that their own time was one such revolutionary era, with industrialization, urbanization, and rapid technological innovation requiring challenges to the social, political, and economic status quo. Pragmatist administrative theory pointed to the administrator as the agent capable of engineering and directing these changes. Such administrators, professionally trained in their areas of technical competence and unconstrained by partisan political considerations, grounded their framework in analytical and scientific reason and the dictates of logic (Stever 1986, 1990).

The attitudes of the pragmatic progressives eventually prevailed and left the most definitive imprint on the practice of public administration. In particular, Dewey's public-oriented philosophy aimed to bring reflective analysis to the problems of society, defining ends and identifying means by which to achieve them. But pragmatism was also inherently utilitarian. Peirce observed that pragmatism was less about solving problems than about showing that supposed problems were not problems after all (cited in Diggins 1994). The reformist impulses of men like Dewey came less from the philosophical impulse of pragmatism than from the fact that they were fundamentally decent individuals (Bronner 1994).

[a] Stever contends that Follett began as an organic idealist but later adopted the pragmatic position when it became more prominent (Stever 1986).

[b] Several good sources on pragmatism and its political influences include Stever (1986), Murphy (1990), Diggins (1994), and Kettner (1995).

[c] See, in particular, the essays "The Fixation of Belief" and "Critical Commonsensism" in Buchler (1955).

[d] "The Fixation of Belief" (Buchler 1955).

---

**From the Conversation: Becoming an Expert**

**Mary:** The reason I went to graduate school in the first place was a public decision which affected the neighborhood in which I lived. The community planner, working in concert with a property interest, approved building a fire substation at the entrance to our subdivision. There was ample land for this facility directly across the road from the subdivision, but the deputy fire chief happened to own the land selected. Two hundred residents attended a borough council meeting to protest, not because of NIMBY but because of genuine concern for the safety of our children. The firehouse would have been located beside the school bus stop. The borough manager and the mayor listened to our concerns and then informed us that we didn't have any standing because the planner was an expert and he had assured the council there was no problem. I decided to go to graduate school so I could be an expert too.

---

reaucrat and professional expert. In conventional administrative situations, the administrator is the agent of the government, working in the in-between of systems and structures that link their profession, the government, and the citizenry. For many public servants, the first duty is to the profession and to norms of autonomy, hierarchy, and brotherhood (Stivers 1993); the second duty is to the state or the agency; the third is to the citizenry. This framework, ironically, limits the administrator's exercise of authority by restricting the range of possible relationships between citizens and administration (Forester 1989).

In the minds of citizen-taxpayers, there is a simple linear relationship between elected officials and civil servants. If they distinguish at all between the politician and the career government worker, it is to see the latter as simply carrying out orders determined politically. When government is viewed as the enemy, as it frequently is these days, administrators and street-level workers become the enemy's foot soldiers. As Tolchin (1996) states,

> No wonder the public now believes the worst about government; why shouldn't it, since it's been lied to so much in the recent past? But the public has gotten the facts mixed up. The lies have come more often from political leaders than from the bureaucracy for whom the public has reserved most of its contempt. Alas, however, the bureaucrats have all too often served as handmaidens to leaders. (35)

Front-line government workers, those who have most frequent and direct contact with ordinary citizens, are most likely to be the target of citizen wrath. For example, in the western United States, park rangers and foresters have begun traveling in teams and not wearing official uniforms for fear of attack. In the infamous bombing of the Alfred P. Murrah Building in Oklahoma City, street-level workers and their children bore the brunt of brutal violence against government. Because angry citizens can neither see nor get near those in command who craft the policies they disagree with, they vent their feelings on the infantry.

For their own part, practitioners have sought to couch their role in professional terms: to see themselves not as foot soldiers simply following orders, but as experts with a certain level of autonomy, who, in the context of vague statutes and conflicting mandates, have to make tough decisions about how to get things done. In this line of thinking, the administrator is the agent of the government, working in the in-between that links legislative initiatives and citizen recipients. Much of the recent theoretical work in public administration focuses on this "discretionary" role, seeking to justify the power it allocates to tenured, unelected bureaucrats (see Rohr 1986; Wamsley et al. 1990). Although this professionalized view imparts dignity to what administrators do, it does little to encourage collaboration with citizens. In fact, when administrators think of themselves as professionals, their relationships with citizens tend to be instrumental, if not inauthentic and conflictual, with participation taking place too late in the process to make any real difference—that is, after issues have already been framed and most decisions made. Administrators become territorial and parochial, guarding information closely and relying on their technical and professional expertise to see them through the challenges of their work. The power that citizens exert, in response, is aimed at blocking or redirecting administrative efforts rather than at working as partners to establish parameters, set agendas, develop methods of investigation, and select approaches and techniques.

As many . . . stories . . . attest, sometimes administrators are able to see their own roles and their relationships with citizens more creatively, to forge ahead to build tangible partnerships with citizens, either working within existing organizational structures and processes or revamping them as they are able to. Building partnerships requires resigning exclusive reliance on professional expertise as the ground of administrative action, a difficult thing to do given the tenacity of the systems and structures in place and their tendency to reify relationships. In fact, building partnerships with citizens requires *redefining* expertise: seeing what citizens know as useful and relevant . . . . Such dissenting from administration-as-usual in fact requires that one swim against the stream. Unfortunately, dissenters are likely to burn out, make promises to the citizenry that can't be kept, or end up having to "whistle-blow," or quit on principle (e.g., the recent high-level resignations in the federal Department of Health and Human Services over the terms of the new welfare-reform measures). These reactions, laudable as some are, do little to improve the administrator-citizen relationship.

## HOW ADMINISTRATORS FEEL ABOUT THEIR WORK

Whereas much has been written, in this [reading] and elsewhere, about how citizens feel about government and their relationship to it, little attempt has been made elsewhere to discover how administrators feel about practicing in such an anti-government climate and in what ways this climate has affected their work. Symptoms of a "debilitated public service" are said to be ubiquitous (Ban & Riccuci 1991, 8), but there have been few studies to find out exactly how demoralized government workers actually are and how they are coping in what they might well feel is a hostile environment. One exception is the National Commission on the Public Service, popularly known as the Volcker Commission (Volcker 1989). The Commission was convened to recommend action to the President and Congress on what was called "the quiet crisis," that is, the erosion of the public service at all levels, as the result of public attitudes, political leadership (or lack thereof), and internal management systems.

When the Commission turned its attention to the impact of bureaucracy bashing on the morale of public employees, not only did it discover low morale and declining self-respect, but also that (a) an increasing majority of upper-level public employees said that they would not recommend government service as a career to young people, particularly their own children; (b) bureaucracy bashing appeared to be taking its toll on the ability to recruit new talent into the ranks of the civil service; and (c) public employees expressed frustration over their inability to meet productivity and efficiency standards because so many factors were out of their control, such as shrinking budgets and regulations governing procurement and hiring.

**Experts and Expertise**
(From the acceptance speech of Terri Swearingen, winner of the Goldman Environmental Prize, April 14, 1997.)

We have to reappraise what expertise is and who qualifies as an expert. There are two kinds of experts. There are the experts who are working in the corporate interest, who often serve to obscure the obvious and challenge common sense; and there are experts and nonexperts who are working in the public interest. From my experience, I am distrusting more and more the professional experts, not because they are not clever, but because they do not ask the right questions. And that's the difference between being clever and being wise. Einstein said, "A clever person solves a problem; a wise person avoids it." This lesson is extremely relevant to the nation, and to other countries as well, especially in developing economies. We have learned that the difference between being clever and being wise is the difference between working at the front end of the problem or working at the back end. Government that truly represents the best interests of its people must not be seduced by corporations that work at the back end of the problem—with chemicals, pesticides, incinerators, air pollution control equipment, etc. The corporate value system is threatening our health, our planet, and our very existence.

. . . We have become the real experts, not because of our title or the university we attended, but because we have been threatened and we have a different way of seeing the world. We know what is at stake. We have been forced to educate ourselves, and the final exam represents our children's future. WE know we have to ace the test because when it comes to our children, we cannot afford to fail. Because of this, we approach the problem with common sense and with passion. We don't buy into the notion that all it takes is better regulations and standards, better air pollution control devices and more bells and whistles. We don't believe that technology will solve all our problems. We know that we must get the front end of the problems, and that prevention is what it needed. We are leading the way to survival in the 21st century.

. . . Even after seeing so much abuse of the system that I have believed in, I still hold on to the slender hope that my government could once again return to representing citizens like me rather than rapacious corporate interests. If they do, then perhaps there is a future for our species; if they don't, we are doomed.

The Volcker Commission recommended three major areas of focus for improving the situation in public agencies: (1) improving leadership, including reforming the federal Senior Executive Service (SES) and limiting the practice of replacing SES positions with political appointments; (2) working to attract and retrain talented people; and (3) focusing on improving performance and productivity (Nigro & Nigro 1994).

Unfortunately, little has been accomplished, to date, in any of these areas for a variety of reasons. On the contrary, instead of building government service, current political leaders at all levels are reducing the ranks through downsizing and "reinvention." In the last four years, 1 in every 10 jobs in the executive branch of the federal government was eliminated. In a 17–year period, mostly within the last four years, the Office of Personnel Management and the General Services Administration both experienced 42% reductions in force (Causey 1997).

In preparation for [the book where this reading appears], we asked a small group of colleagues around the country to distribute an open-ended survey to students in their MPA programs, the bulk of whom were practitioners. Although we are not claiming that our 120 respondents are statistically representative of public administrators, they do come from programs in six widely scattered states (Colorado, Florida, Hawaii, Ohio, Oklahoma, and Washington). The majority are currently working in a variety of local and state government positions, are almost evenly split between males and females, and closely match proportions of the various racial and ethnic groups in the country. Whereas it is not possible to generalize their responses to all public workers, they do give food for thought with regard to how the current anti-government climate is affecting people who work in government agencies.

One would expect anti-government rhetoric to be affecting public administrators in a negative way, and indeed our respondents showed some signs of this. Having come to public sector work with, as many of them said, the desire to make things better, many are not only aware of the criticisms about unresponsiveness and red tape, but they tended to agree with them. Instead of being energized by the criticism, they tended to feel alienated, ineffective, cynical, and even frightened—and with reason, because workplace violence is a significant threat to government workers (Nigro & Waugh 1996). Anti-government attacks, then, have a tendency to become

> a self-fulfilling prophecy, for not very many of the best people in the public services are likely to want to stay there when they are paid far below their peers in the private sector and are the butt of repeated charges of incompetence, dishonesty, and laziness by their neighbors, the media, and indeed their own bosses. (Mosher 1982, xii)

Beneath the surface of these negative reactions, however, were some interesting positive attitudes. Beneath the gloom about the anti-government climate was a continuing commitment to service. In addition to recognizing the extent to which anti-government feelings constitute a dilemma and a threat, our respondents also saw them as a challenge.

Some were quite worried over what one called "near constant harassment from malcontents and political wannabees," found their working conditions "frustrating," or declared, "I resent it. I work hard and the people I work with work hard." Another commented, "It is scary—it makes me want to stay out of the limelight and not make waves."

Some seemed to have internalized public views about bureaucracy, agreeing that government is the problem. For example, one said, "Having been part of the process, I can relate to the public's attitudes." Another commented, "The negative attitude is deserved. There is a lot of waste." Many others, however, looked beyond the immediacy of the criticism, seeing an opportunity to improve public service, as the following sample of quotes illustrate:

- "It can be depressing, but employees who take a positive approach often find that the public is not as negative as one is led to expect, particularly on a one-to-one basis."
- "My understanding of the animus toward government is that this is an extension of non-directed anger and anxiety in the public that they are manipulated by powerful elites who do not have their interests in mind. I wouldn't take it personally."
- "It tends to decrease morale but repeated contacts with the same 'customers' help both of us feel better about what I do."
- "It is challenging/empowering—I'm proud to have the opportunity to turn negatives into positives."
- "I feel like a rebel for the public good."
- "I am a good bureaucrat doing good things for the public."

When we asked how the anti-government environment affected their organizations, respondents cited reduced funding, lowered morale, limits on risk taking, delays, pay freezes, and an excessive focus on accountability, which perversely leads to even more red tape. They also felt that criticism makes working with the public more difficult, reducing public involvement, and creating a challenging need for partnerships with the community and a focus on "customer service." As funds are cut, administrators find themselves having to do more with less, and creativity and risk taking are strangled by the need to maintain or increase visible accountability. It becomes difficult to do one's job, maintain high morale, and recruit and retain the best and brightest. As our respondents' comments suggested, however, the silver lining is that criticism points up the necessity of improving relationships between government agencies and the public. As we argued earlier, whether customer service or other market-modeled strategies improve relationships is questionable, and our respondents agree. Certainly customer service soothes some of the irritation citizens feel at long lines and cumbersome processes, but it does little to engender collaborative efforts by citizens and government workers or to raise the sights of citizens beyond their own immediate needs and wants and give them a sense of connection to government.

When we asked what respondents were doing to counteract anti-government feelings, most mentioned improving interactions with the public, one person at a time. They did not couch these efforts in terms of customer service but rather in terms of one-to-one relationships and bringing citizens into administrative processes. They do "simple things that allow you to get to know people as people," as one person said, like being accessible, responding, educating people about what government does, disseminating information, developing interpersonal skills, listening, and showing genuine concern and empathy.

These comments fall roughly in two interrelated areas: improving one-on-one interactions with citizens, and working to break down barriers inherent in the bureaucratic apparatus. Hannah Arendt argues that bureaucratic management paralyzes action by turning people into "behaving citizens" (cited in Ventriss 1995, 576)—by extension, one could say "behaving public employees" as well. Ralph Hummel (1994) agrees that bureaucratization acts to conceal, deny, and shape the political experience; he warns of two dangers when bureaucratic structures prevail in public life: The bureaucracy converts public problems into administrative and technical issues, and it constrains possibilities for "full, human politics" (241). As Tauxe (1995) states, public organizations must "not only democratize formal institutions and procedures, but also make room for nonbureaucratic discourse and organizational forms" (489). Clearly, as we move to address problems raised by trying to do public work in an anti-government climate, both relational and organizational strategies are needed, especially given the extent to which the two are intertwined.

## REFERENCES

Adams, Guy B. 1992. "Enthralled with Modernity: The Historical Context of Knowledge and Theory Development in Public Administration." *Public Administration Review* 52(4): 363–373.

Ban, Carolyn, and Norma M. Riccuci. 1991. *Public Personnel Management: Current Concerns—Future Challenges.* New York: Longman.

Berry, Jeffrey M., Kent E. Portney, and Ken Thomson. 1993. *The Rebirth of Urban Democracy.* Washington, DC: The Brookings Institution.

Bezich, Louis S. 1997. "N.J. County Shows Citizens What's in Store." *PA Times* 20(3): 1–2.

Bronner, Steven E. 1994. *Of Critical Theory and Its Theorists.* Cambridge, MA: Blackwell.

Buchler, Justus, ed. 1955. *Philosophical Writings of Peirce.* New York: Dover.

Causey, Mike. 1997. "The Vanishing Bureaucrats." *Washington Post,* April 27, B2.

Cooke, Jacob, ed. 1961. *The Federalist Papers.* Middletown, CT: Wesleyan University Press.

Diggins, John Patrick. 1994. *The Promise of Pragmatism: Modernism and the Crisis of Knowledge and Authority.* Chicago: University of Chicago Press.

Fishkin, James S. 1995. *We the People: Public Opinion and Democracy.* New Haven, CT: Yale University Press.

Flanigan, William H., and Nancy H. Zingale. 1994. *Political Behavior of the American Electorate.* Washington, DC: CQ Press.

Forester, John. 1989. *Planning in the Face of Power.* Berkeley: University of California Press.

Hollinger, David A., and Charles Capper. 1989. *The American Intellectual Tradition, vol. II.* New York: Oxford University Press.

Hummel, Ralph P. 1994. *The Bureaucratic Experience.* 4th ed. New York: St. Martin's Press.

Jones, Brian. 1981. "Party and Bureaucracy: The Influence of Intermediary Groups on Urban Public Service Delivery." *American Political Science Review* 75 (November): 688–700.

Kearney, Richard J., and Chandan Sinha. 1988. "Professional and Bureaucratic Responsiveness: Conflict or Compatibility." *Public Administration Review* 48(5): 571–579.

Kettering Foundation. 1991. *Citizens and Politics: A View from Main Street America.* Report prepared for the Kettering Foundation by the Harwood Group. Dayton, OH: Kettering Foundation.

Kettner, Kenneth Laine, ed. 1005. *Peirce and Contemporary Thought.* New York: Fordham University Press.

King, Cheryl Simrell, Kathryn M. Feltey, and Bridget O'Neill Susel. 1998. "The Question of Participation: Toward Authentic Public Participation in Public Decisions." *Public Administration Review* 58(4): 317–326.

Lustig, R. Jeffery. 1982. *Corporate Liberalism: The Origin of Modern American Political Theory.* Berkeley: University of California Press.

MacNair, Ray H., Russell Caldwell, and Leonard Pollane. 1983. "Citizen Participation in Public Bureaucracies: Foul-Weather Friends." *Administration & Society* 14: 507–523.

Mills, C. Wright. 1959. *The Sociological Imagination.* London: Oxford University.

Mladenka, Kenneth R. 1981. "Citizen Demands and Urban Services; The Distribution of Bureaucratic Response in Chicago and Houston." *American Journal of Political Science* 25 (November): 693–714.

Morone, James A. 1990. *The Democratic Wish: Popular Participation and the Limits of American Government.* New York: Basic Books.

Mosher, Frederick C. 1982. *Democracy and the Public Service.* New York: Oxford University Press.

Murphy, John P. 1990. *Pragmatism: From Peirce to Davidson.* Boulder, CO: Westview.

Nelson, Michael A. 1982. "A Short, Ironic History of American National Bureaucracy." *Journal of Politics* 44: 749–778.

Nigro, Lloyd C., and Felix A. Nigro. 1994. *The New Public Personnel Administration.* Itasca, IL: F.E. Peacock.

Nigro, Lloyd C., and William L. Waugh. 1996. "Violence in the American Workplace: Challenges to the Public Employee." *Public Administration Review* 56(4): 326–333.

Parr, John, and Christopher Gates. 1989. "Assessing Community Interest and Gathering Community Support." In *Partnerships in Local Governance: Effective Council-Manager Relations*

(ICMA Handbook), ed. International City Management Association. Washington, DC: International City Management Association.

Putnam, Robert D. 1995. "Bowling Alone: America's Declining Social Capital." *Journal of Democracy* 6(1): 65–78.

Rohr, John A. 1986. *To Run a Constitution: The Legitimacy of the Administrative State.* Lawrence: University Press of Kansas.

Sabine, George H., and Thomas L. Thorson. 1973. *A History of Political Theory.* 4th ed. Hinsdale, IL: Dryden.

Selznick, Phillip. 1949. *TVA and the Grass Roots.* Berkeley: University of California Press.

Sharp, Elaine B. 1980. "Citizen Participation: The Co-Production Concept." *Midwest Review of Public Administration* 14 (June): 105–119.

Skowronek, Stephen. 1982. *Building A New American State: The Expansion of National Administrative Capacities, 1877–1920.* Cambridge, UK: Cambridge University Press.

Stever, James A. 1986. "Mary Parker Follett and the Quest for Pragmatic Administration." *Administration & Society* 18(2): 159–177.

———. 1990. "The Dual Image of the Administrator in Progressive Administrative Theory." *Administration & Society* 22(1): 39–57.

Stewart, Thomas. R., Robert L Dennis, and David W. Ely. 1984. "Citizen Participation and Judgment in Policy Analysis: A Case Study of Urban Air Quality Policy." *Policy Science* 77 (May): 67–87.

Stivers, Camilla. 1993. *Gender Images in Public Administration: Legitimacy and the Administrative State.* Newbury Park, CA: Sage.

Tauxe, C.S. 1995. "Marginalizing Public Participation in Local Planning: An Ethnographic Account." *Journal of the American Planning Association* 61(4): 471–481.

Thomas, John Clayton. 1995. *Public Participation in Public Decisions.* San Francisco: Jossey-Bass.

Thompson, Dennis. 1970. *The Democratic Citizen.* Cambridge: Cambridge University Press.

Timney, Mary M. 1996, July. *Overcoming NIMBY: Using Citizen Participation Effectively.* Paper presented at the 57th National Conference of the American Society for Public Administration, Atlanta, GA.

Tolchin, Susan J. 1996. *The Angry American: How Voter Rage Is Changing the Nation.* Boulder, CO: Westview Press.

Van Riper, Paul P. 1958. *History of the United States Civil Service.* New York: Harper and Row.

Ventriss, Curtis. 1995. "Modern Thought and Bureaucracy." *Public Administration Review* 55(6): 575–580.

Volcker, Paul A. 1989. *Leadership for America: Rebuilding the Public Service. Report of the National Commission on the Public Service.* Lexington, MA: Lexington Books.

Waldo, Dwight. 1948. *The Administrative State.* New York: Ronald Press.

———.1984. Response. *Public Administration Review* 44(6): 107–109.

Wamsley, Gary I., Robert N. Bacher, Charles T. Goodsell, Philip S. Kronenberg, John A. Rohr, Camilla M. Stivers, Orion F. White, and James E. Wolf, eds. 1990. *Refounding Public Administration.* Newbury Park, CA: Sage.

Weinstein, James. 1968. *The Corporate Ideal in the Liberal State.* Boston: Beacon.

Wiebe, Robert H. 1967. *The Search for Order, 1877–1920.* New York: Hill and Wang.

Wilson, Woodrow. 1887. "The Study of Administration." *Political Science Quarterly* 2 (June): 197–222.

# INDEX

# ABOUT THE AUTHOR

**Richard C. Box** is a professor in the School of Public Administration, University of Nebraska at Omaha. He served for thirteen years as a land-use planner, department head, and city administrator in local governments in Oregon and California before completing his doctorate at the University of Southern California. His research focuses on democracy and the relationship of public administration to the broader society. It has been published in journals such as *American Review of Public Administration*, *Public Administration Review*, *Administration & Society*, *Administrative Theory & Praxis*, *International Journal of Public Administration*, and others, as well as in his book, *Citizen Governance: Leading American Communities into the 21st Century* (1998).